ELEPHANTS UNDER HU

ELEPHANTS UNDER HUMAN CARE
The Behaviour, Ecology, and Welfare of Elephants in Captivity

PAUL A. REES
School of Science, Engineering and Environment, University of Salford, United Kingdom

Academic Press is an imprint of Elsevier
125 London Wall, London EC2Y 5AS, United Kingdom
525 B Street, Suite 1650, San Diego, CA 92101, United States
50 Hampshire Street, 5th Floor, Cambridge, MA 02139, United States
The Boulevard, Langford Lane, Kidlington, Oxford OX5 1GB, United Kingdom

Copyright © 2021 Elsevier Inc. All rights reserved.

No part of this publication may be reproduced or transmitted in any form or by any means, electronic or mechanical, including photocopying, recording, or any information storage and retrieval system, without permission in writing from the publisher. Details on how to seek permission, further information about the Publisher's permissions policies and our arrangements with organizations such as the Copyright Clearance Center and the Copyright Licensing Agency, can be found at our website: www.elsevier.com/permissions.

This book and the individual contributions contained in it are protected under copyright by the Publisher (other than as may be noted herein).

Notices

Knowledge and best practice in this field are constantly changing. As new research and experience broaden our understanding, changes in research methods, professional practices, or medical treatment may become necessary.

Practitioners and researchers must always rely on their own experience and knowledge in evaluating and using any information, methods, compounds, or experiments described herein. In using such information or methods they should be mindful of their own safety and the safety of others, including parties for whom they have a professional responsibility.

To the fullest extent of the law, neither the Publisher nor the authors, contributors, or editors, assume any liability for any injury and/or damage to persons or property as a matter of products liability, negligence or otherwise, or from any use or operation of any methods, products, instructions, or ideas contained in the material herein.

British Library Cataloguing-in-Publication Data
A catalogue record for this book is available from the British Library

Library of Congress Cataloging-in-Publication Data
A catalog record for this book is available from the Library of Congress

ISBN: 978-0-12-816208-8

For Information on all Academic Press publications
visit our website at https://www.elsevier.com/books-and-journals

Publisher: Charlotte Cockle
Acquisitions Editor: Anna Valutkevich
Editorial Project Manager: Billie Jean Fernandez
Production Project Manager: Joy Christel Neumarin Honest Thangiah
Cover Designer: Matthew Limbert

Typeset by MPS Limited, Chennai, India

For Katy, Clara and Elliot
— my Little Toomai

But it was all for the sake of Little Toomai, who had seen what never man had seen before — the dance of the elephants at night and alone in the heart of the Garo hills!

Toomai of the Elephants,
The Jungle Book,
Rudyard Kipling (1894)

Contents

Acknowledgements xiii
Preface xv
Who is this book for? xvii
'Zoo elephant' or 'elephant living in a zoo': a note on terminology xix

1. Elephants and their relationship with humans

1.1 Another book about elephants 1
1.2 What is an elephant? 3
    1.2.1 Elephant taxonomy 3
1.3 Conservation status 7
    1.3.1 The status of Asian elephants in the wild 7
    1.3.2 The status of African elephants in the wild 7
1.4 The human use of elephants 8
    1.4.1 Elephants in ancient times 8
    1.4.2 Have elephants been domesticated? 9
    1.4.3 Ceremonial and religious use of elephants; elephants as gifts 9
    1.4.4 The use of elephants for transportation and as weapons of war 10
    1.4.5 Elephants and forestry operations 13
    1.4.6 Elephants as entertainers: circuses, sports and tourism 13
    1.4.7 Elephants as ambassadors for conservation 14
1.5 The beginning of elephant research 15
    1.5.1 Aristotle and elephants 15
    1.5.2 Anatomical research 18
    1.5.3 Anecdotes as a source of knowledge about elephants 24
    1.5.4 Papers in academic journals 25
    1.5.5 Early physiological research 25
1.6 Unacceptable elephant science 26
1.7 Captive elephants as proxies for wild elephants 27

2. Ethological data collection and elephant activity budgets

2.1 Introduction 29
2.2 Methodology 29
    2.2.1 Identifying individuals 29
    2.2.2 Studying elephants in zoos 31
    2.2.3 Ethograms 34
    2.2.4 Methodological difficulties 36
    2.2.5 Data collection by caregivers 38
2.3 Activity budgets 39
    2.3.1 Introduction 39
    2.3.2 Feeding 43
    2.3.3 Dusting 45
    2.3.4 Walking 46
    2.3.5 Resting, sleeping and nocturnal behaviour 46
2.4 The 24-hour needs of elephants in zoos 50

3. Elephant social structure, behaviour and complexity

3.1 Introduction 51
3.2 The structure of elephant societies 51
    3.2.1 Elephant societies in the wild 51
    3.2.2 Elephant societies in captivity 53
    3.2.3 Social behaviour and breeding 59
3.3 Associations between individuals and friendships 59
3.4 Introductions into an elephant group 63
3.5 Protective formations 64
3.6 Dominance hierarchies 66
3.7 Aggression, appeasement and chastisement 68
3.8 Personality 76

4. Elephant reproductive biology

4.1 Introduction 81
4.2 Historical accounts of sexual behaviour 81
4.3 Courtship and mating behaviour 83
4.4 Chemical control of reproduction 88

    4.4.1 Musth 88
    4.4.2 Endocrine monitoring of females 89
4.5 Behavioural indicators of oestrus 90
4.6 Gestation, pregnancy management and birth 91
4.7 Parenting and calf development 93
    4.7.1 Developmental milestones and birth statistics 93
    4.7.2 Parenting and allomothering 94
    4.7.3 The effect of a calf on social interactions in the herd 99
4.8 Early sexual behaviour 100
    4.8.1 Early male sexual behaviour 100
    4.8.2 Juvenile mounting 100
4.9 Reproductive challenges and solutions 106
    4.9.1 Acyclicity and sperm quality 106
    4.9.2 Obstetrics and birthing problems 109
    4.9.3 New techniques in reproductive physiology 109

5. Elephant cognition, communication and tool use

5.1 Introduction 113
5.2 Cognition 113
    5.2.1 Historical perspectives 113
    5.2.2 Self-awareness: do elephants know they exist? 114
    5.2.3 Discrimination between objects and between quantities 116
    5.2.4 Insightful behaviour 116
    5.2.5 Pointing 116
    5.2.6 Memory 117
5.3 Communication 118
    5.3.1 Introduction 118
    5.3.2 Vocal communication 119
    5.3.3 Human speech imitation 120
    5.3.4 Chemical communication 120
    5.3.5 Tactile and seismic communication 123
5.4 Visual acuity and visual discrimination 123
5.5 Tool use 124
5.6 Knowing when to cooperate 127

6. Elephant ecology and genetics

6.1 Introduction 129
6.2 Ecophysiology 129
    6.2.1 Introduction 129
    6.2.2 Acclimatisation to new environments 130
    6.2.3 Thermoregulation 130
6.3 Feeding ecology and energetics 133
    6.3.1 Food preferences 133
    6.3.2 Feeding methods 134
    6.3.3 Calculating food consumption 134
    6.3.4 Digestibility 137
    6.3.5 Food passage time 137
    6.3.6 Defaecation 139
    6.3.7 Food supplementation 139
6.4 Energetics 140
6.5 Exhibit design and enclosure use 141
    6.5.1 Introduction 141
    6.5.2 Enclosure use 142
    6.5.3 Substrate and indoor versus outdoor preferences 146
    6.5.4 Multispecies exhibits 148
    6.5.5 Rotational exhibits 149
    6.5.6 Elephants as agents of landscape change in zoos 150
6.6 Population ecology 153
    6.6.1 Introduction 153
    6.6.2 Age determination from teeth 154
    6.6.3 Longevity and life expectancy in zoos 154
    6.6.4 Birth rates and calf survival 155
    6.6.5 Sexual maturity and mean calving interval 157
    6.6.6 Reproductive performance of Asian camp elephants 158
    6.6.7 Reproductive cessation and the 'mother hypothesis' 161
    6.6.8 Sustainability of zoo populations 161
    6.6.9 Importation of elephants from range states 162
6.7 Genetics 163
    6.7.1 Introduction 163
    6.7.2 Interspecific hybridisation 164
    6.7.3 Intraspecific hybridisation 164
    6.7.4 Genetic diversity 164

7. Elephant welfare

7.1 Historical perspectives 169
7.2 Measuring elephant welfare 173
    7.2.1 What is welfare? 173
    7.2.2 How can welfare be measured? 173
    7.2.3 Population-level welfare indices 175

- 7.2.4 Body weight and condition scoring 176
- 7.2.5 The welfare of elephants working in tourism 179
- 7.2.6 Stress and distress 180
- 7.2.7 Behaviour as a welfare indicator 181
- 7.2.8 Stereotypic behaviours 183
- 7.3 Environmental enrichment 194
  - 7.3.1 Defining environmental enrichment 194
  - 7.3.2 Food and foraging as enrichment 196
  - 7.3.3 Substratum and trees as enrichment 199
  - 7.3.4 Water as enrichment 202
  - 7.3.5 Sleep, rest and enrichment 202
  - 7.3.6 Sound, music and art as enrichment 205
  - 7.3.7 Interactive toys 206
  - 7.3.8 Improving elephant welfare through breeding 206
  - 7.3.9 Social contact: the ultimate in enrichment 207
- 7.4 Training 207
- 7.5 Locomotion and gait 208
- 7.6 Obesity 210
- 7.7 Disease 211
  - 7.7.1 Introduction 211
  - 7.7.2 Histology 213
  - 7.7.3 Foot health 214
  - 7.7.4 Tuberculosis 216
  - 7.7.5 Elephant endotheliotropic herpesviruses 217

8. Housing and handling elephants

- 8.1 Introduction 219
- 8.2 Wild elephant decline and the establishment of ex situ breeding programmes 219
- 8.3 Elephant enclosures 222
  - 8.3.1 Housing and containment 222
  - 8.3.2 Early elephant houses 223
  - 8.3.3 Enclosure size and substratum 229
  - 8.3.4 New enclosures 237
- 8.4 The cost of keeping elephants under good welfare conditions 238
- 8.5 Elephants and their caretakers 243
  - 8.5.1 Keeper–elephant bonds 243
  - 8.5.2 Traditional elephant expertise versus zoo husbandry 244
  - 8.5.3 Keeper and mahout deaths 246
  - 8.5.4 Free versus protected contact 248
- 8.6 Transportation 250

9. Ethics, pressure groups and the law

- 9.1 Introduction 259
- 9.2 Is it ethical to keep elephants in captivity? 259
- 9.3 Pressure groups 264
- 9.4 Law 266
  - 9.4.1 Introduction 266
  - 9.4.2 Ivory in human ownership 266
  - 9.4.3 The UN Convention on Biological Diversity and the IUCN 268
  - 9.4.4 Elephants, zoos and the law 270
  - 9.4.5 The law and elephants in private ownership 271
  - 9.4.6 Elephants, entertainment and the law 272
  - 9.4.7 Legal personhood and *habeas corpus* 274
  - 9.4.8 Elephant cruelty and cruel methods 275
  - 9.4.9 Zoos and the wildlife trade 278
  - 9.4.10 A right to companionship and retirement 282
  - 9.4.11 Financial support for in situ conservation 284

10. The conservation value of captive elephants

- 10.1 Introduction 287
- 10.2 The popularity of elephants in zoos 290
- 10.3 Zoo elephants as insurance populations 294
  - 10.3.1 Introduction 294
  - 10.3.2 The Species Survival Plan in North America 295
  - 10.3.3 The European Endangered species Programme 295
  - 10.3.4 Are zoo elephant populations sustainable? 296
- 10.4 Scientific research 297
- 10.5 The development of technologies relevant to field conservation 298
- 10.6 Educational function 300
- 10.7 Professional training of local conservationists and associated technology transfer 306
- 10.8 Fundraising for in situ conservation 307

- 10.9 Do zoo and conservation authorities support captive breeding in zoos? 308
- 10.10 Captive breeding in range states 309
- 10.11 'Domestication' of African elephants 310
- 10.12 Conclusion 310

11. The future of elephants in captivity

- 11.1 Introduction 313
- 11.2 Elephant ranching 313
- 11.3 Rewilding — shades of Jurassic Park 315
- 11.4 Release to the wild 316
- 11.5 Welfare concerns 317
- 11.6 Sanctuaries 317
- 11.7 A repository of useful genes 318
- 11.8 Cloning 320
- 11.9 Elephants as therapy 320
- 11.10 Climate change 321
- 11.11 A role for zoos? 321
- 11.12 Consumptive use or intensive protection zones? 324
- 11.13 The court of public opinion 326
- 11.14 Predictions 326

Postscript 327

Appendix 329
Glossary 333
References 341
Index 379

Acknowledgements

A number of individuals and organisations have assisted in the production of this book. At Elsevier I would like to thank Anna Valutkevich (Acquisitions Editor) for believing that the world needs another book about elephants, as well as all the people who were involved in its production, especially Billie Jean Fernandez.

I am indebted to Dr Alan Woodward, who kindly drew Figs 3.17, 3.19B, 4.1 and 6.5. Thanks are also due to a number of my colleagues at the University of Salford. Prof. Alaric Searle kindly provided me with a copy of his paper comparing the roles of war elephants and tanks, and some of the information contained therein has been used in Table 1.1. Dr Simon Hutchinson kindly allowed me to use Fig. 1.16A and also provided me with information about a number of events of which I would not otherwise have been aware. My colleague Dr Robert Jehle kindly translated the text in Fig. 1.2 from German into English. The text in Figs 1.1 and 1.2 was made available by the Biodiversity Heritage Library (www.biodiversitylibrary.org).

I have reproduced a number of images of historical interest from the US Library of Congress (Figs 1.4, 1.6–1.8, 1.13, 1.14, 7.25, 8.6–8.8), the New York Public Library (Fig. 8.3) and the University of Queensland (Fig. 8.1). These organisations have indicated that there is no known restriction on the publication of these images.

My friend the late Ivor Rosaire kindly supplied me with data on the food eaten by the elephants kept at Knowsley Safari Park, United Kingdom (Fig. 6.4) when I worked there as an elephant keeper in 1976–77. The staff of the Elephant House at Chester Zoo and the zoo management allowed me to study their elephants and I am most grateful to have had this opportunity. I am particularly grateful to Mick Jones and Alan Littlehales for sharing their vast knowledge of elephants with me. Louise Bell, formerly a research officer at Blackpool Zoo, generously provided Fig. 4.14 that was previously published in Rees (2011). I am most grateful to her for this and to Kevin Williams, of Reaseheath College, for kindly providing me with data on zoo elephant populations held in ZIMS. In India, Prof. Madhyastha, formerly of the University of Mangalore, kindly arranged for me to visit elephant camps in Kushnalagar and the elephant at Kateel Shri Durgaparameshwari Temple in Karnataka State during a visit funded by the British Council.

A number of the figures and tables used in this book have previously been published elsewhere in my own papers and books, and the papers of others. I am grateful to Elsevier for permission to use Figs 2.1B, 5.3, 5.4, 5.6, 6.2, 7.9, 7.10 and 11.2, and to Wiley-Blackwell for permission to reproduce Figs 1.15, 2.1A, 2.4–2.6, 2.8, 3.1–3.3, 3.10, 4.1, 4.5, 4.8, 4.12, 4.14, 4.15, 5.8, 6.3, 7.4, 7.11, 8.2C, 8.5B, 8.18 and 8.20, and Table 2.2. Figs 3.19 and 3.20 were first published by the Bombay Natural History Society, and Fig. 6.8 is based on a figure first published by the Association of British and Irish Wild Animal Keepers (in *Ratel*). Fig. 3.6 was first published in *International Zoo News* and Table 6.2 by the Zoological

Society of London (in the *Journal of Zoology*). The images used in Figs 3.7 and 4.13 and Tables 4.6–4.9 were originally published by Taylor & Francis.

Fig. 1.16B was made available by Chetham's Library, Manchester, United Kingdom, from its archive of materials from the former Belle Vue Zoo in Manchester. Stephen Fritz (Stephen Fritz Enterprises, Inc.) generously provided me with information about, and photographs of, his elephant transportation operations, and I am most grateful for this (Figs 8.24–8.26). Sections of UK legislation and excerpts from the *Secretary of State's Standards of Modern Zoo Practice* are reproduced under the Open Government Licence v3.0.

My brother Les Rees very kindly supplied Figs 8.12 and 8.14. My daughter Clara Clark provided me with Fig. 5.2 (and the information accompanying it) taken during a visit to Blackpool Zoo with my grandson Elliot. This book represents the culmination of a lifelong interest in elephants into which I have dragged my wife, daughter and other family members kicking and screaming. I hope that when they see this book, they will think it was a life well spent and, at the very least, Elliot will enjoy looking at his grandpa's pictures.

Preface

In his enlightening book, *The Naked Ape*, the zoologist Dr Desmond Morris made a zoological study of the behaviour of *Homo sapiens*, arguing that such an approach was scientifically justified because humans are simply another species of animal, even though many live in largely human-made environments (Morris, 1967). The biology of elephants has been extensively studied in the wild, but elephants have another biology: their biology when under human care. Captive elephants are deserving of scientific study because they can tell us things about elephants that would be difficult to investigate in the wild, and also because they have a distribution, ecology, population biology, social structure, repertoire of behaviour and other characteristics that are different from their wild conspecifics. This book is about the biology of elephants under human care, their welfare and their unique relationship with humans. It is my attempt to draw together the many studies conducted on elephants living in zoos, circuses, logging camps and other captive situations, many (if not most) of which are ignored by the existing books on elephants. These animals have a story to tell that, although it may not be as exciting as those of elephants living wild in Africa and Asia, nevertheless deserves to be told.

If we were to ask, 'What do we know about the biology of takins in captivity?' the answer would be, 'Not very much'. The same cannot be said of elephants. Elephants have been kept in captivity for thousands of years. In 350 BCE Aristotle described the training of elephants in his *History of Animals*. By the 1970s field biologists were beginning to describe the ecology and behaviour of wild elephants in Africa and Asia in detail. Alongside these studies of elephants in their natural habitats, other scientists have been studying the biology of elephants living in captivity.

Some of the early captive studies were concerned with basic biology (including postmortem anatomical studies), but in recent years there has been considerable interest in the ecology, behaviour and welfare of elephants living in zoos and other captive environments. Their potential role in elephant conservation and the difficulties associated with providing elephants with good welfare have attracted the interest of scientists, conservationists, animal welfare campaigners and politicians.

The published research on captive elephant biology is scattered throughout the academic literature in a wide range of journals and within husbandry manuals, reports commissioned by governments and the publications of animal welfare organisations. The purpose of this book is to draw together the results of published research conducted on the ecology, behaviour and welfare of elephants living in zoos and other captive environments and, where useful, to compare these studies with what is known about elephant biology in the wild. While it is not intended to be a husbandry manual, some of the studies discussed have clear husbandry implications. Although I have referred to some published research concerned with the health and veterinary care of elephants, I have not attempted to replicate the available books on this subject.

If we are to continue to manage elephants under human care, we have a duty to use all of the available scientific knowledge to produce evidence-based management systems that provide elephants with the best possible welfare. The existing elephant husbandry guidelines barely scratch the surface of what we know about elephants in captivity, and much of what has been published is scattered within a myriad of journals and unpublished reports. I hope this book goes some way towards gathering this information together in a way that is accessible to scientists and nonscientists alike.

Paul A. Rees
**University of Salford, Greater Manchester,
United Kingdom**

Who is this book for?

This book should be useful for students studying zoology, zoo biology, animal behaviour, veterinary science, animal welfare and other related disciplines, as well as zoo professionals who work with elephants and anyone who has a serious interest in the Elephantidae. I hope it will also inform the debate about whether the potential conservation role of captive elephants can justify keeping them in zoos.

Over the past few years there has been a considerable increase in the number of papers published on captive elephants, especially those living in zoos. I do not expect this interest to wane any time soon, with the consequence that by the time anyone reads this, science will have moved on and this work will be somewhat out of date. This, of course, is inevitable and the fate of all textbooks. Nevertheless, I hope it will act as a starting point for anyone interested in the subject and save them the chore of starting their research too far back in time.

'Zoo elephant' or 'elephant living in a zoo': a note on terminology

Some journals concerned with animal welfare or animal ethics discourage the use of terms such as 'zoo animals' and 'wild animals', preferring authors to refer to 'animals living in zoos' and 'free-ranging animals', respectively. Some authors refer to 'zoo-housed' elephants, while others use 'in situ' populations to refer to elephants living within their range states. The style guide for the *Journal of Applied Animal Welfare Science* requires authors to: 'Use "animal in the laboratory" (zoo or wild) at first mention in both abstract and text.' I have tried to use terms such as 'elephants living in zoos' and 'elephants working in logging camps' where appropriate. However, I have also used terms such as 'zoo elephants', 'circus elephants', 'wild elephants', and 'captive elephants' to avoid writing sentences that are unnecessarily long and clumsy, and to avoid repetition. I have used the personal pronouns 'he' and 'she' rather than 'it' when referring to individual elephants whose sex is known, and the relative pronoun 'who' when referring to specific individual elephants, rather than 'that'.

CHAPTER 1

Elephants and their relationship with humans

> [The elephant] *is the most significant terrestrial mammal survival, one that we might have expected to have disappeared in the Pleistocene.*
>
> Clive Spinage (2019)

1.1 Another book about elephants

Elephants are probably the most recognisable animals on Earth. They have a long and complex association with humans and have been studied for thousands of years.

People and elephants can form extraordinary relationships, sometimes even life-long relationships. Some mahouts and elephant keepers form bonds with their elephants that last for many decades. Some people care for elephants in general and devote their lives to elephant welfare, either in the wild or in captivity. Others devote their lives to the scientific study of elephants or to their conservation. Some people who have no interest in elephants per se have had their lives blighted by herds that raid their crops and destroy their homes. Others die at the hand of wild or captive elephants, perhaps because they trusted individual animals too much, or because they came between elephants and their food.

In the developed world, human perceptions of elephants are largely based on encounters at circuses and zoos, and from information presented in books, on television and in films. Few people living in the range states of elephants or in developed countries have had the opportunity to see elephants living in the wild.

Zoo and circus elephants are generally slow-moving and subservient individuals. Often visitors see keepers working closely with them. On television we see Asian elephants working with mahouts, again in a subservient relationship. In the developed world, human perceptions of elephants are generally of animals that are endangered by human activity and no real threat to humans.

In *Tarzan*, *The Jungle Book*, *Dumbo*, *Daktari* and other films, TV programmes and cartoons, elephants are usually portrayed sympathetically, as a friend and companion of humans. But where they live alongside people who do not have the luxury of caring a great deal about wildlife conservation, elephants are often perceived as competitors for food and a threat to human life and property.

Why do we need another book about elephants? Many books have been written about elephants, but most have paid little attention to the biology of elephants under human care. Major scientific texts on wild elephants first appeared in the 1970s and were concerned with the behaviour and ecology of particular populations, such as those in southeastern Ceylon (now Sri Lanka) (McKay, 1973), North Bunyoro, Uganda (Laws et al., 1975), Lake Manyara in Tanzania (Douglas-Hamilton and Douglas-Hamilton, 1975) and southern India (Sukumar, 1989). The *Natural History of the African Elephant* (Sykes, 1971) drew together much of what was known about the species at the time, but again focussed on wild elephants, particularly their diseases (including anatomy) and ecology.

In his book *Elephants*, Eltringham (1982) devoted just four pages to 'Elephants in Zoos' in a work of 262 pages, and did not mention any of the research conducted on elephants living in zoos. Most of this section of the book is an account of the life of an elephant called *Jumbo* at London Zoo. Sukumar's *The Asian Elephant: Ecology and Management* is based mainly on a study he conducted on wild elephants in southern India (Sukumar, 1989). It does not concern itself with captive elephants and devotes just half a page of its 255 pages to captive breeding within range states.

Spinage's book *Elephants* (Spinage, 1994) focuses on wild individuals, largely ignoring studies of captive elephants, except for Benedict's study of an Asian circus elephant: *The Physiology of the Elephant* (Benedict, 1936). In Sukumar's *The Living Elephants: Evolutionary Ecology, Behavior, and Conservation*, the section devoted to the 'Management of elephants in captivity' consists of just five pages — out of a total of 478 — and largely ignores the studies of elephants conducted in zoos (Sukumar, 2003). More recently, the behaviour and ecology of African elephants have been described in detail by Moss et al. (2011), but this book is based on the lives of wild elephants living in Amboseli in Kenya and the work of the Amboseli Elephant Research Project that began in 1972.

Early books on captive Asian elephants predate works on wild elephants and began to appear at the end of the 19th century. An early source of information on Asian elephants was published in Madras by Steel (1885), *A Manual of the Diseases of the Elephant and His Management and Uses*. In 1910 Lt. Colonel G.H. Evans, Superintendent of the Civil Veterinary Department in Burma (Myanmar), published a similar book on elephant diseases (Evans, 1910). This was followed by a work entitled *The Care and Management of Elephants* (Ferrier, 1947). Lt. Colonel J.H. Williams's book *Elephant Bill* was published in 1951 and described his exploits as a soldier and elephant expert in Burma working for the Bombay Burma Trading Corporation extracting teak from the forests, and during the Burma Campaign of the Second World War (Williams, 1951). Although he was not a scientist, the work contains much useful anecdotal information about the lives, behaviour and health of working elephants at the time.

Captive elephants suffer from a range of foot problems, some of which may ultimately be fatal. Csuti et al. (2001) published *The Elephant's Foot: Prevention and Care of Foot Conditions in Captive Asian and African Elephants* following the first North American conference on elephant foot care and pathology that was held in 1998 at Oregon Zoo.

Kurt and Garaï (2007) have drawn together captive studies in *The Asian Elephant in Captivity*, largely focussing on working animals kept within the range states and at the

Pinnawala Elephant Orphanage in Sri Lanka. This book predates a great many of the recent studies conducted on elephants in zoos, and there is no equivalent for the African species for the obvious reason that so few individuals have been kept in captivity in Africa.

The major work by Fowler and Mikota (2006) entitled *Biology, Medicine, and Surgery of Elephants* focuses largely on veterinary science. The chapter on 'Behavior and Social Life' extends to just nine of the 565 pages. In 2008, Wemmer and Christen published *Elephants and Ethics: Towards a Morality of Coexistence*. Although the title promised an ethical — that is to say, a philosophical — consideration of the issues surrounding the relationship between people and elephants, unfortunately few of the authors are ethicists and consequently the quality of the chapters varies and many of the articles are narrative in nature rather than academic (Wemmer and Christen, 2008).

This book is concerned with elephants in human care — living in zoos, circuses, and logging and tourist camps — and focuses primarily on the work published in peer-reviewed journals concerned with their behaviour, ecology and welfare. Most of the academic, conservation and popular interest in elephants is in wild elephants, not elephants living in captivity. Nevertheless, elephants living under human care are legitimate subjects of study for a number of reasons. They are interesting subjects in their own right — as are other species kept in zoos — because their biology in captivity is different from the biology of their wild conspecifics; they may help us to understand wild elephants; they may be of conservation value; and they have particular welfare needs that can only be adequately addressed if we find out more about them.

Mankind's relationship with elephants is longstanding and complex. It does not depend on how we classify elephants from a zoological point of view, but before examining the biology of elephants living under human care, and the different types of relationships that people have with them, we should first consider the different 'types' of elephants and the people who have discovered and described them.

1.2 What is an elephant?

For most people an elephant is an elephant. The majority of the general public probably cannot tell Asian elephants from African elephants, and would consider the distinction between savannah and forest elephants of academic interest only. In many old films set in Africa, Asian elephants were used instead of the African species because they were more readily available, and it was no doubt assumed that the film-going public would be oblivious to the error. In the 1960s television series about a vet running a fictional animal behaviour study centre in East Africa, *Daktari*, an Asian elephant called *Modoc* wore false ears to make her look African (Carwardine, 1995).

1.2.1 Elephant taxonomy

Most of what we know about the origin and evolution of elephants is because of one American scientist: Henry Fairfield Osborn. After a lifetime of research devoted to elephant fossils at the American Museum of Natural History, Osborn published a two-volume monograph on the Proboscidea (Osborn, 1936, 1942). Both volumes were published posthumously. A detailed biographical memoir of Obsorn's life, career and publications was published shortly after his death (Gregory, 1937). Most of the species of proboscideans that have existed are now extinct and we have been left with just three species.

All elephants belong to the mammalian order Proboscidea and the family Elephantidae. The African elephant (*Loxodonta africana*) and the Asian elephant (*Elephas maximus*) have long been recognised as the only extant members of these taxa. Authorities disagree over the number of subspecies within each species, and there has not been a convincing scientific argument for the recognition of any other full species until very recently.

The description of the Asian elephant provided by Linnaeus (1758) in his *Systema Naturae* is reproduced in Fig. 1.1. Before considering the elephant specimen described by Linnaeus, I need first to clarify some terminology.

The holotype, or type specimen, of a species is the specimen upon which its original scientific description is based. A syntype is any one of two or more specimens of equal status upon which such a description is based. A lectotype is a specimen that is retrospectively designated as the holotype when a type specimen was not originally designated.

In 1753, Linnaeus persuaded King Adolf Frederick of Sweden to purchase an elephant foetus preserved in ethanol for his natural history collection, and designated it as one of the syntypes for the Asian elephant (*E. maximus*) along with a complete skeleton described by John Ray in 1693. Linnaeus only recognised one species of elephant: the one that originated in Asia. Cappellini et al. (2013) have since shown that the foetus purchased by the King of Sweden was that of an African elephant using evidence from morphology, analysis of ancient DNA and high-throughput ancient proteomic analyses. The same group examined the syntype described by Ray at the Natural History Museum of the University of Florence, and concluded that it was an Asian elephant, and designated this specimen as the lectotype (retrospectively designated holotype).

Johann Friedrich Blumenbach was a German physiologist, naturalist, anthropologist and comparative anatomist who was appointed as Professor of Medicine in the

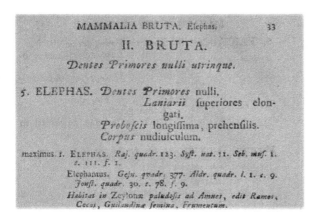

FIGURE 1.1 Part of the entry for the Asian elephant (*Elephas maximus*) in Linnaeus's *Systema Naturae*. Source: *From Linnaeus, C., 1758. Systema Naturae per Regna Tria Naturae, Secundum Classes, Ordines, Genera, Species, cum Characteribus, Differentiis, Synonymis, Locis, vol. 1, tenth ed. Impensis Direct, Holmiae, Laurentii Salvii.*

University of Göttingen in 1776. The description of the African elephant given by Blumenbach (1797) in his *Handbuch der Naturgeschichte (Handbook of Natural History)* is reproduced in Fig. 1.2.

The original range of the African species extended from the Mediterranean and Red Sea coasts to the Cape of Good Hope (apart from deserts). Today it is extinct north of 13°N. Zoologists have traditionally recognised the existence of two distinct subspecies of African elephant: the savannah or bush elephant (*L. a. africana*) and the forest elephant (*L. a. cyclotis*). Eisentraut and Böhme (1989) proposed a new species of African elephant, *L. pumilio* (reclassifying a third subspecies *L. a. pumilio*), based on morphological and behavioural differences between this form and the forest elephant (*L. a. cyclotis*). Spinage (1994) has pointed out that the pygmy elephant (*L. a. pumilio*) has been recognised in Gabonese law, since a lower fee was charged for hunting this form than for the forest elephant, *cyclotis*. However, Kingdon (2004) uses the term 'pygmy elephant' as a synonym for the forest elephant (*L. a. cyclotis*). He suggests that there may be as many as 25 subspecies of *Loxodonta*, based on subpopulations, which show consistent characteristics in size, ear shape, limb proportions, skull and tusk shape, number of nails, skin texture and colour.

After reviewing DNA-based studies of extant elephants in Africa, Roca et al. (2015) concluded that there has been little or no nuclear gene flow between African savannah elephants (*L. africana*) and African forest elephants (*L. cyclotis*), confirming that they comprise two separate species. There is a deep genetic separation between these two species, but Mondol et al. (2015) have identified hybrid zones in West and Central Africa. There is DNA evidence of ancient episodes of hybridisation between forest elephant females and savannah elephant males (Roca et al., 2004), the latter being reproductively dominant to either forest or hybrid males.

Although Grubb et al. (2000) concluded that living African elephants belong to two species — *L. africana* (Blumenbach, 1797) and *L. cyclotis* (Matschie, 1900) — two decades ago,

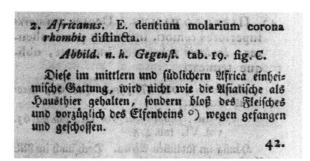

FIGURE 1.2 Part of the original description of the African elephant (*Loxodonta africana*) by Blumenbach (1797) when the species had been assigned the name *Elephas africanus*. Source: *From Blumenbach, J.F., 1797. Handbuch der Naturgeschichte, fifth ed. J.C. Dieterich, Göttingen. Translation courtesy Dr Robert Jehle, University of Salford: 'This genus ('Gattung' is the word for genus, although species is probably referred to) is at home in central and southern Africa, and is not kept as a pet like the Asian (genus/species), but is captured and shot because of its meat and primarily because of ivory.'*

the International Union for the Conservation of Nature and Natural Resources (IUCN) recognises the single species, *L. africana*, in Africa.

Relatively little is known about the extent to which *cyclotis* have been kept in zoos. Schürer (2017) attempted to survey African forest elephants kept in zoos since 1882, but found sources of information to be of mixed quality and reliability. The main sources of these elephants were the elephant training camps in what are now Gabon and the Democratic Republic of the Congo. Some exported animals were determined to be hybrids between forest and savannah elephants. Most forest elephants kept in zoos were short-lived — unlike those kept at training centres — and they were generally kept alone or together with African savannah or Asian elephants and so did not breed. Schürer found no evidence for the existence of pygmy elephants as a taxonomic unit

The taxonomic status of Asian elephants is also complex. Up to nine subspecies have been suggested in the past (Spinage, 1994), but the IUCN recognises only three subspecies based on geographical distribution and following the work of Shoshani and Eisenberg (1982):

E. m. indicus on the Asian mainland,

E. m. maximus in Sri Lanka, and

E. m. sumatranus in Sumatra.

Some authorities (e.g. Barnes, 1995), however, have recognised a separate Malaysian subspecies (*E. m. hirsutus*).

Fleischer et al. (2001) reconstructed the phylogenies of Asian elephants from across much of their range by analysing mitochondrial DNA to study their phylogeography. Their analysis of 57 animals from seven countries (Sri Lanka, India, Nepal, Myanmar, Thailand, Malaysia, Indonesia) revealed two major clades (A and B). All locations contained individuals from both clades except Indonesia and Malaysia, supporting the status of these two populations as evolutionarily significant units (ESUs). Human trade in elephants among India, Myanmar and Sri Lanka appeared to have had an impact on the distribution of clade A individuals. Consequently, the status of Sri Lankan elephants as a separate subspecies and an ESU were not supported by the genetic evidence.

DNA analysis of elephants from Borneo revealed that they are genetically distinct from populations elsewhere in Asia, with molecular divergence indicating colonisation of Borneo during the Pleistocene, and subsequent genetic isolation (Fernando et al., 2003). It was previously thought that this was a feral population descended from imported elephants, and thus of low-conservation value. In the absence of any deleterious effects from inbreeding, Fernando et al. called for elephants from Borneo to be managed separately from other Asian elephants. Further discussion of genetic analyses of wild Asian elephant populations can be found in Choudhury et al. (2008).

It should be noted that the recognition of ESUs below the level of species can cause problems for captive breeding programmes because some individuals — hybrids between individuals from existing or newly recognised ESUs, and individuals of unknown origin — become unsuitable for breeding. Such difficulties have previously been identified in, for example tigers (*Panthera tigris*; Luo et al., 2008) and chimpanzees (*Pan troglodytes*; Hvilsom et al., 2013) living in zoos.

1.3 Conservation status

Much of the discussion around the importance of keeping elephants in zoos centres on the conservation status of wild populations in Africa and Asia. The most recent IUCN global assessments of the status of African and Asian elephants produced for IUCN's Red List were published in 2008 (Blanc, 2008; Choudhury et al., 2008) and need updating. More recent population estimates have been made in 2017 for the Asian elephant (AsERSM, 2017) and 2016 for the African elephant (Thouless et al., 2016).

1.3.1 The status of Asian elephants in the wild

The Asian elephant is classified as 'endangered' by the IUCN. It has been placed in category endangered A2c:

...considered to be facing a very high risk of extinction in the wild [due to]:

A. Reduction in population size based on...

...2. An observed, estimated, inferred or suspected population size reduction of $\geq$ 50% over the last 10 years or three generations, whichever is the longer, where the reduction or its causes may not have ceased OR may not be understood OR may not be reversible, based on...

...(c) a decline in area of occupancy, extent of occurrence and/or quality of habitat.

Under the Convention on International Trade in Endangered Species of Wild Fauna and Flora 1973 (CITES) all populations of the Asian elephant are classified as Appendix I species, thereby receiving the highest level of international protection from the effects of international commercial trade.

The Asian elephant currently occupies fragments of its former range in 13 countries: Bangladesh, Bhutan, Burma, Cambodia, China, India, Indonesia, Laos, Malaysia, Nepal, Sri Lanka, Thailand and Vietnam (Santiapillai and Jackson, 1990). About a quarter of a century ago, it was estimated by the IUCN that there were between 38,000 and 51,000 Asian elephants in the wild (Kemf and Jackson, 1995) along with approximately 13,000 trained domestic elephants, which rarely bred. The population of Asian elephants in 2017 was estimated to be 47,602–50,324 wild individuals and 14,022–14,222 captive elephants within 12 range states (i.e. excluding Nepal where there were around 100 animals) (AsERSM, 2017).

The IUCN Species Survival Commission Asian Elephant Specialist Group (AESG) was formed in 1976 and in 1989 produced an action plan for the conservation of the species (Santiapillai and Jackson, 1990). In 1991–92 the Indian Government launched Project Elephant to provide elephant range states with financial, technical and scientific assistance to ensure the long-term survival of viable populations in their natural habitats. This stimulated a considerable amount of research on wild Asian elephants (Daniel and Datye, 1995).

1.3.2 The status of African elephants in the wild

The IUCN assessment treats African elephants as a single species and does not consider subspecies. While it recognises the genetic evidence for at least two species of elephant in

Africa — and possibly a third (the West African elephant) — the IUCN Species Survival Commission African Elephant Specialist Group has been reluctant to re-classify *L. africana*, believing that premature division into one or more species may leave hybrids with an uncertain conservation status (Blanc, 2008).

The African elephant is classified by the IUCN as vulnerable A2a:

...considered to be facing a high risk of extinction in the wild [due to]:

A. Reduction in population size based on any of the following:

...2. An observed, estimated, inferred or suspected population size reduction of ≥ 30% over the last 10 years or three generations, whichever is the longer, where the reduction or its causes may not have ceased OR may not be understood OR may not be reversible, based on ...

(a) direct observation.

Under CITES, African elephants are listed in Appendix I, except the populations of Botswana, Namibia, South Africa and Zimbabwe, which are included in Appendix II (see Table 9.4). From these populations some trade is allowed in government-owned stocks of ivory, hair, hides, leather goods and trophies as well as live animals (to 'appropriate and acceptable destinations') (see Annotation 2 to the Appendices). The term 'appropriate and acceptable destinations' was originally defined in Resolution Conf. 11.20 (Rev. CoP17) but was updated in Resolution 11.20 (Rev. CoP18) in the summer of 2019. This is discussed further in Box 9.7.

The Great Elephant Census (GEC) — a standardised survey across Africa — reported an estimated 352,271 savannah elephants on study sites in 18 countries, representing some 93% of all individuals of this species in those countries (Chase et al., 2016). In areas for which historical data were available, numbers decreased by approximately 144,000 from 2007 to 2014, and across the continent as a whole the GEC determined that populations were shrinking by 8% per year, primarily as a result of poaching.

The IUCN African Elephant Status Report 2016 reported the results of a survey of 37 countries and estimated a population of 415,428 ± 20,111 at the time of the last survey for each area, and an additional 117,127—135,384 elephants in areas that had not been systematically surveyed (Thouless et al., 2016).

1.4 The human use of elephants

Elephants have been used for ceremonial and religious purposes, in 'sport', as a means of transportation for humans, as pack animals, weapons of war, in forestry operations, as tourist attractions, circus entertainers, royal gifts, ambassadors for conservation and treated as curiosities by inquisitive scientists. For the Romans, elephants were important as symbols of power (Mader, 2006).

1.4.1 Elephants in ancient times

Hieroglyphs exist from ancient Egypt that distinguished between wild and trained elephants as early as 3000 BCE (Wylie, 2008). Although elephants have been kept in captivity for some 5000 years, they have never been truly domesticated. Elephants were present in the Nile valley in ancient times, where they were hunted and distributed to Egypt and the Classical Greco-Roman world for amusement, military purposes and ivory (Lobban and de Liedekerke, 2000). In 879 BCE, the King of Assyria, Assurnasirpal II, collected herds of elephants and kept them in a 'zoo'. A stele (stone tablet) from the early 9th century BCE records one of his largest hauls, which included 30 elephants, 450 tigers (*P. tigris*) and 200 ostriches (*Struthio camelus*). Large-scale organised trade in ivory is not a recent phenomenon and is thought to have begun with the Phoenicians (Delort, 1992).

1.4.2 Have elephants been domesticated?

Domestication is the process whereby a species is tamed and then modified over many generations by selective breeding to perform some useful purpose for humans. This is not what has happened with 'domesticated' elephants.

Baker and Manwell (1983) noted that the most usual pattern of elephant exploitation is one of continuous recruitment from wild populations. This effectively means that the process of domestication begins afresh each time an animal is taken from the wild. They proposed the term 'incipient domestication' for the status of elephants and suggested the process was arrested perhaps more than 3000 years ago.

Elephant domestication has a long history in Southeast Asia. In Nepal, the first king of the Shah Dynasty, Prithivi Narayan Shah, provided the East India Company with seven adult elephants each year from 1743 to 1775 for invading Parsa-Mahotari through the Makawanpur Battle. There were once 31 elephant camps throughout the lowlands of Nepal. The capture and training of wild elephants was commonplace, but between 1954 and 1970 only 10 elephants were captured for domestication. In 1903 Nepal's domesticated elephant population numbered around 330, but by 1973 there were fewer than 50. The captive population had risen again to 208 by 2011 (Pradhan et al., 2011). In 1978 the Department of National Parks and Wildlife Conservation was given management responsibility for domesticated elephants and in 1986 an elephant breeding centre was established in the Chitwan National Park to provide elephants for patrolling and undertaking management work in the national parks. The 'domestication' of African elephants is discussed in Section 10.11.

1.4.3 Ceremonial and religious use of elephants; elephants as gifts

There is a long history of large, fierce, exotic animals being given as gifts by one ruler to another. Elephants, of course, were particularly prized. Charles (2007) has suggested a close association between the use of elephants by the Sassanian Persians (CE 224–651) and Persian notions of kingship.

In 1255 an elephant was given to Henry III of England by Louis IX of France and put on display in the menagerie at the Tower of London (Fig. 1.3). This practice of using elephants as a symbol of friendship has continued into modern times. In 1982, Indian Prime Minister Indira Gandhi presented a baby Asian elephant to Honolulu Zoo as a gift from

FIGURE 1.3 A sculpture of an African elephant in the grounds of the Tower of London representing the elephant given to King Henry III of England by King Louis IX of France in 1255.

the children of India to the children of Hawai'i, and in 1984, US President Ronald Reagan was given an elephant by President Jayewardene of Sri Lanka.

The ceremonial use and religious significance of elephants also has a long history (Fig. 1.4). Searle (2018) describes a report by a 19th-century British writer of the Emperor of Peking displaying 60 extravagantly clothed and decorated elephants to visitors in 1598. Presumably the elephants were kept for ceremonial purposes.

The Asian elephant is important in Hinduism as the embodiment of the elephant god *Ganesh* (Fig. 1.5). Elephants are kept in temples across Southeast Asia for use in religious ceremonies. They have been the subject of a number of studies aimed at improving their welfare. The daily routines of temple elephants in Tamil Nadu, India, have been studied by Vanitha et al. (2010) and recommendations for improvements in their management have been made by Gokula and Varadharajan (1996). Factors affecting thermoregulation in temple elephants have been studied by Vanitha and Baskaran (2010).

1.4.4 The use of elephants for transportation and as weapons of war

Many ancient rulers utilised elephants to support their armies in battle. Accounts of their exploits have been provided by a number of authors, for example Spinage (1994), Kistler (2007) — who provided a map of the locations of important battles involving elephants — and Roy (2013). An account of the use of elephants by the Sassanian Persians against the armies of the Later Roman Empire has been provided by Charles (2007).

The strategic use of elephants as weapons of war in ancient times has been likened to the use of tanks in modern warfare (Searle, 2018). Some of the battles in which elephants were used are listed in Table 1.1.

1.4 The human use of elephants

FIGURE 1.4 Government of India elephant in state costume, south of India, c.1890. Source: *Courtesy US Library of Congress. Reproduction No.: LC-DIG-ppmsca-41472 (digital file from original item)*.

FIGURE 1.5 Wooden carvings representing the Hindu god *Ganesh*.

TABLE 1.1 Examples of the use of elephants in war.

Date	Combatants[a]	Battle/location	Number of elephants
331 BCE	Darius Codomannus against Alexander the Great	Arbela	15
326 BCE	Porus against Alexander the Great	Near Chillianwalla	~200
301 BCE	Seleucus (one of Alexander the Great's generals) against Demetrius	Ipsus	~400
280 BCE	Antiochus I against the Gauls	—	—
273 BCE	Antiochus I against the Galatians	The 'elephant victory'	16
202 BCE	Hannibal against the Romans	Zama	—

[a]Note first named combatant used elephants.
Source: *Based on information in Searle, D.A., 2018. War elephants and early tanks: a transepochal comparison of ancient and modern warfare. Militärgeschichtliche Zeitschrift, 37, 77. https://doi.org/10.1515/mgzs-2018-0002.*

FIGURE 1.6 Print of an engraving showing Scipio Africanus on horseback with Roman soldiers engaging Hannibal, riding a war elephant, during the battle of Zama. Source: *Courtesy US Library of Congress Reproduction Number: LC-DIG-pga-00039 (digital file from original print).*

In 331 BCE Darius Codomannus used 15 elephants against Alexander the Great at Arbela but, in some battles, many hundreds of these animals were used. Some 400 elephants were used by Seleucus (one of Alexander the Great's generals) against Demetrius in 301 BCE at Ipsus, and Alexander's expansion of his empire east of the Punjab is thought to have been deterred by the existence of an army of 8000 elephants possessed by Kaiser Nanda in ancient India.

The use of war elephants in Assam, India, is known from as early as 1000 BCE. They were particularly useful in mountainous regions for transporting supplies instead of using horses, which were believed to be difficult to breed in the prevailing climate, whereas elephants were available in abundance. Little is known of the use of war elephants in ancient and medieval China, but they appear to have been used prior to 1000 BCE (Searle, 2018).

Without a doubt, the best known expedition involving war elephants was that led by Hannibal across the Alps in which he took 37 African elephants into Italy during the Second Punic War between Rome and Carthage (Edwards, 2001) (Fig. 1.6). In an ambitious exercise in experimental archaeology, John Hoyte, a British engineer, successfully led a small party — including Richard Jolly and a veterinary surgeon, Colonel John Hickman, who had worked with elephants in Burma in Second World War — across the Alps from France to Italy in 1959, accompanied by a female Asian elephant called *Jumbo* loaned to him by Turin Zoo (Hoyte, 1960). The British Alpine Hannibal Expedition, as it was called, walked from Montmelian in France, via Col du Mont Cenis, to Susa in Italy in 10 days, thus proving that an elephant — albeit a former circus elephant of the Asian species — was physically capable of traversing the Alps.

Elephants were as used as transportation by the British Army in the Indian subcontinent during the 19th century, often to move heavy artillery pieces (Fig. 1.7). They

FIGURE 1.7 A British Army elephant battery of heavy artillery travelling along the Khyber Pass at Campbellpur (now Attock in Pakistan) in 1895. Source: *Courtesy US Library of Congress. Reproduction Number LC-USZ62-78857 (b&w film copy neg.).*

continued to be used by the British during the Second World War, and a compelling account of their role has been provided by Lt. Colonel J.H. Williams in *Elephant Bill* (Williams, 1951).

1.4.5 Elephants and forestry operations

Around the end of the 19th century there were some 100,000 working elephants in the forests of Thailand. By the end of the 20th century their numbers had fallen to around 4000 (Gröning and Saller, 1999). The decline of logging in Southeast Asia has led to a surplus of working elephants which, in some areas, have no work to do. In places accessible to tourists, some of these animals are being used to carry tourists on short walks or longer safaris into the local jungle. Other elephants are being kept in 'sanctuaries', some of which charge volunteers to work with, and care for, the elephants.

In 1999 the Food and Agriculture Organisation of the United Nations (FAO) published a case study on the use of elephants in logging operations with the objective of assisting developing countries in promoting user- and environment-friendly forest harvesting systems, techniques and methods (FAO, 1999). It considers the training of elephants, the advantages and disadvantages of using them in logging operations and their clinical biology.

1.4.6 Elephants as entertainers: circuses, sports and tourism

Elephants have been used in entertainment for over 2000 years and some countries still allow the use of elephants in circuses and in 'sport'. Elephant football matches take place in Thailand and in Nepal and elephant polo is played in Thailand, Nepal, India and Sri Lanka. The World Elephant Polo Association holds its world tournament in the Royal Chitwan National Park in Nepal each year. The association is based in Tiger Tops Jungle Lodge in the park and established the governing rules for elephant polo in 1982. In several countries, elephants are used to carry tourists on short treks or longer safaris looking for wildlife.

Elephants have been used in circuses in many parts of the world as performers and as beasts of burden, carrying equipment and helping to erect circus tents, especially in North America (Fig. 1.8). The lives of these animals, and the men who worked with them in the heyday of the American circus, have been chronicled by Lewis and Fish (1978) in their book *I Loved Rogues: The Life of an Elephant Tramp*.

Some of the elephants used in entertainment and tourism have been the subject of academic studies on subjects such as the relationship been mahouts and their charges (e.g. Hart and Sundar, 2000), the injuries caused by elephant saddles (Magda et al., 2015), elephant welfare (Kontogeorgopoulos, 2009), social behaviour (Poole et al., 1997) and abnormal behaviour (Friend and Parker, 1999) (Fig. 1.9). Accounts of these studies are distributed across many of the chapters that follow.

1.4.7 Elephants as ambassadors for conservation

Elephants are widely considered to be flagship species by conservationists (Barua et al., 2010; Blake and Hedges, 2004). A flagship species may be defined as:

A charismatic species which is popular with the public and serves as a symbol and focus for raising awareness about conservation issues and stimulates action.

Rees (2013)

There is no doubt that elephants are popular with the public (Carr, 2016) and they are widely acknowledged as a keystone species by ecologists by virtue of the disproportionate role they play in determining community structure, for example by facilitating browsing by mesoherbivores by generating 'browsing lawns' (Makhabu et al., 2006).

FIGURE 1.8 Circus elephants and performers from the Ringling Brothers Barnum and Bailey Combined Shows in front of the US Capitol in 1995. Source: *Courtesy US Library of Congress: Reproduction Number LC-DIG-ppmsca-38849.1.3.*

FIGURE 1.9 An elephant tourist camp at Habarana, Sri Lanka (2003).

Elephants under human care, especially those living in zoos, may have a role to play in attracting support for conservation in general, and elephant conservation in particular. This potential role is explored in Chapter 10, The conservation value of captive elephants.

1.5 The beginning of elephant research

1.5.1 Aristotle and elephants

Studies of elephants kept in captivity have a long history. In ancient Greece and the Roman Empire there was a keen interest in animals. Many species, including elephants, were kept by Roman emperors in private collections for study and for use in the Roman games, where they perished in large numbers in amphitheatres throughout the Empire (Hughes, 2003). Thousands of wild animals were slaughtered when the Colosseum was inaugurated in CE 80. These animals were supplied by the Roman military and acquired from the rulers of the countries where they were endemic (Epplett, 2013). Keeping these elephants in captivity provided an opportunity for them to be studied.

The Greek philosopher Aristotle wrote his *History of Animals* in 350 BCE, almost two and a half millennia ago. Alexander the Great captured many exotic animals on his military expeditions and sent them back to Greece where they were kept in menageries. Most of the Greek city states had such menageries and would have given Aristotle the opportunity to study animal species that he would not have been able to observe in the wild. *The History of Animals* contains descriptions of various aspects of elephant biology, many of which can only relate to captive animals (Cresswell, 1883). Aristotle refers to the anatomy and physiology of elephants, their behaviour, diseases, capture and training (Box 1.1).

BOX 1.1 Observations on elephants made by Aristotle in his *History of Animals* (350 BCE)

On the trunk, toes, and legs...

In the elephant, the nostril is very large and strong, and it answers to the purpose of a hand, for the animal can extend it, and with it take its food, and convey it to its mouth, whether the food is moist or dry. This is the only animal that can do so...

...and all that have toes have nails, and those that have feet have toes, except the elephant, which has its toes undivided, and scarcely distinguished, and no nails at all...

...[the elephant] has the toes less perfectly jointed, and its fore-legs much larger than the hind one; it has five toes, and short ankles to its hind legs. It has a trunk of such a nature and length as to be able to use it for a hand, and it drinks and eats by stretching this into its mouth; this also it lifts up to its driver, and pulls up trees with it; with this organ it breathes as it walks through the water. The extremity of the proboscis is curved, but without joints, for it is cartilaginous.

On the limbs and lying down...

All animals, except the elephant, bend both their fore and hind legs in contrary directions, and also contrary to the way in which man's limbs are bent. For in viviparous quadruped, except the elephant, the joints of the forelegs are bent forwards, and those of the hind-legs backwards, and they have the hollow part of their circumference opposite each other: the elephant is not constructed as some have said, but is able to sit down, and bend his legs, but, from this great weight, is unable to bend them on both sides at once, but leans either to the right side or the left, and sleeps in this position, but its hind legs are bent like a man's.

On body hair...

...the elephant is the least hairy of all quadrupeds.

On mammae...

...the elephant has two mammae near the armpits; in the female they are small, and do not bear any proportion to the size of the animal, so that they are scarcely visible in a side view; the males also have mammae as well as the females, but they are exceedingly small.

On the teeth, tusks and tongue...

The elephant has four teeth on each side, with which he grinds food, for he reduces his food very small, like meal. Besides these, he has two tusks: in the male these are large, and turned upwards; in the female they are small, and bent in the contrary direction. The elephant has teeth as soon as it is born; but the tusks are small, and therefore inconspicuous at first. It has so small a tongue within it mouth, that it is difficult to see it.

On the gut and liver...

There are enlargements in the bowels of the elephant, which give it the appearance of having four stomachs; in these the food is detained, and apart from these there is no receptacle for the food. Its intestines are very like those of the hog, except that the liver is four times greater than that of the ox, and other parts also; the spleen is small in proportion to its size...

The elephant...has a liver without a gall, but when the part where the gall is attached in other animals, is cut open, a quantity of fluid like bile, more or less abundant, runs out.

On vocalisations...

The elephant utters a voice by breathing through its mouth, making no use of its nose, as when a man breathes forth a sigh; but with its nose it makes a noise like the hoarse sound of a trumpet.

On reproduction...

The elephant arrives at puberty, the earliest at ten years of age, the latest at fifteen, and the male at five or six years old. The season for the intercourse of the sexes is in the spring: and the male is ready again at the end of three years, but he never touches again a female whom he has once impregnated. Her period of gestation is two years, and then she produces one calf, for the elephant belongs to the class of animals which have but one young one at any time. The young one is as large as a calf of two or three months old...

The elephant has a penis like a horse, but small and less in proportion to the size of its body; its testicles are not external but internal, and near the kidneys, wherefore also the work of copulation is quickly performed. The female has the pudendum in the same position as the udder of the sheep, and when excited with desire, it is lifted up outwards, so as to be ready for copulation with the male; and the orifice of the pudendum is very wide...

Elephants begin to copulate at twenty years old. When the female is impregnated, her period of gestation, some persons say, is a year and a half; other people make it three years. The difficulty of seeing their copulation causes this difference of opinion respecting the period of gestation. The female produces her young bending upon her haunches. Her pain is evident. The calf, when it is born, sucks with its mouth, and not with its proboscis. It can walk and see as soon as it is born...

Elephants...retire into desert places for intercourse, especially by the sides of rivers which they usually frequent. The female bends down and divides her legs, and the male mounts upon her.

On the breeding season and 'madness'...

Elephants also become wild at this period [the breeding season]. Wherefore they say that in India those who have the acre of them do not permit them to have sexual intercourse with the females; for they become mad at such season and overturn the houses, which are badly built, and do many other violent acts. They say also that abundance of food will render them more gentle. They also bring others among them which are directed to beat them, and so they punish them and reduce them to a state of discipline.

On longevity...

Some say that the elephant lives two hundred, and other three hundred years...

...Some persons say that the elephant will live for two hundred years, others an hundred and twenty, and the female lives nearly as long as the male. They arrive at perfection when sixty years old.

On food and drink, feeding and drinking...

The elephant can eat more than nine Macedonian medimni at one meal, but so much food at once is dangerous; it should not have altogether more than six or seven medimni, or five medimni of bread, and five mares of wine, the maris measures six cotylae. An elephant has been known to drink as much as fourteen Macedonian measures at once, and eight more again in the evening.

On flatulence...

The elephant does not appear to suffer from any other infirmity except flatulency...Elephants suffer from flatulent diseases, for which reason they can neither evacuate their fluid or solid excrements. If they eat earth they become weak, unless used to such food. If it is accustomed to it, it does no harm. Sometimes the elephant swallows stones. It also suffers from diarrhoea. When attacked with this complaint, they are cured by giving them warm water to drink, and hay dipped in honey to eat; and either of these remedies will stop the disease. When fatigued for want of sleep, they are cured by being rubbed on the shoulders with salt and oil, and warm water. When they suffer from pain in the shoulders, they are relieved by the application of roasted swine's flesh. Some elephants will drink oil, and some will not; and if any iron weapon is struck into their body, the oil which they drink assists in its expulsion; and to those which will not drink it, they give wine of rice cooked with oil.

On adaptation and ecology...

They bear winter and cold weather badly. It is an animal that lives in the neighbourhood of rivers, though not in them. It can also walk through rivers, and will advance as long as it can keep its proboscis above the surface; for it blows and breathes through this organ, but it cannot swim on account of the weight of its body.

On elephant fighting, courage and war elephants...

Elephants...fight fiercely with each other, and strike with their tusks; the conquered submits entirely, and cannot endure the voice of the victor: and elephants differ much in the courage they exhibit. The Indians use both male and female elephants in war, though the females are smaller and far less courageous. The elephant can overthrow walls by striking them with its large tusks; it throws down palm trees by striking them with its head, and afterwards putting its feet upon them, stretches them on the ground.

On elephant hunting...

Elephant-hunting is conducted in the following way: men mount upon some tame courageous animals; when they have seized upon the wild animals they command the others to beat them till they fail from fatigue. The elephant-driver then leaps upon its back and directs it with a lance; very soon after this they become tame and obedient. When the elephant-drivers mount upon them they all become obedient, but when they have no driver, some are tame and others not so, and they bind the fore legs of those that are wild with chains, in order to keep them quiet. They hunt both full-grown animals and young ones. Such is the friendship and enmity of these wild animals originating in the supply of food, and the mode of life.

On taming...

Of all wild animals the elephant is the most tame and gentle; for any of them are capable of instruction and intelligence, and they have been taught to worship the king. It is a very sensitive creature, and abounding in intellect.

From the translation by Cresswell, R., 1883. Aristotle's History of Animals in Ten Books. George Bell & Sons, Covent Garden, London. Some of the material quoted is clearly based on information from other sources.

1.5.2 Anatomical research

Captive elephants have been an important source of research material for scientists since the animals began living — and dying — in logging camps, menageries and zoos.

The first dissection of an elephant performed in the British Isles was undertaken by an Irish physician, Allen Mullen, in 1681 (Isaac, 2017). It was a bull Asian elephant who arrived in London in 1675 — probably via one of the ships belonging to the East India Company — and was exhibited at Garraway's Coffee House in the City. The entrance fee was three shillings. In 1681 the elephant was taken to Dublin by a showman called Mr Wilkins, where he was exhibited in a wooden booth near the Custom House on Essex Street. On 17 June, 1681, the booth caught fire and the elephant perished. Allen Mullen oversaw the dissection of the elephant and sent his observations to the Royal Society (Mullen, 1682). The animal's skeleton was put on display some three months after his death (Fig. 1.10).

On 19 June 1847, a drawing of a dead Asian elephant appeared on the front page of *The Illustrated London News* (Fig. 1.11). He was *Jack* and had been living at the Gardens of the Zoological Society in Regent's Park. According to the article that accompanied the picture:

...His remains were hastily dissected under the direction of Professor Owen, who has, we are most glad to learn, perfectly recovered from a slight wound received during the operation. Various portions of the vast frame have been distributed to the Royal College of Surgeons, the Anatomical School of Oxford; King's College, London; the Royal Veterinary College, &c.; there to undergo an examination, which will give important additions to our knowledge of the anatomy of this mightiest of those that walk the earth.

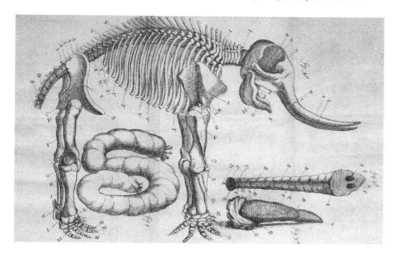

FIGURE 1.10 Anatomical drawings of the bull Asian elephant dissected by Allen Mullen after dying in a fire in 1682. Source: *From Isaac, S. 2017. Treasures from the collections. An anatomical account of the elephant accidentally burnt in Dublin, on Fryday, June 17 in the year 1681 — Allen Mullen (1682). Bulletin of The Royal College of Surgeons of England 99 (3), 127 https://publishing.rcseng.ac.uk/doi/pdfplus/10.1308/rcsbull.2017.127. https://doi.org/10.1308/rcsbull.2017.127.*

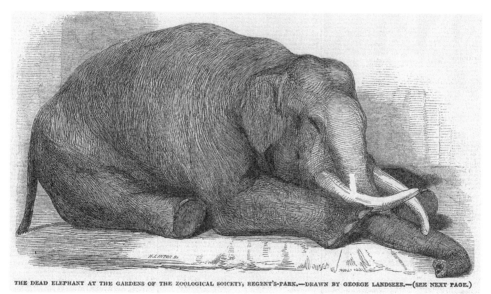

FIGURE 1.11 The death of *Jack*, the Asian elephant at London Zoo, drawn by George Landseer. Source: *From the front page of* The Illustrated London News, *for the week ending Saturday, June 19, 1847.*

At this time, Richard Owen was Hunterian Professor at the Royal College of Surgeons, but he became the superintendent of the natural history department of the British Museum in 1856.

In 1871 Dr Morrison Watson, Demonstrator of Anatomy at the University of Edinburgh, published a paper entitled *Contribution to the anatomy of the Indian elephant. Part I. The thoracic viscera*. The viscera had been part of the collection of Prof. John Goodsir, Professor of Anatomy at Edinburgh, who had purchased the elephant from a travelling menagerie in

FIGURE 1.12 Skeleton of one of a pair of Asian elephants on display at the Old Medical School, University of Edinburgh.

1856 (Watson, 1871). The University purchased the material after his death. To this day two elephant skeletons stand guarding a doorway in the Old Medical School at the University (Fig. 1.12).

Many of the elephant specimens found in museums around the world were taken from the wild in Africa and Asia when big game hunting was a popular and ethically acceptable pastime and wild animals were viewed by hunters as an inexhaustible resource. During his 1909 expedition from Mombasa to Khartoum, President Theodore Roosevelt and his son Kermit — accompanied by the English hunter Frederick Courteney Selous — killed over 500 large animals destined for the American Museum of Natural History, including 11 elephants (Roosevelt, 1910) (Fig. 1.13). The President carried with him a double elephant rifle presented to him by over 50 of his English friends.

Captive elephants from zoos and circuses have been important sources of anatomical specimens for museums. The preserved body of a male Indian elephant called *Don Pedro* stood as the centrepiece of the Upper Horseshoe Gallery of the Liverpool Museum until the night of 3 May 1941, when the building was destroyed by fire during a bombing raid by the German air force. He had formerly been owned by the Barnum and Bailey Circus which toured Europe between 1897 and 1898. When the circus arrived in Liverpool it was

FIGURE 1.13 President Theodore Roosevelt with an elephant he shot at Meru, Kenya, for the American Museum of Natural History. Source: *Courtesy US Library of Congress. LC-DIG-ppmsca-36545 (digital file from original item) LC-USZ62-998 (b&w film copy neg.).*

decided that *Don Pedro* had become too aggressive to continue with the show. He was donated to the museum after being euthanised in the presence of the museum's director Prof. Henry Ogg Forbes. James William Cutmore led a team of taxidermists who prepared the body for display after it was delivered to the museum on a wooden cart pulled by a traction engine (Anon., 2018a).

The African elephant *Jumbo* spent time in zoos (in Paris and London) and Barnum and Bailey's circus in the United States before he was hit by a train and killed while touring in Canada. Even after the elephant's death, Barnum continued to make money out of *Jumbo* by displaying his skeleton (Box 1.2).

BOX 1.2 *Jumbo*: the most famous elephant that has ever lived

Jumbo the African elephant is undoubtedly the most famous zoo animal of all time. He was born in eastern Sudan, around Christmas 1860. In 1862 he was captured by Taher Sheriff and walked to Kassala, where he was delivered to Johan Schmidt, and then walked across the Sahara by Casanova. *Jumbo* was first sold to the Jardins des Plantes in Paris. Then, in 1865 he was traded with the Zoological Society of London for £450 (£30,000 today) and an Indian rhinoceros (*Rhinoceros unicornis*), two dingoes (*Canis familiaris*), a black-backed jackal (*Canis mesomelas*), a possum (Didelphidae), a kangaroo (Macropodidae) and a pair of wedge-tailed eagles (*Aquila audax*).

In 1882 *Jumbo* was sold by London Zoo to Barnum & Bailey Circus in the United States for £2,000 (£138,000 today). The British public protested and pleaded with the zoo to keep *Jumbo*. In March that year Matthew Berkley-Hill brought a lawsuit in the Court of Chancery against the Zoological Society, its Council and Barnum, claiming that the society had no right to sell any of its animals without the consent of its fellows. The suit failed and the sale of *Jumbo* went ahead.

Jumbo left for America after 16 years and nine months at London Zoo. On 15 September 1885, he was hit by a train and killed while crossing a railway line in St. Thomas, Ontario, Canada. Barnum continued to make money out of the elephant by touring America with his stuffed body. On 4 April 1889, *Jumbo*'s mounted skin was delivered to Barnum's museum at Tufts College where it remained until 14 April 1975, when it was destroyed by fire (Chambers, 2007) (Fig. 1.14).

FIGURE 1.14 Poster advertising the display of *Jumbo*'s skeleton by P. T. Barnum & Co.'s Greatest Show on Earth and The Great London Circus combined with Sanger's Royal British Menagerie and Grand International Shows in 1888. Source: *Courtesy US Library of Congress. Reproduction Number: LC-DIG-ppmsca-32620 (digital file from original print).*

Jumbo's skeleton is now held by the American Museum of Natural History in New York. There is no doubt that *Jumbo* was in considerable pain for much of his life. Documentary evidence suggested that he was prone to aggressive behaviour, especially at night, most of which was exhibited as attacks on his 'den'. *Jumbo* broke off both his tusks and examination of his skull showed that his teeth were soft and impacted (because new teeth erupted before old teeth had been worn away), probably due to a poor diet. Some of the damage to his skeleton was consistent with having carried heavy weights (zoo visitors) on his back. Other damage to the long bones was indicative of a much older animal, even though *Jumbo* died when he was 24 years old. Towards the end of his life, he appeared to have been suffering from some kind of wasting disease. When his dead body was examined, his stomach was found to contain several hundred coins (probably given to him to pass to his keeper in payment for buns) along with other metal trinkets. An account of *Jumbo*'s fate in North America can be found in McClellan (2012).

The Manchester Museum in the United Kingdom owns the mounted skeleton of *Maharajah*, a male Asian elephant who was purchased by Belle Vue Zoo in Manchester from Wombwell's Royal Number One Menagerie when it closed down in 1872 (Figs. 1.15 and 1.16A). After he destroyed a railway wagon, the elephant was walked from Edinburgh to Manchester by his keeper, Lorenzo Lawrence (Fig. 1.16B), although this may have been a publicity stunt on the part of Lawrence who intended to secure his future employment with the zoo (Barnaby, 1988).

1.5 The beginning of elephant research

FIGURE 1.15 The visit of Bostock and Wombwell's Royal Number 1 Menagerie to Windsor Castle in 1847 from the cover of the menagerie's catalogue of animals on display. Source: *Reproduced with permission from Wiley-Blackwell. First published in Rees, P.A., 2013. Dictionary of Zoo Biology and Animal Management. Wiley-Blackwell, Chichester, West Sussex.*

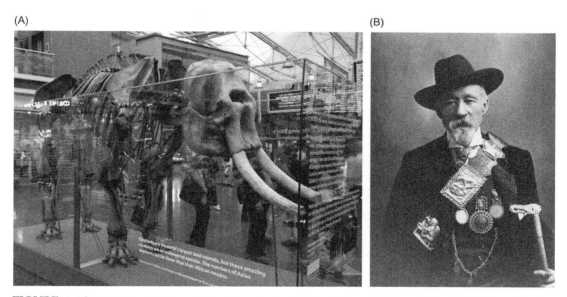

FIGURE 1.16 (A) The skeleton of the Asian elephant *Maharajah* on display at Piccadilly Railway Station, Manchester, United Kingdom. The skeleton is normally on display at Manchester Museum. (B) Lorenzo Lawrence, *Maharajah*'s keeper. Source: *(A) Courtesy Dr Simon Hutchinson, University of Salford. (B) Courtesy Chetham's Library, Manchester, United Kingdom.*

1.5.3 Anecdotes as a source of knowledge about elephants

The writings of hunters and colonial era elephant managers in Southeast Asia are an interesting source of knowledge about elephant biology (e.g. Williams, 1951). Scientists sometimes claim to discover phenomena that are well known to zookeepers, professional hunters and others who have years of experience working with animals. For example Poole and Moss (1981) claimed to have provided the first description of *musth* in African elephants.

> *The phenomenon of musth in male Asian elephants,* Elephas maximus, *has long been recognized... The most obvious manifestations are a sharp rise in aggressive behaviour, copious secretions from and enlargement of the temporal glands, and the continuous discharge of urine. It has been speculated that a similar phenomenon occurs in males of the African genus,* Loxodonta africana, *but most workers have concluded that it does not exist... Here we show that musth does occur in the African elephant and that its manifestations are similar to those in the Asian elephant.*

However, some 18 years earlier, in 1963, the professional hunter T. Murray Smith described musth in African elephants in his book *The Nature of the Beast* (Smith, 1963: pp. 49–50):

> *It is* [the bull] *that comes into season and that rampant rumbustious period is called musth. He is very dangerous then and is in this amative and lethal condition three or four times a year. His desires draw him to the herd... A musth elephant's sex gland — which is in his head — opens and exudes a sticky strong-smelling fluid.*

Some 2300 years earlier in 350 BCE, Aristotle recorded the aggression exhibited by bull elephants during the breeding season (Box 1.1).

In 1922, writing from Maihongsong, Siam (now Thailand), J.C.C. Wilson published a letter entitled 'The breeding of elephants in captivity' in the *Journal of the Bombay Natural History Society* in which he discussed musth, reproduction and gestation in Asian elephants living in timber camps:

> *I have kept a record extending over a number of years and put the period of gestation at 22 months.*
>
> Wilson (1922)

Writing about elephants in timber extraction camps in Burma (now Myanmar), Gordon Hundley of Moulmein, Burma, referred to the behaviour that we now call 'allomothering':

> *...a calf usually adopts a foster mother, who assists the dam in protecting and caring for the calf.*
>
> Hundley (1923)

Hundley also referred to the fact that Asian elephants in logging camps bred successfully and supplied the camps with new stock:

> *Numbers of calves born amongst the herds of elephants employed and owned by the large timber working firms are to be seen nowadays, born and bred, trained, worked and growing aged in captivity.*

The value of anecdotal evidence in the study of elephant cognition is discussed in Section 5.2.

1.5.4 Papers in academic journals

Elephant groups living in zoos are typically small, and widely dispersed around the world, mostly in Europe and North America. Consequently, most studies of zoo elephants have few subjects, typically fewer than 10. Some studies have been based on a single individual (e.g. Benedict, 1936; Elzanowski and Sergiel, 2006) while others have considered hundreds of animals [Meehan et al., 2016a ($n = 255$); Rees, 2009a ($n = 831$)].

The journal *Zoo Biology* was first published in 1982. It was the first journal to concern itself exclusively with scientific studies of zoos and zoo animals, although this type of work had previously been published elsewhere.

The first paper on elephants published in this journal was written by Michael J. Schmidt of Washington Park Zoo, Portland, Oregon, entitled *Studies on Asian elephant reproduction at the Washington Park Zoo* (Schmidt, 1982). The abstract of the paper predicted the imminent success of artificial insemination (AI) in elephants:

> Studies of Asian elephant reproduction at the Washington Park Zoo in the areas of the estrous cycle, semen collection, bull elephant management and health care, pregnancy, and pheromones have yielded sufficient information to attempt repeated artificial insemination (AI) at the time of ovulation. While no pregnancy has been achieved to date, with the information now at hand AI can be expected to become a practical technique for breeding elephants in captivity.

In fact, the first successful AI birth was not achieved for another 17 years, in November 1999 at by Dickerson Park Zoo, Springfield, Missouri (Schmitt, 1998).

The earliest academic papers on the behaviour of elephants in zoos appear to be by Wolfdietrich Kühme of the Max Plank Institute in Germany in 1961:

Kühme, W., 1961. Beobachtungen am afrikanischen Elefanten (*Loxodonta africana* Blumenbach 1797) in Gefangenschaft. [Observations on the African elephant (*Loxodonta africana* Blumenbach 1797) in captivity]. Zeitschrift für Tierpsychologie 18, 285−296. doi:10.1111/j.1439-0310.1961.tb00420.x.

This paper was published in the journal now known as *Ethology* and was followed by a chapter on the same subject in the *International Zoo Yearbook*:

Kühme, W., 1963. Ethology of the African elephant (*Loxodonta africana* Blumenbach 1797) in captivity. International Zoo Yearbook 4, 113−121.

This work was conducted on three 11-year-old African elephants (1.2) at Kronberg Zoo in Germany.

1.5.5 Early physiological research

There have been few physiological studies conducted on elephants until relatively recently. Their large size and relative rarity (in zoos) have made elephants challenging subjects for physiologists and even modern texts still refer to the work of Benedict, *The Physiology of the Elephant*, which was conducted in the United States and published in 1936. Francis Benedict undertook a detailed study of the physiology of a female Asian elephant named *Jap* who was borrowed from a circus (Benedict, 1936) (Fig. 1.17).

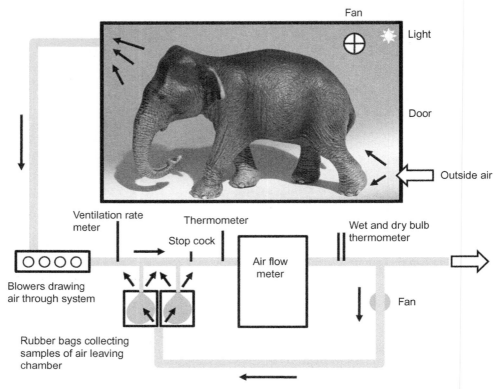

FIGURE 1.17 The apparatus used by Benedict (1936) to study the energetics of the female Asian elephant *Jap*. The arrows indicate the direction of air flow. Source: *Based on a drawing in Eltringham, S.K., 1982. Elephants. Blandford Press, Poole.*

Other studies include reports of observations on the hearts of two captive elephants in the United States (King et al., 1938) and a study of the sensorimotor specialisations of the trunk tip of the Asian elephant using material obtained when the tip of an elephant's trunk was severed in an accident (Rasmussen and Munger, 1987).

1.6 Unacceptable elephant science

The welfare of captive elephants is now seen as such a high priority that some governments have been financing research and while some zoos have invested in new elephant enclosures, others have transferred their animals to zoos with better facilities or to sanctuaries. Two historical incidents serve to illustrate the extent to which attitudes towards elephants have changed.

The first incident occurred in January 1903 when an adult female Asian elephant called *Topsy* was executed by electrocution in the United States. *Topsy* was a member of a herd kept at Coney Island's Luna Park. She was considered dangerous because she had reportedly killed three men, although the first two deaths may have been fabricated to justify her execution (Scigliano, 2002). The last of these was that of an abusive trainer whom she killed after he had put a lit cigarette in her mouth. *Topsy*'s owners decided to dispose of her, but their initial attempt at poisoning her with potassium cyanide failed. They eventually decided to kill her by electrocution after a proposal to hang her was opposed by the American Society for the Prevention of Cruelty to Animals (ASPCA).

At this time, Thomas Edison was attempting to establish an electricity transmission system in the United States using direct current (DC). He considered the competing Westinghouse alternating current (AC) system to be much more dangerous and had staged a number of public electrocutions of dogs and cats in an attempt to prove his point. Edison suggested that *Topsy* should be killed by electrocution using AC, and the ASPCA considered the method to be humane. The electric chair was already in use for human executions in New York and used AC. However, it was reported at the time that the equipment used to kill *Topsy* was wired to the DC supply at Coney Island and not an AC supply. *Topsy* was killed by electrocution within a few seconds. Edison filmed the execution and showed his film to audiences around the United States as part of an unsuccessful attempt to discredit AC. The circumstances surrounding *Topsy*'s death and the killing of other elephants by zoos and circuses by poisoning, hanging and shooting are described by Scigliano (2002) in his book *Love, War and Circuses*.

The second incident took place in 1962 when scientists at the University of Oklahoma administered a large dose of the drug LSD to *Tusko*, a 14-year-old male elephant from Lincoln Park Zoo, Chicago, in an attempt to study the animal in musth (West et al., 1962). It has been suggested that the dose of 297 mg represented an overdose of almost 750 times, based on the amount required to have a behavioural effect in humans (Spinage, 1994). *Tusko* collapsed five minutes after receiving the drug and was declared dead one hour and 40 minutes later. The scientists concluded that elephants are extremely sensitive to the effects of LSD and suggested that this might prove valuable in elephant control work in Africa.

Although many zoo and circus elephants still have less-than-perfect lives today, it is inconceivable that they could now be exposed to such barbaric practices in the name of 'humane destruction' or scientific experimentation.

1.7 Captive elephants as proxies for wild elephants

Elephants living in zoos, logging camps and other captive environments are studied so that we may learn more about their biology in captivity. However, in some circumstances, they may also be used as proxies for wild elephants. For example Seltmann et al. (2018) have evaluated the personality structure of semi-captive Asian elephants living in their natural habitat in Myanmar. They worked in the timber industry during the day, but foraged freely in the forest at night, received no human assistance in determining their diet and consorted with wild elephants who consequently sired their offspring. For these reasons the authors considered that their study elephants expressed natural behaviours. Such studies make it possible to discover new knowledge about elephants that would be extremely difficult – and probably impossible – to obtain from free-ranging animals.

CHAPTER 2

Ethological data collection and elephant activity budgets

Watching an elephant drink, one is reminded of someone filling a cistern with a bucket.
Richard Carrington (1958)

2.1 Introduction

This chapter discusses the scientific methods used to study captive elephants and what is known about their basic behaviour, that is their activity budgets during the day and night. Other aspects of the behaviour of elephants are considered elsewhere, especially in Chapters 3, 4, 5 and 7.

2.2 Methodology

2.2.1 Identifying individuals

Most behavioural studies of elephants require the identification of individual animals. This is only likely to be problematic when the group size is large or when several people who are unfamiliar with the animals are making recordings over an extended period of time. Obvious differences among individual animals might include size, the length or thickness of tusks (or their absence), body condition and distinguishing features, such as a damaged tail. The shape and size of the ears are particularly useful (Fig. 2.1). However, care must be taken to record both ears and update any drawings made or photographs taken to record any changes with time (such as new tears or increased folding) (Fig. 2.1B). This is especially important if research is to be conducted on the same animals over a long period of time by different researchers. With more experience, it is possible to determine identification from more subtle cues such as gait and characteristic behaviours.

Broom (2013) argues that it is not necessary to use human names for elephants to convey the idea that they are individuals and suggests that these trivialise the animals

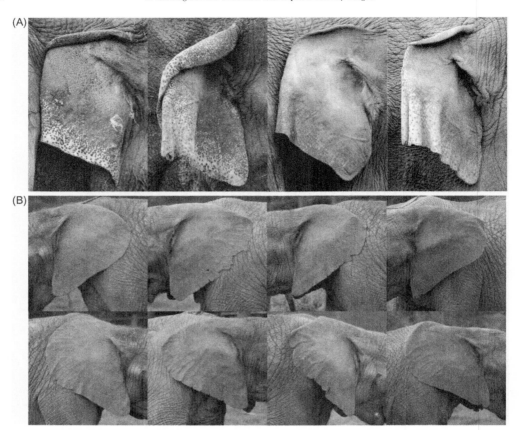

FIGURE 2.1 Elephants may be identified by the shape and size of their ears, along with any holes or tears. It is important to record both ears because only one may be visible when behaviour recordings are being made: (A) Asian and (B) African. Source: *(A) Reproduced with permission from Wiley-Blackwell. First published in Rees, P.A., 2015. Studying Captive Animals: A Workbook of Methods in Behaviour, Welfare and Ecology. Wiley-Blackwell, Chichester, West Sussex; (B) Reproduced with permission from Elsevier. First published in Rees, P.A. 2018. Examining Ecology. Exercises in Environmental Biology and Conservation. Elsevier, London.*

because the names are not concepts initiated by the animals and their use implies that they are in some way human. Broom concludes that, 'It would be better if human names were not used for nonhuman animals in scientific papers and books'. However, the fact is that many zoo animals are given human names — and sometimes made-up names — and it does not make sense for a researcher to call a particular bull elephant male 4 if he is known as *Chang* by his keepers.

Some zoos have traditionally given their elephants African or Asian names — depending on the species — and it does at least seem more acceptable to call an Asian elephant *Chang* rather than *Richard*. In some papers individual animals — and zoos — are anonymised by the authors. Lukacs et al. (2016) refer to their study elephants using letters (T, S, K, R and C) for no obvious reason when the animals were known by their names by everyone else. Furthermore, this practice prevents researchers from determining which studies used the same elephants.

2.2.2 Studying elephants in zoos

Studying wild elephants in the field is often difficult and time-consuming. In contrast, elephants in captive environments are accessible and, one might expect, much easier to study than their wild counterparts. However, if the purpose of the study is to investigate natural behaviour this may be impossible if the animals are not kept in a naturalistic environment and normal social groups, or if individuals have developed stereotypic behaviours. Behaviours may be restricted by opportunity. An elephant cannot wallow if no mud is available, cannot swim if there is no pool, cannot browse if not provided with suitable food. This may make it difficult to compare the behaviours exhibited in different zoos. Elephants in zoos and other captive environments cannot freely choose their associates and may not be kept with relatives, so social groups rarely reflect those found in the wild.

The presence of humans (zookeepers or visitors) may affect elephant behaviour. Keepers may alter the behaviour of elephants when providing food and cleaning enclosures, and elephants may throw objects and squirt water at visitors – or researchers – within their reach. Although these interactions may be of interest for some studies, for many others it interferes with data collection. In zoos, access to elephants may be restricted due to safety considerations, so it may only be possible to make recordings during certain times of the day, for example, only during zoo opening hours.

In the wild, there may be many elephants living within a relatively small area. A researcher studying elephants living in zoos, however, may have to travel long distances to collect data on elephants living in different facilities. This may affect the amount of data collected if funds for travelling are in short supply. Elephants from a study group may be moved to another zoo for breeding purposes or taken off-show due to illness or for some other reason, with resultant disruption to the collection of data on social behaviour.

In addition to the problems of data collection, it may also be difficult to interpret behaviour exhibited in a zoo environment. When an elephant exhibits an abnormal behaviour, is this a response to its current enclosure, current husbandry, current social relationships, or has the animal brought this behaviour from elsewhere?

The authors of some peer-reviewed papers about elephants living in zoos did not identify where their work was undertaken, and the elephants and zoos concerned have been anonymised, presumably to protect the zoo owners and management from any adverse publicity that might result after publication. For example the zoos holding the Asian elephants studied by Williams et al. (2015) were identified as Zoos A, B and C and those studied by Harvey et al. (2018) as A and B. Although this may have been a condition of allowing the research to take place, nevertheless, in my view this approach is counterproductive as I feel that zoos would benefit from being seen to engage actively with researchers, especially in relation to animal welfare issues. Zoos cannot receive any credit for their participation in research if they hide behind the cloak of anonymity. In any event, it is often possible for anyone familiar with the zoos that hold elephants to identify them from the social and age structure of the group.

Some of the information presented in this book is derived from a study of the Asian elephant herd at Chester Zoo, United Kingdom, that I began in 1999. The results of various aspects of this study appear in the relevant chapters and a brief description of the herd and the enclosure is provided in Box 2.1.

BOX 2.1 The Asian elephant herd at Chester Zoo.

In January 1999 I began a study of the Asian elephant herd living at Chester Zoo in the United Kingdom. At that time it consisted of one adult bull, five adult cows, one juvenile bull and a young calf (Table 2.1; Fig. 2.2).

TABLE 2.1 The composition of the Asian elephant herd at Chester Zoo in 1999.

Elephant	Sex	Order of arrival at Chester	Approximate age at start of study (years)	Approximate number of years already spent at Chester at start of study
Sheba	♀	1	43	33
Chang	♂	2	17	10
Kumara	♀	3	32	9
Maya	♀	4	30	8
Thi[a]	♀	5	17	7
Jangoli	♀	6	30	4
Upali	♂	7	4	1
Sithami[b]	♀	8	1	1

[a]Mother of Sithami.
[b]Born in the group.

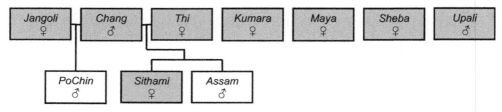

FIGURE 2.2 The relationships within the Asian elephant herd at Chester Zoo, United Kingdom. Those present at the beginning of the study are coloured grey and those born during the study are coloured white.

During the study, the elephants were usually released into their outdoor enclosure at around 1000 (where food was immediately available) and returned to the elephant house at around 1600 (when food was waiting for them inside). During the day they usually received at least one scatter feed and sometimes browse. The outdoor enclosure was flat with a very shallow pool and surrounded by a dry moat. By the end of 1999, work began on a new elephant exhibit. When finished this contained a waterfall, a deep pool, a landscaped substratum and a feeder wall, and elephants were contained with a ha-ha with high walls and a line of artificial tree trunks. Both the old and the new exhibits feature in various photographs in this book (Fig. 2.3A and B).

(Continued)

BOX 2.1 (Continued)

FIGURE 2.3 The elephant exhibit at Chester Zoo: (A) old (c.1999) and (B) new (c.2019).

The study examined the response of the elephants to a novel area (a new bull pen) (Rees, 2000); the effect of temperature on the frequency of dust bathing (Rees, 2002a) and the frequency of stereotypic behaviour (Rees, 2004a); early experience of sexual behaviour (Rees, 2003a); the social facilitation of mounting behaviour (Rees, 2004b); appeasement behaviours (Rees, 2004c); and activity budgets (Rees, 2009b). The results of these studies are discussed in the relevant sections below. In addition, data were collected on enclosure use, social relationships and aggression, and the results of these studies are reported here for the first time.

2.2.3 Ethograms

An ethogram is a species-specific library of behaviours that describes the elements and functions of each behaviour. A species ethogram is a master list of all the known behaviours exhibited by a species. For the purposes of conducting a particular experiment an experimental ethogram is used that contains only those behaviours relevant to the hypothesis being tested (Table 2.2). Unfortunately, inconsistencies exist between experimental ethograms used in different studies whereby the same behaviour may have several different definitions, making comparisons between studies difficult (Table 2.3). This is particularly important where activity budgets of animals are compared between studies.

TABLE 2.2 Ethogram of Asian elephant behaviour.

Behaviour	Description
High-frequency behaviours	
Dusting	Collecting soil and throwing it over the body/rubbing it into the skin (while standing still or walking), including digging in soil for this purpose.
Feeder ball	Feeding or attempting to feed at a metal feeder ball containing small quantities of food.
Feeding	Collecting solid food with the trunk and placing it in the mouth while standing or walking. (Does not include suckling or activity at the feeder ball.)
Locomotion	Walking (except while feeding or stereotyping)
Playing	Chasing another elephant or mock-fighting with another elephant (but not as a result of an antagonistic encounter or as part of courtship).
Standing	Standing motionless (not while stereotyping or dusting).
Stereotyping	Repetitive behaviour with no obvious purpose: weaving, head bobbing, pacing backwards and forwards or in an arc, walking in circles.[a]
Suckling	Calf suckling from mother or other female. Measured separately from feeding.
Low-frequency behaviours, grouped as 'other'[b]	
Aggression	Hitting/pushing as a result of an antagonistic encounter (but not as part of play).
Bathing	Standing/laying in pool/squirting water from pool over body with trunk.
Digging	Digging in soil using the foot (but not as part of a dusting behaviour).
Drinking	Collecting water in the trunk and squirting it into the mouth.
Lying down	Lying down on the ground (on its side or prone).
Rolling	Rolling in soil or mud (but not as part of playing with another individual).
Sex	Courting or being courted/mounting another elephant or being mounted by another elephant of either sex.

[a]*Stereotypic behaviours were recorded separately, but concatenated for the purpose of analysis.*
[b]*These behaviours have been concatenated and described as 'other' in the analysis owing to their relatively low frequencies.*
Source: *Reproduced with permission from Wiley. Based on Rees, P.A., 2009. Activity budgets and the relationship between feeding and stereotypic behaviors in Asian elephants* (Elephas maximus) *in a zoo. Zoo Biology 28, 79—97.*

TABLE 2.3 Comparison of the definitions of selected behaviours listed in the elephant ethograms used by Rees (2009b), Posta et al. (2013a) and Lukacs et al. (2016).

Rees (2009b) – Asian elephant ethogram		Posta et al. (2013a) – African elephant ethogram		Lukacs et al. (2016) – Asian elephant ethogram[a] (used during diurnal observations)	
Behaviour	Definition	Behaviour	Definition	Behaviour	Definition
Feeding	Collecting solid food with the trunk and placing it in the mouth while standing or walking. (Does not include suckling or activity at the feeder ball.)	Feed	Seeking or ingesting of food or water. Often involves gathering food with its trunk and lifting it into its mouth. Animal is not engaged in any other behaviour.	Eating (includes enrichment)	Using the trunk to pick up hay or concentrate and place it in the mouth; includes chewing; may include walking while chewing.
Standing	Standing motionless (not while stereotyping or dusting).	Stand	Individual is stationary in an upright position. No other behaviours are occurring simultaneously.	Standing	Standing with no forward or backward momentum.
Lying down (grouped as 'Other')	Lying down on the ground (on its side or prone).	Lie	Individual is in lateral recumbence. Weight is no longer supported by legs. No other behaviours are occurring simultaneously.	Lying down	Lying down on the ground, side or prone
Locomotion	Walking (except while feeding or stereotyping).	Walk	Animal takes 2 or more steps in any direction, but not in a stereotypic pattern. Is not playing, feeding, or exhibiting any other overt behaviour simultaneously.	Walking	Movement of all four feet in forward or backward motion
Stereotyping	Repetitive behaviour with no obvious purpose: weaving, head bobbing, pacing backward and forward or in an arc, walking in circles.	Stereotypy (grouped as 'Other')	Any behaviour that occurs in repetitive pattern including pacing, swaying or head bobbing.	No category described for stereotypic behaviour in this study	
Behaviours grouped as 'other'	Aggression Bathing Digging Drinking Lying down Rolling Sex	Behaviours grouped as 'other'	With keepers Public (interacting with or watching) Stereotypy	Behaviours grouped as 'other'	Drinking Spraying self Spraying environment Hose playing Wading Swimming Urinating Defaecating

[a]Defined as mutually exclusive behaviours.

Where behaviours are grouped together as 'other' (often due to their low frequencies) information is lost in the analysis and the term may mean different things in different studies, depending on the focus of each study (Table 2.3).

Ethograms of particular types of elephant behaviours have been published by a number of authors. For example a detailed ethogram of solitary behaviours, social behaviours and proximity measures used in a study of the nocturnal behaviour of African elephants living in a zoo has been published by Wilson et al. (2006), and Kahl and Armstrong (2000) have produced an ethogram of visual and tactile displays in African elephants focussing particularly on the elephants of north-western Zimbabwe. The organisation *ElephantVoices* plans to publish the first version of its (African) Elephant Ethogram in 2020 combining its existing databases of gestures and calls based on decades of study of elephants in Amboseli, Kenya, and data from other projects conducted in the Maasai Mara, Kenya, and Gorongosa in Mozambique (Anon., 2019a). The ethogram will include behaviour videos.

Stafford et al. (2011) have highlighted the difficulties of analysing activity budget data and developed a statistical method for the multivariate analysis of such data using recordings collected from several species including five adult male African elephants in the Pongola Nature Reserve in South Africa. The authors claim that their technique — which involves principal components analysis — provides a novel method of determining significant differences between the behaviour classifications measured.

2.2.4 Methodological difficulties

Superficially, it would appear straightforward to compare the results of similar studies. However, this assumes that all the data that might be compared have been collected and analysed in the same way, and this is often not the case. Instantaneous scan sampling is frequently used to study activity budgets in small groups of animals in zoos. At regular intervals (say 5 minutes) the behaviour being exhibited by each individual is recorded. So, if an animal is recorded as feeding in 20 of a total of 100 recordings, it was feeding for $20/100 \times 100 = 20\%$ of the time. However, if the animal was out-of-sight (e.g. off-show) at the time when 10 of the recordings were made, the time spent feeding may be calculated in two ways. If 'out-of-sight' is recorded as such, the time spent feeding is 20% and the time spent off-show is 10%. But if 'out-of-sight' recordings are excluded — because we cannot know what the animal was doing if it could not be seen — the percentage time spent feeding now becomes $20/(100-10) \times 100 = 22.2\%$. If there were 30 'out-of-sight' recordings the time spent feeding would increase to 28.6%, and if there were 34 such recordings the time spent feeding would be 30.3%.

Lukacs et al. (2016) measured activity during the day in five Asian elephants by studying focal animals continuously and recording the behaviour performed for the majority of each minute using 15 mutually exclusive categories including 'out-of-view' and 'trainer interaction'. I studied activity budgets in eight Asian elephants during the day using instantaneous scan sampling (at 5-minute intervals) and 15 mutually exclusive categories — which were different from those used by Lukacs et al. — but the percentage time spent on each activity was calculated using the total number of scans made per day minus those when keepers were present or the animals could not be seen (or a recording was not made due to disturbance) (Rees, 2009b). Assuming that the difference in recording methods (continuous vs scan sampling) had no effect on the results, the ethograms had been identical and the elephants in the two groups had behaved identically, the method of treating recordings made when the animals were 'out-of-view', disturbed or with keepers would have resulted in higher values in my study than in that of Lukacs et al.

Other problems can arise when sample periods are combined to represent a pattern of activity throughout the day. Imagine two elephants (A and B) living in the same zoo studied for 2 days between 0900 and 1200 and between 1300 and 1600. On day 1 elephant A is studied in the morning. Food is put out at 1000 and she spends all morning feeding. In the afternoon of day 1 elephant B is studied and spends no time feeding because all of the food had been eaten by both elephants in the morning. On day 2 food is put out at 1230. Elephant B is studied in the morning and spends no time eating because there is no food available. In the afternoon elephant A is studied and spends all of her time eating because food was only provided at 1230. If these data are now combined to provide a picture of what happens on a typical day, it would appear that elephant A eats continuously and elephant B does not eat at all. If both elephants had been studied in the morning and the afternoon on each day it would have been clear that feeding time was affecting behaviour.

A similar difficulty may arise when elephants are studied for a small number of whole days. Imagine that elephant C exhibits stereotypic behaviour all day on days 1, 3, 6 and 8, but not at all on days 2, 4, 5 and 7. If a study of her behaviour is made on all of these days, we would conclude that she spent half of her time stereotyping. If we studied her only on days 1, 6 and 8 we would conclude that she spends all day stereotyping; but if we studied her only on days 2, 5 and 7 we would observe no stereotypic behaviour at all.

I calculated the frequency of stereotypic behaviour recorded in a single adult Asian elephant cow kept at Chester Zoo using scan sampling during five consecutive 7-day sample periods and over the total 35-day study, using data collected for a study of activity budgets (Rees, 2009b) (Fig. 2.4). By calculating a moving average frequency for this behaviour, it is possible to determine the minimum number of sample periods (in this case days) necessary to produce a constant value. In this case the frequency decreased as the number of sample days increased right up until the end of the study (after 35 days). After 3 days of study, the frequency of stereotypic behaviour was around 0.37 (37%) but after 35 days it had fallen to around 0.09 (9%). If the 35 days are treated as five separate 7-day studies the frequencies obtained range from 0.003 (0.3%) to 0.189 (18.9%). Clearly, the length, and timing, of the sample period had a profound effect on the frequency of stereotypic behaviour calculated.

Some data relating to behaviour are almost meaningless. For example statements such as:

> ...the number of elephants with stereotypies was the highest in temple system (49%) followed by private (25%) and the lowest in the forest department (7%).
>
> Varadharajan et al. (2016)

This tells us very little in isolation. We cannot say, from this information alone, how many elephants in each category exhibited stereotypies and how frequently they were exhibited.

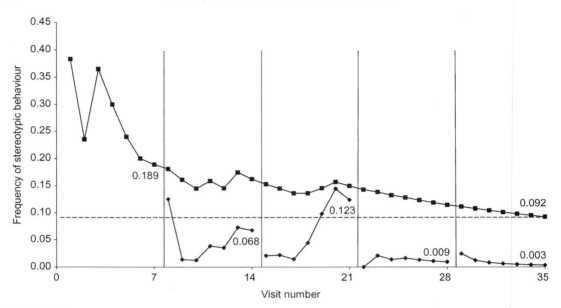

FIGURE 2.4 The effect of calculating a cumulative moving average (from samples taken on consecutive visits) on the estimated frequency of a single category of behaviour: stereotypic behaviour in an adult Asian elephant cow in a zoo. The moving averages have been calculated over the entire 35-day study period and for each of the five 7-day samples (days 1–7, 8–14, 15–21, 22–28 and 29–35). Values shown on the graph are the average after each individual 7-day period and the average of all 35 days (0.092). Source: *Reproduced with permission from Wiley-Blackwell. First published in Rees, P.A., 2015. Studying Captive Animals: A Workbook of Methods in Behaviour, Welfare and Ecology. Chichester, West Sussex.*

There is clearly a significant difference between the welfare of an elephant who spends 50% of every day stereotyping and one that exhibits very occasional swaying of the body.

Some years ago, an animal welfare organisation asked me to comment on the welfare of an elephant shown exhibiting stereotypic behaviour in an undercover video they had made. I declined to comment on the basis that the recording was just a few minutes long and I had no way of knowing if this was an isolated event or if the animal behaved like this all day, every day. I discovered some time later that a vet had been prepared to comment authoritatively from the same evidence.

2.2.5 Data collection by caregivers

Collecting data on animal behaviour and welfare is time-consuming and scientists are increasingly turning to caregivers to supply information on their animals. Keepers were asked to evaluate the social rank of female African elephants living in zoos in a study of the impact of rank and age on social behaviour (Freeman et al., 2010a). Information on elephant personalities has been collected from mahouts in Myanmar (Seltmann et al., 2018) and from keepers in zoos (Yasui et al., 2013); on possible elephant welfare indicators from zoo professionals (Chadwick et al., 2017); and on reproductive status in North American zoos (Brown et al., 2004a) and the relationship between foot problems and floor type from institutions holding elephants (Haspeslagh et al., 2013). However, Posta (2011) concluded that routinely kept keeper records were inadequate to assess the effectiveness of enrichment.

This method of data collection has the advantage that it is inexpensive compared with the cost of collecting data first hand, but it relies on the cooperation — and competence — of others, and this may not always be forthcoming. An attempt by the Elephant TAG/SSP (Taxon Advisory Group/Species Survival Plan Program) to survey the reproductive status of female elephants in North America received responses from nearly 100% of the SSP facilities, but for the studbook populations only 81% of facilities holding Asian elephants and 71% holding African elephants responded.

When Clubb and Mason (2002) attempted to use a questionnaire to collect data on the elephant population in the United Kingdom for a report commissioned by the Royal Society for the Prevention of Cruelty to Animals (RSPCA) they failed spectacularly:

A questionnaire was designed to gather data on the current U.K. captive elephant population, including housing and husbandry conditions, and information on various welfare indicators...

...A copy of this questionnaire was sent to the director or head curator of each zoo that held elephants in the U.K., totalling 18 facilities. Unfortunately, at this time the Zoo Federation of Great Britain and Ireland [now BIAZA], which had previously supported the preparation of this review, withdrew their support. The result was that no questionnaires were returned from Federation member zoos (plus none of the four non-member zoos sent back the questionnaire), and so no data were collected.

2.3 Activity budgets

2.3.1 Introduction

The most basic question we can ask about an animal's behaviour is: 'What does it spend its time doing?' To answer this question, ethologists construct an activity budget: a description of the amount of time an animal spends engaging in behaviours that have been previously described in a suitable ethogram. Activity budgets allow a comparison to be made of animals of different ages or sexes, animals in the same social group or animals in different social groups. They may also be used to determine whether or not behaviour alters after a change to an enclosure, such as the introduction of an enrichment device, or if it varies between daytime and night time. Schulte (2000) has noted that captive elephants display a similar behavioural repertoire to elephants observed in the wild, in spite of the fact that the environmental pressures associated with wild living are largely absent.

Most studies of captive elephant behaviour have been small scale, with short sampling periods taken from a small number of days during a limited period of the year. The subjects have usually been small groups of adult females, and sometimes single individuals (e.g. Elzanowski and Sergiel, 2006).

Detailed studies of activity budgets in zoo elephants were rare until relatively recently. In their review of elephant welfare in European zoos, Clubb and Mason (2002) identified just 10 studies that quantified elephant behaviour, four of which were of zoo elephants, and only one of these was published in a peer-reviewed journal. Studies of activity patterns in elephants living in zoos help us to understand the extent to which they behave in captivity as they do in the wild. The amount of stereotypic behaviour exhibited is of particular interest because it is generally considered to be an indicator of poor welfare (Mason, 1991a,b). This behaviour is considered in detail in Chapter 7, Elephant welfare.

The activity budgets of captive elephants vary between individuals (e.g. depending upon age, sex and temperament), at different times of the day and seasonally. They also vary between institutions due to differences in the physical environment and husbandry regimes. Comparing activity budgets between different studies made in captivity, different studies undertaken in the wild, and between studies of captive and free-ranging elephants is problematic. The methods of collecting data vary between studies, as do the length of the studies, the seasons when they were undertaken and the time of day recordings were made. Most have only collected data during the daylight hours and some of those undertaken in captivity have been constrained by access difficulties.

I studied the activity budgets of eight Asian elephants while they were in their outdoor enclosure at Chester Zoo for 35 days, between January and November 1999 using instantaneous scan sampling between 1000 and 1600 (with most behaviour frequencies calculated between 1000 and 1400 because the animals were taken inside early when the weather was inclement) (Rees, 2009b). The amount of time elephants spent on various activities varied between individuals (Fig. 2.5) and depended on their age, sex, the time of day and the time of year (Fig. 2.6A and B). As the day progressed the frequency of 'active' behaviours (walking, dusting, feeding) decreased and the frequency of 'inactive' behaviours (stereotyping, standing, other) increased in the five adult female elephants (Fig. 2.7). Seasonal patterns within these five animals showed an increase in dusting behaviour and a decrease in stereotypic behaviour in the warmer months (Fig. 2.8). These patterns have

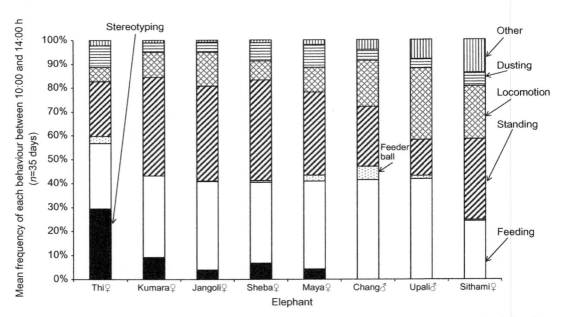

FIGURE 2.5   Individual variation in the activity budgets of eight Asian elephants at Chester Zoo, United Kingdom. Source: *Reproduced with permission from Wiley. First published in Rees, P.A., 2009. Activity budgets and the relationship between feeding and stereotypic behaviors in Asian elephants* (Elephas maximus) *in a zoo. Zoo Biology 28, 79—97.*

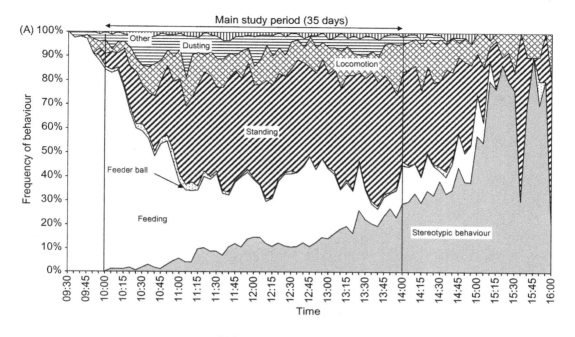
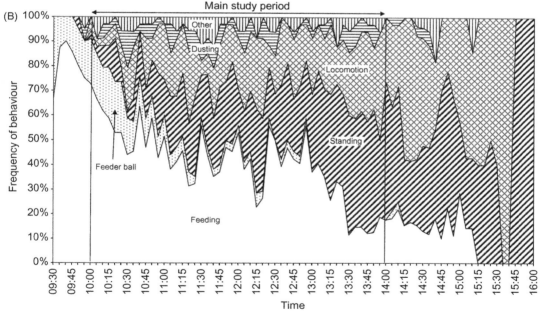

FIGURE 2.6 Diurnal activity in Asian elephants at Chester Zoo, United Kingdom: (A) five adult females; (B) one adult bull. Source: *Reproduced with permission from Wiley. First published in Rees, P.A., 2009. Activity budgets and the relationship between feeding and stereotypic behaviors in Asian elephants* (Elephas maximus) *in a zoo. Zoo Biology 28, 79–97.*

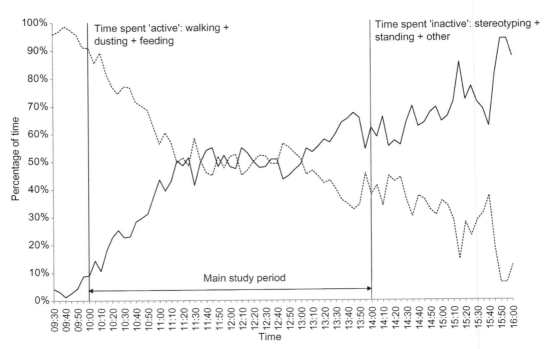

FIGURE 2.7 The change in the frequency of 'active' behaviours (walking, dusting, feeding) and 'inactive' behaviours (stereotyping, standing, other) in five adult Asian cow elephants at Chester Zoo, United Kingdom. Between 1000 and 1400 $n = 35$ days. Before and after these times $n < 35$ days. Source: *Based on data in Rees, P.A., 2009. Activity budgets and the relationship between feeding and stereotypic behaviors in Asian elephants* (Elephas maximus) *in a zoo. Zoo Biology 28, 79–97.*

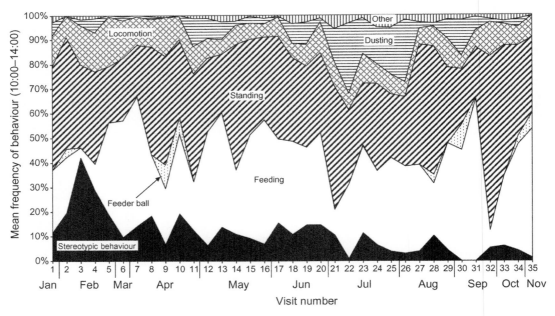

FIGURE 2.8 Seasonal variation in activity budgets in five adult Asian cow elephants at Chester Zoo, United Kingdom. Source: *Reproduced with permission from Wiley. First published in Rees, P.A., 2009. Activity budgets and the relationship between feeding and stereotypic behaviors in Asian elephants* (Elephas maximus) *in a zoo. Zoo Biology 28, 79–97.*

implications for studies that use short sampling periods and may make comparisons of data collected by different studies at different times of the day or year invalid.

2.3.2 Feeding

Wild elephants spend much of their time feeding. The time spent on this activity in zoos is often reduced due to a lack of continuous access to food.

McKay (1973) studied wild elephants in Sri Lanka and found that bulls spent almost 94% of the daylight hours feeding; the corresponding figure for cows was almost 91%. He estimated that this represented 80% of any 24-hour period. However, Eltringham (1982) noted that this only allowed for 2 hours' sleep, so was probably an overestimate. In the same area, Vancuylenberg (1977) reported that about 75% of his daytime observations of elephants were of feeding. A more recent study from Sri Lanka found that wild bulls fed for 44.4% and cows for 46.7% of the time between 1000 and 1600 (Ahamed, 2015). Mohapatra et al. (2013) observed wild elephants in Odisha, India, during the 12 hours of daylight and found that they fed for 8.1 hours (67.5%) in winter, 7.0 hours (58.3%) in summer and 6.1 hours (50.8%) during the monsoon.

Adult Asian elephants at Chester Zoo spent on average approximately 27.4%–41.4% of their time feeding (between 1000 and 1400) in an enclosure where the substratum was sand/soil and almost all food was provided by keepers (Rees, 2009b). Food was made available from first release into the outside enclosure in the morning and, although this was usually supplemented by scatter feeds at other times, feeding activity reduced progressively through the day (Fig. 2.6A and B).

In a study of five adult Asian elephants at Busch Gardens Tampa Bay, Florida, Lukacs et al. (2016) found that they were feeding for 19%–44% of the time during daylight hours but only 1%–19% of the time during the night.

Time spent feeding is likely to increase where vegetation is freely available. The mean percentage of time spent feeding by seven African elephants kept on grass at Knowsley Safari Park, United Kingdom, between 1030 and 1745 on 1 day in summer was 74.9% (range 63.3%–83.3%) (Rees, 1977). This agrees well with data from wild elephants in Uganda studied over a 24-hour period or more who spent about 75% of their time feeding (Wyatt and Eltringham, 1974).

A preliminary study of six elephants at the Pinnawala Elephant Orphanage in Sri Lanka (840 minutes) found that they spent most of their time feeding (males: 46.8%; females: 50.6%), but data were only recorded from 0800 to 0900 and from 1500 to 1600, and the extent to which they had access to food was unclear (Samarasinghe and Ahamed, 2016) (Fig. 2.9). A much more extensive study of free-foraging translocated captive elephants at the Seblat Elephant Conservation Center in Sumatra (4496 daytime hours) found that they spent on average 82.2% of the time between 0700 and 1700 feeding, with the two males feeding for longer (86.8%) than the 12 females (81.6%) (Sitompul et al., 2013).

Elephants living in zoos often engage in contrafreeloading: working for food when it is already freely available (Fig. 2.10A and B). Similar behaviour may be observed in the wild when elephants expend additional energy gathering food that is out of reach when apparently surrounded by other suitable food (Fig. 7.17A). Elephant managers may make use of

FIGURE 2.9 Elephants tethered by chains to trees while feeding at the Pinnawala Elephant Orphanage in Sri Lanka.

FIGURE 2.10 Contrafreeloading: (A) *Sheba* at Chester Zoo, United Kingdom; (B) *Victor* at Berlin Zoo, Germany.

FIGURE 2.11 An Asian elephant cow (*Maya*) dusting at Chester Zoo, United Kingdom. Sequence of images from top left to bottom right taken within a 3-second period.

this tendency by providing food at a high level as an enrichment, thereby extending the time spent feeding (Figs 7.17B and 7.18). When the *Elephants of the Asian Forest* exhibit was created at Chester Zoo, the elephants helped themselves to decorative plants around the edge of their new enclosure, above the wall delimiting the ha-ha, prior to the erection of a hot wire.

2.3.3 Dusting

When bare soil or sand is available to elephants in zoos they will spend a considerable amount of time dust bathing by throwing material over their bodies with their trunks (Fig. 2.11). The function of this activity is unclear, but its frequency increases with temperature (Fig. 6.2) so this must be considered when comparing data from studies made in different locations (latitudes) or in different seasons (Rees, 2002a).

At Chester Zoo, the six adult Asian elephants spent 3.9%–9.6% dusting (between 1000 and 1400) (Rees, 2008). At Knowsley Safari Park, seven African elephants spent between 0% and 26.7% (mean = 12.1%) of the time between 1030 and 1745 dusting on a summer's day (Rees, 1977). Wild male and female elephants in Sri Lanka spent 3.65% and 1.33% (respectively) dusting between 1000 and 1600 (Ahamed, 2015). Wyatt and Eltringham (1974) did not record dusting as a separate activity in their study of wild elephants in

Uganda and neither did Sitompul et al. (2013) in their work on the activity budgets of elephants in Sumatra. Lukacs et al. (2016) recorded dust bathing as a 'behavioral context' rather than a 'mutually exclusive behavior' in their study of five female Asian elephants at Busch Gardens Tampa Bay, Florida, so did not measure the amount of time their study elephants spent on this activity. This is curious considering that other studies have found this behaviour to occupy around 10% of daytime activity in some elephants.

2.3.4 Walking

Within the herd of Asian elephants kept at the Chester Zoo I found that the six adults spent 6.1%–19.2% of their time walking (Rees, 2009b). The adult bull and calves (of both sexes) spent considerably more time walking than did adult females (Fig. 2.5). Lukacs et al. (2016) found that the Asian elephants at Busch Gardens Tampa Bay, Florida, spent 1%–9% of their time walking during the daylight hours, but recorded no walking at night.

Wild elephants in Odisha, India, spent around 2 hours 'moving' in the 12 hours of daylight (16.7%), with little seasonal variation (Mohapatra et al., 2013). Similar results were obtained by Ahamed (2015) who found that male wild elephants in Lahugala Kitulana National Park in Sri Lanka spent 16.7% of the time between 0600 and 1800 walking while females engaged in walking for 13.8% of the time. However, elephants in a Sumatran conservation centre moved on average for only 9.5% of the day (between 0700 and 1700), with the two males moving much less than the 12 females (4.7% and 10.2%, respectively) (Sitompul et al., 2013).

Wild African elephants in Uganda followed by Wyatt and Eltringham (1974) for at least 24 continuous hours spent about 9.3% of the time walking. In Sri Lanka the comparable figures for wild elephants were 5.4% (McKay, 1973) and 16.1% (Vancuylenberg, 1977) measured from intermittent daytime-only observations.

2.3.5 Resting, sleeping and nocturnal behaviour

The six adult Asian elephants at Chester Zoo spent 22.9%–42.0% of their time standing/resting (between 1000 and 1400) (Figs 2.5 and 2.12). The Asian elephants studied by Lukacs et al. (2016) were recorded as standing for 17%–49% of the time during the day and 29%–87% of the time at night. These animals were recorded lying down for between 0% and 45% of the time at night, but during the day only one elephant was observed lying down on a single occasion for just 6 minutes inside the elephant barn.

Free-foraging captive elephants at a conservation centre in Sumatra spent an average of 6.6% of the time between 0700 and 1700 resting (Sitompul et al., 2013). Wild elephants in Odisha, India, studied during 12 hours of daylight spent 0.8 hours (6.7%) resting in winter, 1.5 hours (12.5%) in summer, and 1.1 hours (9.2%) during the monsoon. Wild bull and cow elephants in Sri Lanka were observed standing for 7.4% and 4.0%, respectively, during the time between 0600 and 1800 (Ahamed, 2015).

An early study of sleep examined four female elephants and noted that the presence of human observers resulted in the two older animals (15 and 35 years old) lying down for only short periods at night as they were suspicious of them (Hartmann et al., 1968). The

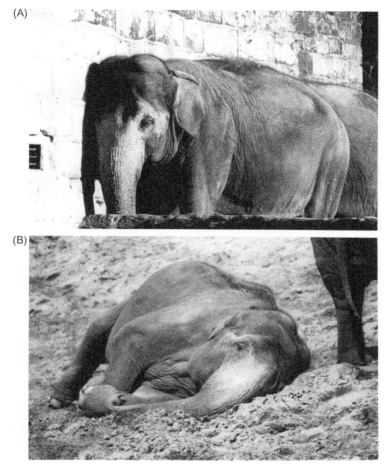

FIGURE 2.12 Female Asian elephant (*Sheba*) sleeping while standing (A) and lying down (B) at Chester Zoo, United Kingdom.

two younger elephants (5 and 12 years old) spent 2–4 hours lying down. D-state (desynchronised or dreaming) sleep was observed while elephants were lying down. This was characterised by a relaxed stretched-out position, rapid eye movements (REM) easily visible, heavy irregular respiration and twitching of the muscles in the trunk, face and tail. The 5-year-old elephant exhibited a mean D-period of 21 minutes' REM and a sleep–dream cycle length of 96 minutes, while the corresponding periods in the 12-year-old were 40 and 124 minutes respectively.

Hartmann et al. conducted their study using visual observation and minute-by-minute behavioural flow sheets for just three nights. Tobler (1992) investigated sleep in 12 female Asian circus ($n = 7$) and zoo elephants ($n = 5$), including one infant, using continuous time lapse video recording. The elephants slept while standing and while recumbent. Sleep onset began after 2100 and increased progressively with a maximum between 0100 and 0400. In the adults, the total sleep time was 4.0–6.5 hours/night – including 13.8–130.9 minutes of standing sleep – and was similar in the circus and zoo groups. Recumbent sleep episodes lasted for 72 minutes. The elephants slept for longer periods in winter than in summer, and seasonal differences were recorded in the distribution of sleep during the night. REM sleep was observed, but its duration could not be accurately determined. During the course of a 15-month period of recording, the total time the infant spent sleeping fell from 8.1 to 5.1 hours, with standing sleep recorded for the first time at the age of 9 months.

A study of the nocturnal behaviour of six African elephants at Schönbrunn Zoo in Austria showed that the animals were active for at least half of the night, based on data recorded using a time lapse video recorder between 1800 and 0700 each night between 11 December 1998 and 10 April 1999 (Weisz et al., 2000). This amounted to at least 6 hours of nocturnal activity. When this is added to activity during the day the total active period is 18 hours in every 24. The maximum sleeping time recorded was 5 hours/night, but there was considerable individual variation. Young animals generally slept for longer than older animals and were more likely to lie down. In order to enrich night time activity the zoo installed automatic feeding machines that provided small quantities of fruit and hay in the early morning.

Brockett et al. (1999) studied the nocturnal behaviour of three unchained female African elephants at Zoo Atlanta in 1992 and 1994. They found that the animals were most active between 1800 and 2400 and between 0600 and 0700. The elephants spent half of their time within one body length of another and utilised all three of the available enclosures. The authors concluded that zoos could permit increased activity and more social interaction between elephants by extending the period when elephants are unchained.

Wilson et al. (2006) studied the same group of elephants 8 years later at night for 10 weeks between 1700 and 1800. They found that affiliative behaviours accounted for 57% of all social behaviours and agonistic behaviours were infrequent and caused no injuries. The elephants used all the areas to which they had access. The authors concluded that unrestricted social access during the night was the appropriate management strategy for this elephant group.

Elephants can sleep while standing or lying, and time spent resting may be an important indicator of welfare in these animals. Williams et al. (2015) studied overnight standing and lying rest behaviour in 14 Asian elephants (2.12) living in three zoos in the United Kingdom using video footage of indoor accommodation recorded between 1600 and 0830. The elephants rested for a mean of 58–337 minutes/night. The longest bouts of lying rest occurred between 2201 and 0600. Lying rest was not observed on concrete or tiled floors. A sand floor was available to 11 of the elephants and all these animals engaged in lying rest on this. Only two of the eleven elephants who had access to rubber flooring engaged in lying rest on it. The mean length of resting bouts was longer when a conspecific was within two body lengths than when conspecifics were further away. This study demonstrates that elephants show individual substrate preferences and engage in more rest – and hence may experience better welfare – when in close proximity to conspecifics.

Walsh (2017) measured sleep behaviour in eight Asian elephants at Dublin Zoo, Ireland, for 704 nights extending over 33 months. The herd consisted of three adult cows, a juvenile cow, one adult bull, two bull calves and a cow calf. Behaviour was recorded using infrared CCTV cameras from 1900 to 0800 each night. Adults slept for an average of

3 hours 33 m/night while calves slept for an average of 5 hours and 8 m. Old elephants slept less than younger elephants. The three adult cows each gave birth during the study. One slept 68% less in the 9 months after giving birth; the second slept 13% less after parturition; but the third slept up to 10% more after giving birth. Walsh suggested that the reduction in sleep may have been due to calf-guarding and also the establishment of successful suckling. Night time observations of four (3.1) Asian elephants at Berlin Zoo, Germany, found that a 25-year-old pregnant female slept for only 1.5 hours, while the two older cows (42 and 38 years old) slept lying down for almost 4 hours (Ibler and Pankow, 2012).

Elephant calves are sometimes orphaned when the mother has been killed by poachers or during human−elephant conflict. Orphans may be captured and taken to rehabilitation facilities with the intention of eventually returning them to the wild. High postsurvival rates depend upon calves developing natural behaviours. In an attempt to determine the extent to which orphaned Asian elephant calves exhibit normal patterns of nocturnal behaviour, Stokes et al. (2017) studied 34 calves at the Elephant Transit Home, a rehabilitation centre in Udawalawi National Park in Sri Lanka (0−24 months $n = 9.4$; 25−36 months $n = 12.4$; >36 months $n = 3.2$). At night they were confined in a 4 ha paddock and their behaviour was recorded for 18 nights between 1830 and 0600 using an infrared camera. Lying rest and feeding were the two most frequently recorded behaviours, accounting for 46.2% and 28.4% of the time respectively; the calves were fed a milk formula through a tube three times during the night. Calves spent more time in lying rest than has been observed in adults, and standing rest accounted for just 1.2% of the time. There was no significant difference in the time spent lying in the three age groups, but two distinct lying resting periods were noted: 2300−0100 and 0330−0530. Calves spent most of their time during the night within 5 m of their nearest neighbour; the youngest calves remaining closer to other calves than did older individuals. It is common for adults to stand guard over sleeping calves in the wild and in captivity, but this 'sentinel' behaviour was rarely observed among the calves. Stokes et al. suggested that this may have been because there was no perceived predation threat requiring vigilance behaviour − and therefore no behavioural need for it − or because the individuals were all young, had a high requirement for lying rest, and no adult−calf relationships existed. The authors of this study concluded that the calves exhibited behaviours reported from wild and captive (zoo) elephants and appeared to be consistent with 'natural' behaviour patterns and indicative of good welfare.

Schiffmann et al. (2018) have reviewed the scientific literature on quantitative studies of elephant resting and sleeping behaviour, along with a discussion of case studies on elephant falling bouts resulting from insufficient lying rest. The welfare implications of sleep are discussed in Chapter 7, Elephant welfare.

Yawning, or oral gaping, is known to occur in a range of taxa, such as rats, dogs, chimpanzees, bird and reptiles. Its function is uncertain, but contagious yawning − the onset of a yawn that results from seeing or hearing another individual yawn − may be part of a neural network involved in empathy (Platek et al., 2005), however the evidence for this is inconsistent (Massen and Gallup, 2017). Yawning in elephants has only recently been recorded by scientists. Rossman et al. (2017) studied seven African elephants kept at the Knysna Elephant Park in South Africa for 'guided interaction with visitors', and found that yawning was particularly associated with arousal from night time recumbencies. The same study examined video recordings of Asian and African elephants kept in North

American zoos and yawning was recorded in some individuals of both species. Of 133 yawns associated with arousal from recumbencies, over half were associated with arousal from the final recumbency of the night, just before morning. Yawning did not appear to be associated with vocalisation, but some instances of contagious yawning were recorded.

2.4 The 24-hour needs of elephants in zoos

Traditionally, elephants living in zoos have been returned to their indoor accommodation at the end of the day and kept inside until the next morning. For some elephants this may mean being outside between 1000 and 1600 (when zoos tend to be open to the public in northern latitudes in winter) and inside for the remaining 18 hours. In theory, provided that enclosures are secure, there is no reason why elephants should not have access to their outdoor enclosure during the night, provided that the climate is suitable.

A study of three female Asian elephants at New York's Bronx Zoo compared their activity budgets overnight when they were housed indoors with those exhibited when they were allowed free access to an outdoor yard (Powell and Vitale, 2016). Data were recorded between 1900 and 0700 from July to September using infrared video cameras. Two of the elephants showed a slight preference for being outdoors, while the third spent most of her time indoors. When the elephants had outdoor access, they exhibited increases in standing and play behaviour and decreases in feeding behaviour and lying down. There was a significant decrease in stereotypical swaying behaviour when the animals had outdoor access. The mean frequency of swaying was low when housed indoors (2.7 ± 0.7), but the behaviour was almost eliminated when the elephants had outdoor access (0.4 ± 0.2). The sand floor stall was little used in either condition. This study shows that the wellbeing of elephants may be promoted by giving them access to alternative areas overnight as it allows them to express their preferences for different locations, while also reducing stereotypic behaviour.

The 24-hour behavioural needs of 15 African elephants were studied during two summers (2010 and 2011) at San Diego Zoo Safari Park, California, by Horback et al. (2014). They found that foraging was the most common behaviour during the day, followed by resting and walking. In the evening, the elephants spent most of their time foraging, resting and sleeping. Negative social events occurred at low rates throughout the day and night. Most positive behaviour events occurred between mother−calf pairs during daylight hours and at night time. Calves and juveniles initiated approximately 60% of all social interactions during the daytime and 57% at night. This study highlighted the importance of managing elephants to meet their 24-hour needs.

Although it is interesting to study the way captive elephants spend their time during the day, we must be careful not to assume that their activity budgets should necessarily reflect those of free-ranging elephants if they are to experience good welfare (Hutchins, 2006). Wild elephants exhibit variation in their behaviour, as do elephants in zoos and logging camps. The extent to which captive elephants exhibit abnormal behaviour is discussed in Chapter 7, Elephant welfare.

CHAPTER 3

Elephant social structure, behaviour and complexity

Strong bonds between cow elephants are the cement of elephant society on which the survival of the younger animals depends.
Iain Douglas-Hamilton (Douglas-Hamilton and Douglas-Hamilton, 1975)

3.1 Introduction

Maintaining captive elephants in natural — or at least naturalistic — social groupings is important for their welfare and breeding prospects. This chapter considers the structure of elephant society in free-ranging populations and compares this with that of elephants living in captivity. It goes on to consider the extent to which elephants living in zoos are able to exhibit normal social behaviour, including establishing friendships and cooperating in the protection of calves. To conclude, the chapter examines dominance in elephant society, aggression, appeasement and elephant personality.

3.2 The structure of elephant societies

3.2.1 Elephant societies in the wild

In the wild, elephant social structure is complex. Laws et al. (1975) suggested that the largest stable population unit (or the smallest discrete unit) averaged 5–6 elephants in Murchison Falls National Park, South, Uganda. The mean group size recorded was 11.6 (range 2–29). The simplest family group has an average of three members consisting of a cow and her 1–2 offspring (Spinage, 1994) (Fig. 3.1).

Asian elephants live in family groups which typically contain one adult female and between one and five immature offspring (Sukumar, 1994). Larger groups may contain two or more adult females and sometimes groups contain three generations (Fig. 3.2).

FIGURE 3.1 A family unit of African elephants (adult females and their young) in Tarangire National Park, Tanzania. Source: *Reproduced with permission from Wiley-Blackwell. First published in Rees, P.A., 2013. Dictionary of Zoo Biology and Animal Management. Wiley-Blackwell, Chichester, West Sussex.*

FIGURE 3.2 A herd of approximately 15 Asian elephants in Minneriya National Park, Sri Lanka.

A group of elephants with more than one adult female has been called an 'extended family unit' (Douglas-Hamilton and Douglas-Hamilton, 1975) or a 'joint family' (Sukumar, 1994). Larger groups of related animals have been called 'kin groups' (Douglas-Hamilton and Douglas-Hamilton, 1975) or 'bond groups' (Moss, 1988), the latter term not necessarily implying relatedness (Fig. 3.3).

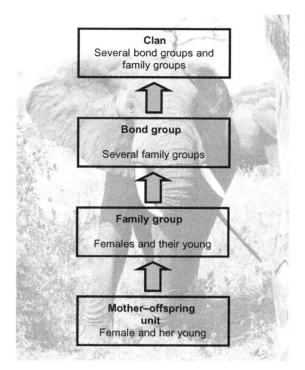

FIGURE 3.3 The social structure of African elephants. Source: *Reproduced with permission from Wiley-Blackwell. First published in Rees, P.A., 2013. Dictionary of Zoo Biology and Animal Management. Wiley-Blackwell, Chichester, West Sussex, based on Moss (1988).*

In a zoo environment, keeping related animals together will be a lower priority than moving them between institutions for breeding purposes. Nevertheless, the formation of naturally sized groupings of unrelated individuals is preferable to keeping elephants alone or in very small groups

3.2.2 Elephant societies in captivity

It was once common to find that zoos kept a single elephant. When the animal died it would be replaced by another, either taken from the wild, a logging camp in Asia or perhaps another zoo. For example Bristol Zoo, in the United Kingdom, kept elephants for over 130 years (from 1868 to 2002) with a period of around 3 years when they had no elephants (Fig. 3.4). For most of this time they kept a single elephant and they never had more than two at the same time (Fig. 3.5). When Bristol kept two elephants, they were of the same sex (female) and, for most of the time, from different species. Consequently no breeding occurred. Nevertheless, in zoos that have kept elephants in larger groups, breeding populations have been difficult to establish. At the time of writing, Chester Zoo, also in the United Kingdom, has a thriving breeding herd of Asian elephants, but this has taken over 80 years to establish (Fig. 3.6). The zoo acquired two Asian elephants (*Manniken* and *Molly*) from Dourley's Tropical Express Revue in 1941, and at one time kept African and Asian species together. It was not until 1977 that the zoo produced the

FIGURE 3.4 The elephant house and the female Asian elephant, *Wendy*, at Bristol Zoo, United Kingdom (1999). Since Wendy's death in 2002, the elephant house has been repurposed to house lowland gorillas. Inset: The primary barrier between the enclosure and the public was a stone wall.

first Asian elephant calf in the United Kingdom, a male called *Jubilee*. I have published a short history of the first 60 years of Chester's elephant herd and this is summarised in Fig. 3.6 (Rees, 2001a).

Schulte (2000) has noted that female elephants are generally housed together — although usually in smaller groups than in the wild — and adult bulls are kept separate from each other and from females most of the time, reflecting their wild behaviour. However, Schulte also noted that captive elephants are kept in a wider range of group sizes, degrees of relatedness and age structures than would be typical of their wild conspecifics. He suggested that this provides opportunities for research on the development of social behaviour, the fundamental significance of allomothering and the ability of elephants to recognise kin and members of their social group. Some 20 years on, many of these areas remain largely uninvestigated.

I examined the global distribution of elephants living in zoos in 2006 using the database of the International Species Information System (ISIS) — now the Zoological Information Management System (ZIMS) — and found that many zoos were still keeping small numbers of elephants that did not reflect their social structure in the wild (Rees, 2009a). Most were kept by zoos in Europe or North America (see Section 8.2).

In these zoos, cows outnumbered bulls four to one (*Loxodonta*) and three to one (*Elephas*). Groups contained seven or fewer individuals: mean, 4.28 ($\sigma = 5.73$). One fifth of elephants lived alone or with one conspecific; 46 elephants (5.5%) had no conspecific. Many zoos ignored the minimum group sizes recommended by regional zoo association guidelines. At that time, in North America, the Association of Zoos and

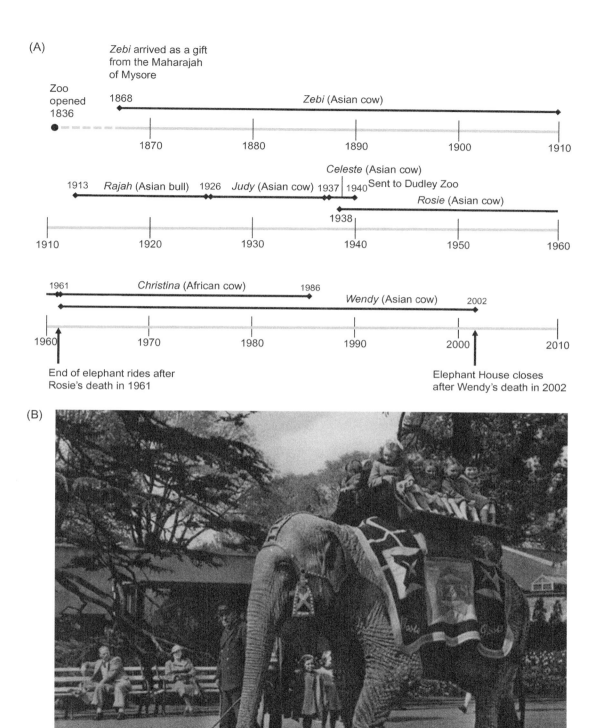

FIGURE 3.5 (A) The elephants kept at Bristol Zoo, United Kingdom, between 1868 and 2002. (B) Rosie giving rides at Bristol Zoo in the c1940s. Source: Courtesy Bristol Archives, Bristol City Council. Image reference number: 40826/ZOO/1/10.

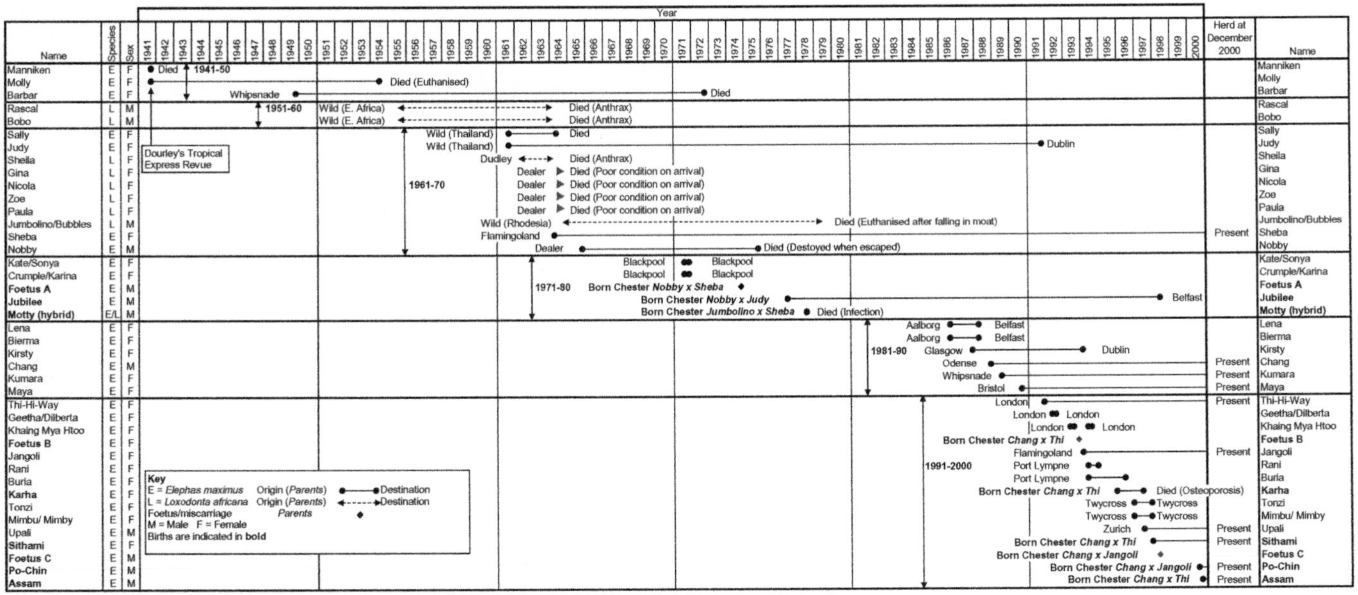

FIGURE 3.6 The elephants kept at Chester Zoo, United Kingdom, from 1941 to 2000. Source: From Rees, P.A., 2001. A history of the National Elephant Centre, Chester Zoo. International Zoo News 48, 170–183.

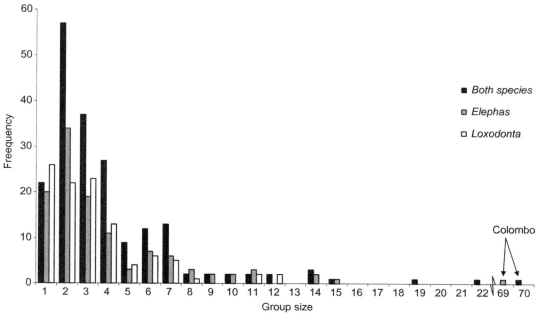

FIGURE 3.7 Distribution of elephant group sizes in zoos in 2006. The column labelled 'Both species' indicates the frequency of group sizes when groups containing *Loxodonta* and *Elephas* are included. The largest groups were held by Colombo Zoo, Sri Lanka, and included elephants at the Pinnawala Elephant Orphanage. Source: Reproduced with permission from Taylor & Francis. First published in Rees, P.A., 2009. The sizes of elephant groups in zoos: implications for elephant welfare. Journal of Applied Animal Welfare Science 12, 44–60.

Aquariums (AZA) recommended that breeding facilities should keep herds of 6–12 elephants. The British and Irish Association of Zoos and Aquariums (BIAZA) recommended keeping together at least four cows over 2 years old. Over 69% of Asian and 80% of African cow groups — including those under 2 years — consisted of fewer than four individuals (Fig. 3.7).

In 2009 I suggested that the welfare of individual elephants should outweigh all other considerations and zoos should urgently seek to integrate small groups into larger herds. The trend towards moving elephants between facilities to increase herd size and closing inadequate facilities is discussed in Section 9.3. To examine recent changes in elephant mean group size, I have compared the ISIS data from 2006 with the data held in ZIMS in 2018 in Table 3.1. These data have been extracted from Tables 8.1 and 8.2.

A basic analysis suggests there has been a small increase in mean group size in both genera between 2006 and 2018, on a global basis. However, only the increase in group size in *Loxodonta* is statistically significant (*Loxodonta*: $t = 1.675$, $d.f. = 205$, one-tailed test, $P < .05$; *Elephas*: $t = 0.314$, $d.f. = 274$, one-tailed test, $P > .10$). There are regional inconsistencies. In Europe, the mean group size has increased very slightly in *Loxodonta*, but in *Elephas* it has fallen by some 9.5%. In North America the mean group size has increased in both genera — 60.6% in *Loxodonta* and 18.1% in *Elephas* — but mean group sizes are still

TABLE 3.1 Changes in elephant mean group sizes in zoos between 2006 and 2018.

	Loxodonta		Elephas	
	2006	2018	2006	2018
Global totals				
Elephants	336	397	495	750
Zoos	104	103	114	162
Mean	3.23	3.85	4.34	4.63
European totals				
Elephants	116	182	239	305
Zoos	42	45	54	76
Mean	3.95	4.04	4.43	4.01
N. American totals				
Elephants	147	178	124	100
Zoos	53	40	44	30
Mean	2.77	4.45	2.82	3.33

Source: *Based on data from ISIS; Anon., 2006a. ISIS Abstracts:* Loxodonta africana. *International Species Information System. https://app.isis.org/abstracts/abs.asp (accessed 27.10.06.); Anon., 2006b. ISIS Abstracts:* Elephas maximus. *International Species Information System. https://app.isis.org/abstracts/abs.asp (accessed 27.10.06.); Rees, P.A., 2009. The sizes of elephant groups in zoos: implications for elephant welfare. Journal of Applied Animal Welfare Science 12, 44–60; ZIMS, 2018a. Species holding report:* Elephas maximus/*Asian elephant. Zoological Information Management System. Species 360. https//www.species360.org (accessed 26.03.18.); ZIMS, 2018b. Species holding report:* Loxodonta africana/*African elephant. Zoological Information Management System. Species 360. https//www.species360.org (accessed 26.03.18.)*

only 4.45 and 3.33, respectively. Based on these data there appears to have been little progress towards increasing the sizes of elephant group in zoos with the exception of *Loxodonta* in North America, but this population started from a low base.

Although group sizes have not altered greatly in zoos, there have been improvements in sex ratio. In 2006 the global ratio of males to females was 1:3.93 in *Loxodonta* and 1:3.14 in *Elephas*. These had changed to 1:2.98 and 1:2.32, respectively, by 2018, probably reflecting the increased success of captive breeding programmes in zoos (Table 7.2). Elephants taken from the wild in the past produced a skewed sex ratio in favour of females as they were preferred by zoos and circuses, probably because they were perceived to be less aggressive and more manageable than bulls. Again, there are regional inconsistencies. While sex ratios improved in both genera in Europe, there was a slight deterioration in *Elephas* in North America, from 1:4.39 in 2006 to 1.4.88 in 2018.

Lewis and Fish (1978) noted that the entertainment magazine *Billboard*'s 1952 census of elephants in the United States listed 264 elephants, only 92 of which were in zoos and only six were bulls. Bulls were routinely executed in these times either because they had killed a handler or because they had become unmanageable. All but 8 of the 264 elephants were Asian.

3.2.3 Social behaviour and breeding

While captive breeding requires that animals are moved from one zoo to another, it is important that social groups are not unnecessarily fragmented (Rees, 2001b). In the wild, elephant society (in both species) is dominated by the relationships formed between females (Douglas-Hamilton and Douglas-Hamilton, 1975; Moss, 1988; Sukumar, 1994). Similar strong bonds have been observed among captive elephants.

These studies illustrate that elephants need to interact with other elephants and, furthermore, they form special bonds with particular individuals. Female friendships may be particularly important in the rearing of calves. Allomothering is an important feature of calf rearing (Rapaport and Haight, 1987) and breaking social bonds between adult cows may interfere with the normal social development of calves.

The development of normal courtship and copulatory behaviour in mammals requires the presence of conspecifics (Estep and Dewsbury, 1996). Juvenile male elephants are stimulated to mount juvenile females by observing adults mating (Rees, 2004b). In some zoos young female calves are reared by their mothers in the absence of adult or juvenile bulls, thus depriving these calves of observing adult sexual behaviour and engaging in play mounting with calves of the opposite sex (see Chapter 4: Elephant reproductive biology). Dorresteyn and Terkel (2000b) reported that zoos which attempted to form cohesive family groups were more successful in producing calves than other facilities, although they presented no supporting data. Research on wild African elephants has suggested that families with older matriarchs are able to rear more calves because of their advanced social awareness (McComb et al., 2001).

3.3 Associations between individuals and friendships

The existence of special relationships (friendships) between female Asian elephants in zoos is well known (Garaï, 1992) and should be taken into account by zoos when they transfer cows to other facilities for breeding purposes.

Garaï recorded the existence of special relationships between female Asian elephants based on an analysis of spatial proximity, partner-specific reactions to arousal and vocalising as well as the absence of agonistic behaviour, in three different zoo groups. While collecting data on the activity budgets of the members of the Asian elephant herd at Chester Zoo (Rees, 2009b), I also recorded their associates at 15-minute intervals (Box 3.1). Clear friendships within the herd emerged from this analysis and it was apparent that the pattern of associations between individuals changed with time.

Some zoos view acyclic cows as 'expendable' and as an unproductive burden on resources. On the contrary, they may be a social asset in a captive herd as partners in social exchanges and even as allomothers. Schmid (1995) has emphasised the importance of elephants having a choice of social partners. She found that 16 out of 23 circus elephants kept in paddocks exchanged social contacts most frequently with an animal they could not reach when they were shackled. It has been suggested that integration into unfamiliar groups may be the reason why only 20% of breeding transfers have resulted in pregnancies (Kurt, 1994; Schmid, 1998).

Koyama et al. (2012) studied the effect of the loss of a sole companion — a conspecific male — on the behaviour of a female African elephant in Higashiyama Zoo in Japan for a year after his death. They concluded that management changes which involved habituating her to a new space — an indoor exhibition room — affected her mental stability in the early stages of social isolation.

Kinship is important in the welfare of zoo elephants. The social interactions within two groups of four female Asian elephants housed in unidentified zoos 'within the United Kingdom and Ireland' were studied by Harvey et al. (2018) through a limited social

---

BOX 3.1 Associations between individuals in the herd of Asian elephants at Chester Zoo

Recordings were made from January to November 1999 on 35 days when all eight elephants were able to associate freely. Recordings were made every 15 minutes while the elephants were in their outside enclosure. Individuals were considered to be associating if they were within one adult elephant length of each other. An association index was calculated for each dyad using the method employed by Schaller with wild African lions (*Panthera leo*) (Schaller, 1972):

$$\text{Association index} = \frac{2N}{n_1 + n_2}$$

where

N = the number of times animal 1 and animal 2 were seen together (including in a group with others)

n_1 = the total number of times animal 1 was seen (alone or as part of a group)

n_2 = the total number of times animal 2 was seen (alone or as part of a group)

An index of 0 indicated that two individuals were never seen together, while an index of 1.0 meant that they were always recorded together.

It was clear that there were strong bonds between certain adult female Asian elephants and that the frequency of dyadic associations changed with time (Fig. 3.8). The strongest associations were seen between *Thi*, her daughter *Sithami* and *Sithami*'s allomother *Sheba*. The most sociable elephant was *Sithami* — as she spent most of her time near her mother and/or *Sheba* — and the least sociable was *Kumara* (Fig. 3.9). The young bull, *Upali*, was very active and moved around the enclosure a great deal, reducing the likelihood that he would be recorded near others. In addition, he was sometimes excluded from the herd by adult females if he annoyed them (see Box 3.2).

In the last 4 months of the study there were fewer strong associations within the herd than in the previous 6 months. This may have been because *Sithami* was straying further from *Thi* and *Sheba* as she grew older — she was 1 year old at the beginning of the study — and because a bull pen was opened in May 1999. By July all of the elephants except *Thi* and her calf were seen using the bull pen, so this provided additional space for the herd. From the third week in August all the elephants used the bull pen.

(*Continued*)

BOX 3.1 (Continued)

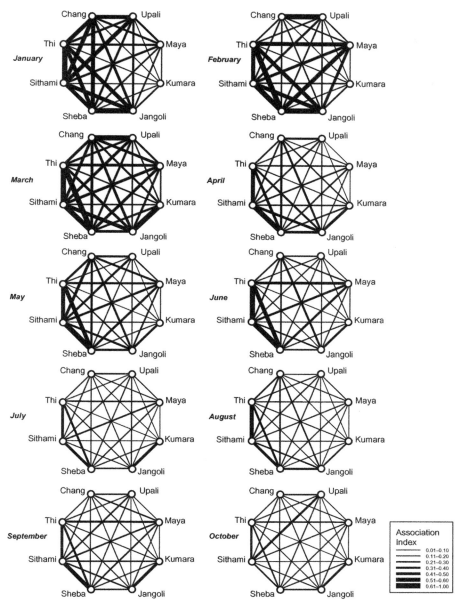

FIGURE 3.8 Sociograms of the relationships between elephants at Chester Zoo, United Kingdom. Individuals were defined as associating when they were closer than one adult elephant body length apart. Source: *Author's unpublished data.*

(Continued)

BOX 3.1 (Continued)

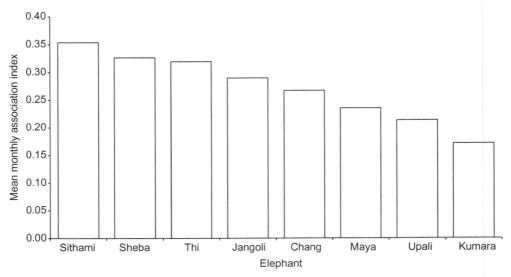

FIGURE 3.9 Mean monthly association indices for individuals in the Chester Zoo herd. Source: *Author's unpublished data.*

network analysis over a period of 8 days using video recordings. One group contained all-related individuals and the other group was a mixture of related and unrelated animals. Unsurprisingly, more affiliative and fewer agonistic interactions were observed in the related group than the other group and a higher frequency of noncontact displacement behaviour was observed in the related group (suggesting an established functioning hierarchy avoiding the need for overt aggression over resources). The authors suggested that this study supported the keeping of elephants in multigenerational family herds.

Moss (1988) described greeting ceremonies performed by wild African elephants who have not seen each other for a long time. Captive elephants who are close friends may perform intense greeting ceremonies after being separated for just a short time. This may occur in some zoo elephants several times a day, even when the individuals have not been out of sight of each other. During such ceremonies, the participants stand close together, urinate, defaecate, vocalise and touch each other with their trunks (Fig. 3.10). Elephants may also greet their keepers in a similar manner (see Fig. 8.18).

Taken together these studies illustrate not only that elephants in captivity need to interact with other elephants, but that they also form special bonds with particular individuals and with keepers.

FIGURE 3.10 A greeting ceremony between adult Asian elephants at Chester Zoo, United Kingdom.

3.4 Introductions into an elephant group

Elephants are routinely moved between facilities for breeding and other purposes, and it is important to ascertain that the new herd members are compatible with the existing group. This requires that individuals are only introduced to each other after a period of association separated by barriers.

In 2018–19, Twycross Zoo, United Kingdom, sent its four female Asian elephants (*Mimbu*, *Tara*, *Noojahan* and her calf *Esha*) to join an adult female (*Kate*) at Blackpool Zoo, United Kingdom. The elephants were sent in two groups and during the process, *Kate* was kept in a separate enclosure where she could still see, smell and touch the newcomers without any risk of exposing any of the elephants to intraspecific aggression (Fig. 3.11).

Wild female elephants are highly social, but young bulls leave their natal herd and join loosely associated all-male groups. Historically, there have been few bulls kept in zoos, and as numbers have increased the bull population has outgrown the bull accommodation available. Zoo Heidelberg was the first zoo in Germany to keep a group of young male Asian elephants and was the first zoo to attempt to integrate another young bull into an established group. Hambrecht and Reichler (2013) studied the social behaviour of this group for the first 4 months after the introduction of the newcomer to the herd. They found that the three members of the original group interacted much more among themselves than with the new member. The behaviour of the new elephant indicated social isolation and an elevated level of stress. However, the existence of a dominance hierarchy within the group and the progressive integration of the newcomer suggested a stable social structure. Hambrecht and Reichler concluded that the results of their study supported the concept of keeping young bulls in same-sex groups in zoos.

FIGURE 3.11 In 2018−19 Twycross Zoo, United Kingdom, transferred its four female Asian elephants − *Mimbu, Tara, Noojahan* and her daughter *Esha* − to Blackpool Zoo, United Kingdom. The resident female Asian elephant (*Kate*) is pictured here in the rear pen at Blackpool prior to her introduction to the new arrivals.

3.5 Protective formations

One major advantage of group living is the increased ability to protect individuals against predators. A number of mammal species, for example musk oxen (*Ovibos moschatus*) (Manning, 1972), exhibit group vigilance behaviour that results in adult animals forming a protective 'circle' around their offspring when they are threatened by a predator. Elephants exhibit similar behaviour.

From his experiences of working with Asian elephants in Burma (now Myanmar) while in the British Army during the Second World War, Williams (1951) wrote that when the elephants are disturbed they do not stampede, but

> '...with an uncanny intelligence, they close up around one animal as though they were drilled, and their leader then decides on the best line of retreat.'

Although Williams was not a scientist, this is more or less what happens.

Spinage (1994) has used the term 'defensive ring' to describe this protective response of African elephants and Moss (1988) describes the same phenomenon using the terms 'defensive formation' or 'defensive circle' in which the elephants form a circle with the large females facing the threat (Fig. 3.12A). These 'circles' are not really circles, but rather very close groupings of elephants. Moss describes the elephants rumbling, flapping their ears, shaking their heads and making mock charges towards the threat. Dugatkin (1997) has referred to 'circular defence formations (with juveniles in the centre)' when discussing

FIGURE 3.12 Protective formations: (A) A herd of African elephants responding to passing trucks in the Serengeti National Park, Tanzania; (B) Asian elephants at Chester Zoo, United Kingdom, responding to the noise from a chain saw being used by a nearby construction worker. The elephants are facing the source of the sound.

the behaviour observed in the wild (Douglas-Hamilton, 1972; Dublin, 1983; Lee, 1987). In African elephants, the matriarch and larger females will take a lead position (or guard the rear, depending on the situation) when threatened. Elephants may also form such circles when a cow is giving birth.

This type of defensive behaviour is well documented in the scientific literature, but some second-hand accounts of the protective responses of elephants do not do these events justice. In his book *Sociobiology*, Edward Wilson describes an incident witnessed by Iain Douglas-Hamilton in Lake Manyara National Park in Tanzania:

> *When Douglas-Hamilton felled a young bull with an anesthetic dart, the adult cows rushed to his aid and tried to raise him to his feet.*
>
> (Wilson, 1980)

Wilson's brief description gives no indication of the scale and extent of the response of this herd. From Wilson's account it is not possible to tell how many animals were involved or whether the incident lasted 10 minutes or an hour. Douglas-Hamilton's first-hand account — which occupies almost three pages of his book (pp. 119−121) — describes the incident in great detail. The incident involved a total of 67 elephants from five family units and lasted over 2 hours (Douglas-Hamilton and Douglas-Hamilton, 1975).

The type of defensive behaviour I have described here depends upon the presence of a social group. In a zoo environment, large social groups are rare. Highly threatening situations are also rare but it is, nevertheless, possible to observe this behaviour in a zoo.

At Chester Zoo I observed the Asian elephant herd form a protective circle within their enclosure on several occasions. The most spectacular was in response to the arrival of a mobile crane that had to drive along the internal road past the edge of the elephant enclosure. When it arrived the adult cows gathered around the calf, *Sithami*, but the adult bull, *Chang*, ignored it and continued to feed. When the 'threat' had passed the group dispersed. Later, when the crane left the zoo, *Chang* was in the bull pen with *Jangoli*. As soon as the crane approached, *Chang* and *Jangoli* walked quickly into the main enclosure (a distance of over 100 m) to join the remainder of the herd who were forming a protective group as before. The circle quickly fragmented as the crane (or the perceived danger) passed.

On a separate occasion the herd formed a protective circle around the two young herd members (*Upali* and *Sithami*) when a construction worker was using a chain saw to cut wood for the decking that was to form part of the public area of the new elephant exhibit. The sound of the saw was almost certainly one they had not previously experienced and they stood facing the source (Fig. 3.12B).

The elephants at Chester Zoo also formed protective groups during heavy rainstorms, often moving out into the middle of the enclosure and dusting in the rain. Spinage (1994) has described the great excitement exhibited by orphan elephants kept by Daphne Sheldrick in Tsavo during rainstorms.

If this behaviour is learned, then it is an important component of the behavioural repertoire of adult elephants that should be learned by young elephants so that they will know how to protect future generations of youngsters if they are ever released into the wild. As this is a social activity, it can only take place in an appropriately sized social group. At the moment, many zoos cannot provide this.

3.6 Dominance hierarchies

The possible existence of dominance hierarchies in elephants has received little scientific attention until relatively recently, although zookeepers often refer to the dominance relationships between animals in their care and at least one study has assumed that keepers provided accurate information about these relationships (Freeman et al., 2010b). Archie et al. (2006) conducted the first quantitative analysis of dominance relationships within 'family' groups of wild African elephants, by recording overt aggressive behaviours [similar to those recorded by Lee (1986)] along with displacements and supplants. Sukumar (2003) previously noted that little attention had been paid to competition among matriarchs, although Dublin (1983) studied the effect of female competition on reproductive success and Moss et al. (2011) more recently provided a detailed discussion of the social dynamics of the elephants of Amboseli National Park, Kenya, including a consideration of aggression.

Dominance hierarchies have been described within and between family groups of female African elephants, and dominance in both males and females is based largely on intrinsic factors relating to age. In free-ranging African elephants, dominance relationships among the matriarchs of different social groups are primarily based on age rather than physical or group size. Furthermore, group matriarch rank influenced dominance relationships among other (nonmatriarchal) females in the population (Wittemyer and Getz, 2007).

O'Connell-Rodwell et al. (2011) quantified the existence of a linear dominance hierarchy for the first time in a group of elephants in Etosha National Park, Namibia. Under normal arid conditions when water was limited and resource competition was high, they found that males formed a stable linear dominance hierarchy. However, there was no linearity to the dominance hierarchy in unusually wet years with increased water availability. Under these conditions there were fewer interactions between individuals and more agonistic behaviours were exhibited, particularly in lower-ranking individuals.

The preservation of dominance systems in elephants may have important implications for the survival of migratory elephants in the wild. Wittemyer et al. (2007) collected data from seven families of free-ranging African elephants of different ranks and showed that dominant groups disproportionately used preferred habitats; they avoided unprotected areas thereby limiting their exposure to predation/conflict with humans and expended less energy than subordinate groups during the dry season. These behaviours were not recorded during the wet season, indicating that the spatial segregation of elephants is related to resource availability. The authors emphasised the importance of protecting the natural dominance systems in elephants for mitigating the ecological impacts of high elephant density.

A study of wild elephants in north-eastern India showed that tuskless bulls in musth were dominant over nonmusth bulls, and that musth and body size were more important than the possession of tusks as a male—male signal of dominance (Chelliah and Sukumar, 2013).

De Silva and Wittemyer (2012) used the simple ratio index of association to quantitatively compare social organisation in wild Asian and African savannah elephants. They compared the African elephants of Samburu (northern Kenya) with Uda Walawe Asian elephants (in Sri Lanka). They found that, compared with the African elephants, the Asian elephants in Uda Walawe were found in smaller groups, did not maintain coherent core groups, exhibited much less social connectivity at the population level and were socially less influenced by differences in seasonal ecology. The African population exhibited a significant increase in the number of breeding females per group and levels of association during wet seasons compared with dry seasons, whereas the Asian population did not.

African elephant society is well known to be hierarchical and multilevelled. De Silva and Wittemyer's study demonstrated that Asian elephants maintain a complex, well-networked society consisting of at least two differentiated types of associates that they termed 'ephemeral' and 'long-term' affiliates. They described the African elephant population as a multitiered society and the Asian elephant population as multilevelled. Elephants under human care are not kept in large enough social groups to develop these complex social structures.

Freeman et al. (2010b) reported that dominance hierarchies in captive elephant herds are common. They compared behavioural assessments of dominance relationships — based on 8 hours of recordings — with keeper assessments determined using questionnaires for 33 nonpregnant female African elephants housed at 14 zoos in North America. The authors found that body weight — and to a lesser extent age — were related to rates and types of 'body movements' (such as back away, displace, push and present) and that these demographic factors dictated the hierarchy, in a similar fashion to that observed in the wild. They concluded that short-term behavioural observations and zookeeper questionnaires produced similar profiles for these elephants.

For African and Asian elephants, dominance status in individuals living in zoos is positively correlated with age, relative size, temperament, disciplinary nature and willingness to share novel objects (Freeman et al., 2004). Social status was the best predictor of ovarian activity in African elephants. Acyclic females ranked higher in the dominance hierarchy than cycling females in the same group, and gave more discipline to herd mates. It was not clear from this study whether these were cause-and-effect relationships.

Coleing (2009) used social network analysis to study group structure and interactions in the herd of 10 Asian elephants (3.7) at Chester Zoo. She analysed 40 hours of video recordings collected over a 2-month period and found that most affiliative (body contact) interactions were between related individuals and that there was no clear dominance hierarchy within the four adult cows (*Sheba, Thi, Maya* and *Jangoli*). This latter finding concurs with my earlier analysis of aggressive interactions between these four cows (see Box 3.3).

3.7 Aggression, appeasement and chastisement

Aggression and dominance relationships in captive elephants are of interest for a number of reasons. First, the extent to which levels of aggression in captive herds are comparable to those seen in the wild may be an indicator of welfare. Second, knowledge of the existence of a dominance hierarchy within an elephant herd may inform management and husbandry decisions. Third, the identification of aggressive individuals, the factors that affect the frequency of aggressive behaviours and knowledge of the position of each animal in any hierarchy that exists, may have implications for the safety of zookeepers and other caretakers.

Elephant management guidelines published by BIAZA (Walter, 2010) state that aggression between zoo elephants is 'relatively common', although it acknowledges that this had not been quantified. A number of risk factors are identified, including a lack of relatedness between group members, mixing of unfamiliar animals and enclosures which do not allow sufficient flight distances or incorporate refuge areas. Walter (2010) suggests that there is a need to quantify aggression in zoo elephants and determine the relationship with risk factors due to welfare implications. In North America, the AZA's *Standards for Elephant Management and Care* (AZA, 2019a) refer to the need for staff to be aware of social compatibility between elephants and dominance hierarchies, and assert that institutions must be able to manage these and aggression in the group.

3.7 Aggression, appeasement and chastisement

Incidents involving aggression by captive elephants against people are not uncommon. People for the Ethical Treatment of Animals (PETA) documented incidents involving captive elephants in North America between January 1987 and June 2019, which resulted in 20 human deaths and more than 140 human injuries (PETA, 2019a). Since the 1980s, many — but not all — zoos have introduced a system of protected contact for managing their elephants (Roocroft, 2007).

The few studies of aggression in elephants living in zoos that exist have considered dominance hierarchies (Coleing, 2009; Gunhold et al., 2006; Lincoln and Ratnasooriya, 1996), 'problem' females (Tresz and Wright, 2006), the effect of oestrous state in females (Slade-Cain et al., 2008) and musth in males (De Oliveira et al., 2004; Niemuller and Liptrap, 1991), appeasement behaviours (Rees, 2004c), managing aggression during captive social group formation (Burks et al., 2004), and injuries and deaths among elephant caretakers (e.g. Benirschke and Roocroft, 1992; Gore et al., 2006).

Most studies of aggression in wild elephants have been conducted on African elephants and levels of aggression are generally considered to be relatively low. It occurs mostly between musth males competing for females (Moss, 1983; Poole, 1989), especially where densities are high (Whitehouse and Schoeman, 2003). In some populations, hyperaggression in bulls appears to have been caused by an inability to regulate stress-reactive aggressive states following a breakdown in social structure resulting from elephant culls (Bradshaw et al., 2005). High aggression levels have been observed in bulls kept within a small fenced area in Addo Elephant National Park, South Africa. This may have been caused by the high frequency of contact between musth and nonmusth bulls (Whitehouse and Schoeman, 2003). Jachowski et al. (2012) have suggested that chronic stress and increased refuge behaviour in African elephants may be linked to an increase in aggression towards humans. Prusty and Singh (1995) described male–male aggression in Similipal Tiger Reserve, Orissa, India, that resulted in the deaths of two tuskers.

In elephants, within family groups, adult aggression is often directed towards pubertal bulls and calves. Lee (1986) refers to aggressive interactions in African elephants consisting of contacts such as pokes, shoves, slaps with the trunk, chases and threats. Adults, especially temporarily associating males, were the most frequent initiators of aggression against calves. Immature females received significantly more aggression from their mothers than did immature males, possibly because females spend more time with their mothers than do males.

Wild elephants have been observed exhibiting interspecific aggression [e.g. towards bovids (Bovidae; Henshaw, 1972); black rhinoceros (*Diceros bicornis*; Berger and Cunningham, 2006); humans (Datye and Bhagwat, 1995)]. Young, orphaned musth African bulls that had been introduced to Pilanesberg National Park, South Africa, killed more than 40 white rhinoceros (*Ceratotherium simum*) between 1992 and 1997. Around 45 juvenile bulls had been translocated from Kruger National Park following culling operations. In 1998, six older bulls were introduced from the Kruger Park population and the killing ceased when musth in the young bulls reduced as a consequence (Slotow et al., 2000).

Aimed object throwing occurs in the wild [e.g. at a white rhinoceros (Wickler and Seibt, 1997)] and in zoos (Kühme, 1963). This behaviour has been observed in zoos and has management implications because of the risk to the public (Fig. 3.13). In 2016, a 7-year-old girl was killed by a stone thrown by an elephant in Rabat Zoo, Morocco (Anon., 2016a; Rothwell, 2016).

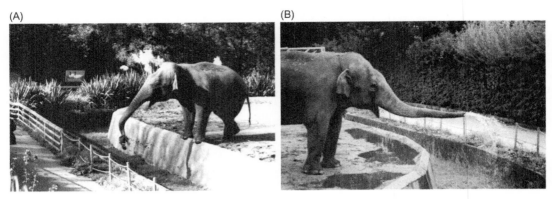

FIGURE 3.13 A bull Asian elephant (*Chang*) at Chester Zoo, United Kingdom, throwing soil (A) and squirting water (B) at visitors.

Historically, most elephant groups kept in zoos have been small, thereby providing little opportunity to study aggression in a group that approximates the size and composition of those found in the wild. A survey of 495 Asian elephants and 336 African elephants kept in 194 zoos found the mean group size to be 4.28 animals (Rees, 2009a).

Since June 2017, zoos that keep elephants in England have been required to meet more exacting welfare standards as a result of amendments to Appendix 8 (Special exhibits, Elephants) to the *Secretary of State's Standards of Modern Zoo Practice* (SSSMZP) published by the Department for Environment, Food and Rural Affairs (Defra, 2017). Section 8.8.3 requires a zoo to produce a long-term management plan for its elephants that must include 'herd compatibility details'. All elephants must have the option to get away from other elephants through the use of space or physical barriers (Section 8.8.12). Section 8.8.4 requires an individual welfare plan to be produced for each elephant that contains a behaviour profile. A 'bull profile' must be drawn up and documented for each bull from the age of 4 years to inform its management (Section 8.8.5). Information about aggressive behaviour should be an important part of these profiles, but little baseline data on aggression in elephants living in zoos is available.

Aggressive tactile behaviours within a group of six female Asian elephants (aged 16–64 years) living at Busch Gardens Tampa Bay, Florida, were studied by Makecha et al. (2012) over a period of 152 hours. One elephant did not initiate any aggressive tactile behaviour during this period. In the remaining five animals, the highest rate of initiation recorded was 3.06/h; the rate for the other four cows was 0.48/h or less. Aggressive tactile behaviours received by each elephant ranged from a single incident for one animal to a maximum rate of 1.95/h. Two pairs of elephants did not direct any aggressive behaviours towards each other during the study. The individuals in these pairs also exhibited low rates of nonaggressive tactile behaviour towards each other. Only 25 aggressive tactile behaviours were directed by low-ranking to high-ranking elephants (none of which involved the trunk), while 427 such behaviours were directed from high-ranking to low-ranking elephants.

Occasionally fighting within a group of elephants living in a zoo may become serious in that it threatens the safety of individual animals. In November 1999, three adult Asian cows — two of which were accompanied by their male calves — were moved from Rotterdam Zoo in The Netherlands to Port Lympne Zoo in the United Kingdom because they were fighting with other members of the herd (Rees, 2001b). It was thought that this aggressive behaviour was caused by insufficient space at Rotterdam. As part of a study of their social behaviour, I recorded aggressive interactions between individuals within the Asian elephant herd kept at Chester Zoo, including the chastisement of juvenile elephants. A summary of this study is presented in Box 3.2.

BOX 3.2 Aggression and dominance within the herd of Asian elephants at Chester Zoo

As part of a larger study of the herd of Asian elephants (2.6) kept at Chester Zoo, I made all occurrence recordings of overt aggressive behaviours (Table 3.2) within the group for a total of 213 hours over 47 days.

In total, 556 aggressive behaviours were recorded. Most (93.6%) consisted of head butts (32.2%), trunk slaps and pushes (25.4%), kicks (19.8%) and chases (16.2%) (Figs 3.14 and 3.15). The average frequency of aggressive behaviours exhibited for all members of the herd was 0.326 incidents per hour, ranging from 0.821/h exhibited by the juvenile bull to just 0.066/h in the youngest adult cow.

Most aggression was directed at the juvenile bull ($>1/h$) largely because of adult cows pushing him out of the herd (Fig. 3.16). He received 40% of the aggression observed; 47.5% of this was from the oldest cow, even though she exhibited almost no aggression towards other adults. There was no evidence of a hierarchy among the five adult cows (Table 3.3).

TABLE 3.2 An ethogram of aggressive behaviours.

Behaviour	Definition
Aggressive restraint	Pushes another to the ground and may subsequently hold it down, preventing it from getting to its feet. Attempting to do this
Bite/tusk	Attacks another with mouth open and bites/tusks it, either breaking the skin or causing no apparent damage. Usually directed at the rear or flank
Chase	Pursues another, who withdraws in response
Chastisement	Behaviour directed by adult at juvenile. Consists of holding the juvenile's trunk and kicking or rubbing it gently with a forefoot (Fig. 3.17A), tapping juvenile's face or head several times with trunk in close succession (Fig. 3.17B, C) or pulling at a juvenile's ear with the trunk
Head butt	Pushes with the head, usually directed at the rear of another
Kick	Strikes another with the foot, usually using the rear leg
Threat	Throws trunk out at another, lunges towards another or charges towards another briefly (but does not pursue)
Trunk slap/push	Briefly slaps or pushes with the dorsal surface of the distal end of the trunk, or spars with the trunk when separated from another elephant by a fence

(Continued)

BOX 3.2 (Continued)

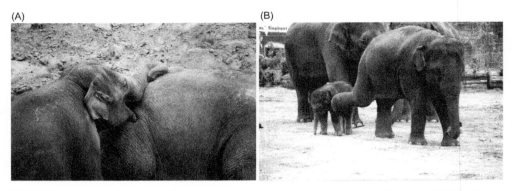

FIGURE 3.14 *Upali*, a young bull, (A) pushing and biting an adult cow (*Maya*) and (B) kicking a young calf.

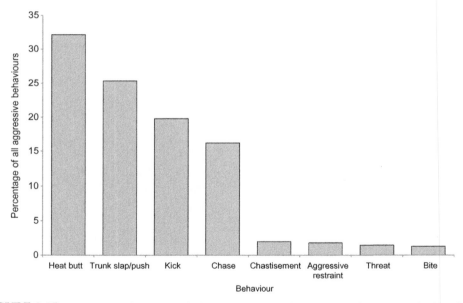

FIGURE 3.15 Frequency of aggressive behaviours exhibited by members of the Asian elephant herd at Chester Zoo, United Kingdom. Source: *Author's unpublished data*.

(*Continued*)

BOX 3.2 (Continued)

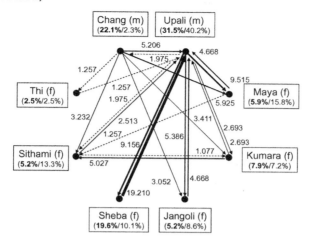

FIGURE 3.16 Aggressive interactions between members of the Asian elephant herd at Chester Zoo, United Kingdom. The diagram shows the percentage of all aggressive behaviours directed at each elephant (only values over 1% of total shown). Figures in parentheses show the total percentage of all aggressive behaviours exhibited (bold) or received by each animal. Note that aggressive interactions between all dyads of adult females were less than 1% of the total within the herd and therefore not represented by arrows (see Table 3.2). Source: *Author's unpublished data*.

TABLE 3.3 Mean rates of aggressive behaviours exchanged between herd members (incidents/h) calculated over 47 days.

Elephant	Aggressive behaviours received/h								Total
	Sithami juv. ♀	Upali juv. ♂	Chang ♂	Thi ♀	Kumara ♀	Maya ♀	Jangoli ♀	Sheba ♀	
Sithami juv. ♀	—	0.066	0.000	0.014	0.028	0.005	0.005	0.019	0.136
Upali juv. ♂	0.052	—	0.052	0.019	0.070	**0.249**	**0.141**	**0.239**	0.821
Chang ♂	0.084	**0.136**	—	0.033	0.089	**0.155**	0.080	0.000	0.577
Thi ♀	0.019	0.033	0.005	—	0	0.005	0	0.005	0.066
Kumara ♀	**0.131**	0.070	0.005	0	—	0	0	0	0.206
Maya ♀	0.033	**0.122**	0	0	0	—	0	0	0.155
Jangoli ♀	0.014	**0.122**	0	0	0	0	—	0	0.136
Sheba ♀	0.014	**0.497**	0	0	0	0	0	—	0.511
Total	0.347	1.046	0.061	0.066	0.188	0.413	0.225	0.263	2.608

Note: The only aggressive interactions between adult cows (enclosed within bold border) were a single kick directed by *Thi* at *Maya* and another by *Thi* at *Sheba*.
Values ≥ 0.1 = light grey; ≥ 0.2 = mid grey; ≥ 0.4 = dark grey.
Source: *Author's unpublished data*.

(Continued)

BOX 3.2 (Continued)

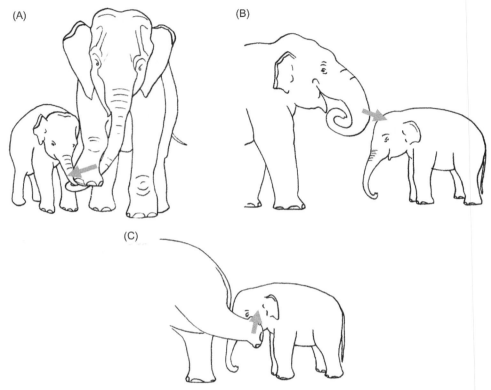

FIGURE 3.17 Chastisement of calves by adults in the herd of Asian elephants at Chester Zoo, United Kingdom: (A) Adult female holding the trunk of a calf with her trunk and tapping the side of the calf's trunk with her forefoot; (B) adult bull tapping the head of a calf with his trunk; (C) adult kicking back gently at a calf. Source: *Courtesy Dr A. J. Woodward, based on photographs taken by the author.*

The levels of aggression were considered low and similar in type and frequency to those found in wild African elephants. Aggression was focused on juveniles (53.4%) — especially in relation to chasing the juvenile bull from the herd in a manner similar to that seen in wild herds — and one particular adult cow (15.8%).

Some of the aggression observed appeared to be related to the chastisement of the young female calf, *Sithami*, or the juvenile bull, *Upali*. It consisted of gently taping the young elephant with the anterior surface of the trunk (Fig. 3.17), a rear foot, or holding the offending youngster with the trunk and hitting them with a forefoot swung laterally.

Upali frequently attempted to mount *Sithami* during this study (see Box 4.3) and often harassed and attacked some of the adult females. As a consequence of this he was occasionally chased from the group by one or more of the adult cows and sometimes took refuge behind the bars or a corral within the enclosure to which he could gain access, but the adults could not (Fig. 3.18).

(*Continued*)

BOX 3.2 (Continued)

FIGURE 3.18 A juvenile bull (*Upali*) being chased from the herd at Chester Zoo, United Kingdom: (A) *Upali* being chased by two adult females – *Jangoli* (left) and *Sheba* (right) – into a corral; (B) *Upali* sparring with *Sheba* from behind a steel fence.

This study provides important baseline data from a stable herd in which there was no perceived aggression or compatibility problem against which other herds may be compared, and provides evidence of the importance of refuges for young bulls.

Ritualised appeasement behaviour is known in a wide range of mammalian species. Its purpose is to avoid harm to a subordinate individual and includes exposure of vulnerable parts of the body, lowering of the head and body, lying down, sexual presentation and submission to mounting (Ewer, 1968).

Early descriptions of aggression in elephants were concerned with the fighting behaviour of adults (Carrington, 1958; Kühme, 1961, 1963). Estes (1991) has categorised defensive/submissive displays in African elephants as: avoidance (turning away, backing up, running away), flattening ears, arching back, raising tail, agitated trunk movements, touching temporal gland, throwing dust, pawing, foot-swinging, swaying and exaggerated feeding behaviour. In an agonistic encounter between two bulls, the smaller animal flattens his ears, keeps his head lowered, moves backwards and sideways and makes writhing trunk movements.

Langbauer (2000) summarised apprehension and submissive behaviour in elephants as jaw out, face check, trunk twitch, trunk curl, swaying, tail up and back in (Douglas-Hamilton, 1972; Kühme, 1961; Payne and Langbauer, 1992; Poole, 1999).

During a study of the behaviour of a herd of Asian elephants at Chester Zoo, United Kingdom, I recorded a number of instances of appeasement behaviour. These are described in Box. 3.3

3.8 Personality

Weinstein et al. (2008) defined personality as 'those characteristics of individuals that describe and account for consistent patterns of feeling, cognition and behaving'. Scientific studies of animal personality often use behaviour measures, behaviour ratings or adjective ratings to describe personality (e.g. Uher and Asendorpf, 2008). Such studies are not new but began in the early part of the 20th century (Pavlov, 1906).

Personality assessment has implications for captive management and reintroduction programmes because differences in behaviours may affect postrelease survival (Bremner-Harrison et al., 2004). The value of animal personality ratings as a tool for improving the breeding, management and welfare of zoo mammals has been discussed by Tetley and O'Hara (2012).

Personality assessment of captive elephants is a relatively new field of study. It has important implications for keeper safety and may be important in the reintroduction, translocation and relocation of elephants.

Seltmann et al. (2018) collected personality data for 150 female and 107 male Asian semicaptive elephants owned by Myanmar Timber Enterprise using questionnaires in which mahouts rated individual animals against 28 behavioural adjectives (translated into Burmese). These included active, affectionate, aggressive, inventive, moody, obedient, playful, solitary, timid and vigilant. Explanatory factor analysis was used to determine the personality structure of the population and this suggested that personality was manifested as three factors, namely attentiveness, sociability and aggressiveness. This structure was the same for both sexes. The authors reported that this three-factor personality structure is similar to that found in other studies of elephants and nonhuman primates and may be the result of common ancestry, but acknowledged that this could be a methodological artefact because personality questionnaires tend to employ similar questions.

Yasui et al. (2013) reported the first discovery of an association between genetic polymorphism and personality in elephants. They collected information on the personalities of elephants from 75 elephant keepers and 196 elephant DNA samples. Three genes that are expressed in the brain, and evidently linked to personality traits, contained polymorphic regions, namely the androgen receptor, fragile x-related mental retardation protein interacting protein (NUFIP2) and acheate-scute homologs 1 (ASH1). Yasui et al. examined the

BOX 3.3 Appeasement behaviours observed in the Asian elephant herd at Chester Zoo

I have described previously unreported submissive behaviours exhibited by Asian elephants at Chester Zoo (Rees, 2004c). These behaviours are rare and were only observed on a few occasions during 420 hours of observations made on 93 days over two and a half years. When approached by an adult bull, some adult cows were observed to decline their head and bow down, lowering the head to ground level (Fig. 3.19). This occurred when the bull was in musth, or when he had been separated from the cows for many days. On other occasions, cows kneeled on their rear legs and, in extreme cases, lay prone (Fig. 3.20). A cow was also observed bowing low to a very young calf soon after birth and when his mother was far away. This behaviour may have been intended to reassure the calf that the cow was not a threat. The appeasement of aggression by submission to ritualised mounting was also observed.

FIGURE 3.19 (A) An adult cow (*Kumara*) 'bowing' to an adult bull (*Chang*) while he was in musth; original photograph. (B) A drawing showing the posture of both elephants based on other images taken of similar behaviour. Source: *From Rees, P.A., 2004. Unreported appeasement behaviours in the Asian elephant (Elephas maximus). Journal of the Bombay Natural History Society 101, 71–78. Drawing courtesy Dr A. J. Woodward.*

(Continued)

BOX 3.3 (Continued)

FIGURE 3.20 Sequence of photographs showing an adult female Asian elephant (*Kumara*) appeasing an adult bull (*Chang*) taken over a period of 14 minutes at Chester Zoo, United Kingdom: (A) *Kumara* (left foreground) bowing to *Chang* (right) (1115); (B) *Kumara* lowering her head and rubbing it against *Chang* (1116); (C) *Kumara* (right) bowing to *Chang* (1124); (D) *Kumara* (left) lying prone briefly as *Chang* (middle) walks behind her from left to right (1129). Source: *From Rees, P.A, 2004. Unreported appeasement behaviours in the Asian elephant* (Elephas maximus). *Journal of the Bombay Natural History Society 101, 71–78.*

Kühme (1963) described a kneeling behaviour in captive African elephants which occurred at the end of a hostile encounter. The elephant would kneel with ears spread wide in front of its partner, a dog or hostile human.

Appeasement behaviours at Chester Zoo were displayed by adult female Asian elephants in situations where they were being attacked or harassed by bulls, or following extended periods of separation from the adult bull. Similar behaviour appeared to be used by an adult cow to signal the absence of threat to a young, unrelated calf.

Lowering the head is an antithreat appeasement behaviour in elephants (Manning, 1972) since aggressive animals generally hold their heads high. Lying down may also be categorised as antithreat behaviour. Crouching down and kneeling on the rear legs may be intended to appease

(*Continued*)

BOX 3.3 (Continued)

an aggressor by arousing a conflicting, sexual, tendency. In the few incidents of submission to mounting that were observed, this appeared to prevent further aggression, at least temporarily.

Appeasement behaviours in Asian elephants may serve the function of allowing subordinate animals to remain within a social group. It is interesting that the adult bull (*Chang*) showed considerable aggression towards *Kumara* at the beginning of the study, but two and a half years later there was little sign of antagonism between them. Keepers reported that when *Chang* was younger (and smaller than her) *Kumara* had been very aggressive towards him.

It is not surprising that the appeasement behaviours reported here exist in Asian elephants as they are similar to behaviours displayed by many ungulate species. Neither is it surprising that appeasement behaviour of this type has not previously been reported from the wild. Studies of captive rhesus monkeys (*Macaca mulatta*) have shown that submission and appeasement gestures increase in crowded conditions (De Waal, 1996). Such behaviour is more likely to be observed in a captive environment because the animals who behave antagonistically towards each other cannot easily avoid contact.

---

association between intraspecific genetic variation and personality in 28 African and 17 Asians elephants, and found that the ASH1 genotype was associated with neuroticism in the Asian species. Individuals with long alleles had higher neuroticism scores than those with short alleles.

The stress responses of eight Asian elephants during relocation from the Cocos (Keeling) Islands – an Australian territory in the Indian Ocean – en route from Thailand to mainland Australia were studied by Fanson et al. (2013). They examined adrenocortical activity in serum, urine and faeces, and found that an individual elephant's adrenocortical response to relocation was correlated with personality traits. Those animals that were identified as 'curious' by keepers exhibited a more prolonged release in faecal glucocorticoid metabolites (FGM) post transfer, while elephants described as 'reclusive' showed a greater increase in FGM values.

CHAPTER 4

Elephant reproductive biology

In the past there has been little incentive to breed elephants in zoos because there has been no problem over obtaining a replacement from the wild for any animal that dies.

S. Keith Eltringham (1984)

4.1 Introduction

This chapter is largely concerned with the behaviour associated with reproduction in elephants. Although it is not my intention to provide a detailed account of the reproductive anatomy and physiology of elephants, it would be remiss of me to ignore the contribution that elephants living in captivity have made in these areas, helping us to understand the problems experienced by captive breeding programmes. Recent studies of elephant reproduction have focused on gathering physiological data from living animals held in captivity in an attempt to improve breeding success. Particular emphasis has been placed on the development of noninvasive methods of monitoring oestrus and the development of artificial insemination (AI).

Birth statistics, such as the length of gestation and birth weights, are considered in this chapter, while the demography of elephant populations is discussed in Chapter 6, Elephant ecology and genetics.

4.2 Historical accounts of sexual behaviour

'A female elephant was observed throughout most of oestrus, and was then shot'. This was the opening sentence in the abstract to a paper by R. V. Short, a Cambridge University scientist, published in 1966, about a study of reproduction in a single wild African elephant (Short, 1966). The study included observations on the behaviour of this elephant and attendant bulls, and investigated the formation of the corpus luteum, hence the need to shoot the female. This work was undertaken at a time when it was not uncommon to kill wild animals to study them. Prior to this study, a number of descriptions of elephant reproductive behaviour had been published that vary considerably in both detail and accuracy, with some later texts perpetuating some elements of earlier accounts, often without reference to any specific authority.

In *Elephant Bill*, Williams (1951) wrote:

In the final mating position the male is standing almost vertically upright, with the forelegs resting gently on the female's hindquarters.

Carrington's subsequent account (1958) says:

...he is standing almost upright, with his fore-feet resting gently on the female's hindquarters.

Carrington (1958) also wrote:

After congress, the male moves off and remains silent; the female sometimes trumpets softly, and shows her excitement by flapping her ears and whisking her tail.

Carrington's account of sexual behaviour is anthropomorphic and quotes no specific authority. A later account by Spinage (1994) is remarkably similar and states that:

...the cow has only been recorded as showing 'some excitement' after mating by trumpeting softly, flapping her ears or whisking her tail. The bull just moves off and remains silent.

Surprisingly, some relatively modern standard texts on mammals make no attempt to describe elephant courtship and mating in detail (e.g. Macdonald, 1984; Nowak, 1999).

In natural conditions animals are free to exhibit normal sexual behaviour, assuming a suitable mate is available. Uninterested females may flee from ardent males, and pubescent males may be driven away from females by older males. In enclosures, some of these natural responses are not possible; unwilling females may be cornered by males, and juvenile males may be able to harass females in the absence of older males who would drive them away – sometimes permanently – in the wild.

All captive elephants have, by definition, access to limited space and a limited choice of mates. Some captive herds are very small and contain juvenile bulls, but no adult bulls. Physical constraints and unnatural social groups may result in sexual behaviours that would not normally occur in the wild.

One early description of courtship and mating in captive Asian elephants was based on a completely unnatural situation that was not apparent from the title of the paper in which it was reported. The frequently quoted paper by Eisenberg et al. (1971) entitled *Reproductive behaviour of the Asiatic elephant* (Elephas maximus maximus L.) was based on a study of three captive elephants in Sri Lanka. Two cows (approximately 18 and 22 years old) were introduced to a bull (approximately 27 years old) who was tethered to a tree by a 60 ft (18 m) chain. If a cow exhibited avoidance behaviour when the bull initiated courtship, she was tethered by a 40 ft (13 m) chain to prevent her escape. Clearly, the practice of chaining the animals severely constrained the opportunity for them to exhibit normal courtship behaviour. Nevertheless, Eisenberg et al. identified a number of component behaviours (Table 4.1).

TABLE 4.1 Reproductive behaviour categories observed in Asian elephants described by Eisenberg et al. (1971).

Behaviour type	Component behaviour
Contact-promoting	Exchange of chemical information using trunks
	Cow investigates bull's temporal gland
	Bull investigates cow's urinogenital sinus
Precopulatory	Wrestling with intertwined trunks
	Bull reaching over cow's back with his trunk
	Driving
	Neck biting
	Attempted mounts
Copulation	Cow remains passive and standing or kneeling
	Bull mounts and achieves intromission through independent movements of penis
	Mount duration often <30 s

Captive elephants, of course, will not mate on demand and this severely limits opportunities to study their reproductive behaviour. In a study of three captive African elephants, only two copulations were observed during 438 hours of observations (Kühme, 1963). However, during a study of the social facilitation of mounting behaviour in Asian elephants, I recorded 60 adult mountings (including attempts) over a period of 230 hours (Rees, 2004b) (see Section 4.8.2).

4.3 Courtship and mating behaviour

Courtship and mating behaviour in Asian elephants are similar to that observed in other mammals, particularly ungulates (Ewer, 1968). During courtship the cow is moved forward by the bull (a behaviour known as 'driving'), but she may stop repeatedly and reverse towards him. She may even exhibit some aggression towards him. Sometimes the cow may initiate courtship by reversing towards the bull and presenting her rear. This simultaneous attraction to, and flight from, the male is sometimes called 'coquettishness' and is well known in other placental mammals (Ewer, 1968) (Box 4.1).

Occasionally when the normal sequence of courtship behaviour is interrupted or 'stalls' for some reason, resulting in both animals standing still, the bull may briefly engage in displacement digging behaviour with one of his forefeet. Displacement behaviour is characterised by its apparent irrelevance to the situation in which it appears (Tinbergen, 1952) and often occurs in conflict situations, especially during courtship (Manning, 1972). Digging is usually associated with dusting behaviour (Rees, 2002a). Courtship is not essential to successful mating; sometimes the bull simply walks up to the cow and mounts her.

Courtship may involve other members of the herd or it may not. If the herd pursues the couple, other adults and young elephants may impede their progress. Calves may be pushed aside by the bull and other adult cows may attempt to place themselves between the bull and his chosen mate. This may be an attempt to solicit the bull. This behaviour by other elephants sometimes allows the prospective mate to 'escape' from the bull so that the courtship sequence comes to an end. Alternatively, the bull may have his attention

BOX 4.1 Courtship and mating in Asian elephants

This ethogram describes the major elements of courtship and mating in the Asian elephant as observed in a zoo environment based on observations I made during a study of the elephants at Chester Zoo in the United Kingdom (Rees, 2004b) (Fig. 4.1). The specific sequence with which the various elements occur is not fixed, but varies from one occasion to another. However, these elements may be divided into a number of categories that occur at the beginning, middle and end of the process. Eisenberg et al. (1971) described behaviours as being:

1. contact-promoting,
2. courtship, and
3. copulatory (Table 4.1).

Elements of reproductive behaviour in the Asian elephant illustrated in Fig. 4.1

1. Bull and cow entwine trunks. Bull and cow butt heads. Cow may investigate bull's temporal gland.
2. Bull inserts his trunk into the cow's anus. This often stimulates the cow to defecate and urinate. This behaviour is not confined to bulls. Cows have been observed inserting their trunk into the anus of other elephants. The sexual significance of this behaviour is therefore unclear.
3. Bull tests the cow's urine and faeces by sniffing with his trunk. He may also sniff the cow's rear legs and feet where they are wet with urine.
4. Bull places small amounts of urine, and possibly faeces, in his mouth. The Jacobson's (vomeronasal) organ in the roof of the bull's mouth is used to determine the cow's hormonal status (Rasmussen and Schulte, 1998). The cow may stand still while he 'urine tests' or she may walk away.
5. Cow walks away and bull follows, touching the cow's genitals with his trunk, with his penis extended from its sheath. The cow may occasionally stop and then walk off again with the bull following. Eisenberg et al. (1971) refer to this behaviour as 'driving'. Driving may proceed at a walking pace or it may develop into a chase. The point at which the penis becomes unsheathed varies. It may be visible from the outset of courtship or it may descend much nearer to the time of mounting.
6. Bull stands alongside the cow and holds one of her forelegs with his trunk (possibly while sniffing her foot).
7. Cow stops. Bull stands alongside and rests his trunk along her back. The bull may grip the cow's back with his mouth. Cow walks off again and is pursued by the bull.
8. Cow walks away with bull pursuing, sniffing and touching her genitals. Cow may stop and reverse towards the bull. (The behaviour occurs repeatedly throughout the sequence.)
9. Cow stops and lowers her rear with her tail held to one side while the bull pushes down on her rump. The bull uses his head to push the cow into position for mounting. The cow may reverse towards the bull.

(Continued)

BOX 4.1 (Continued)

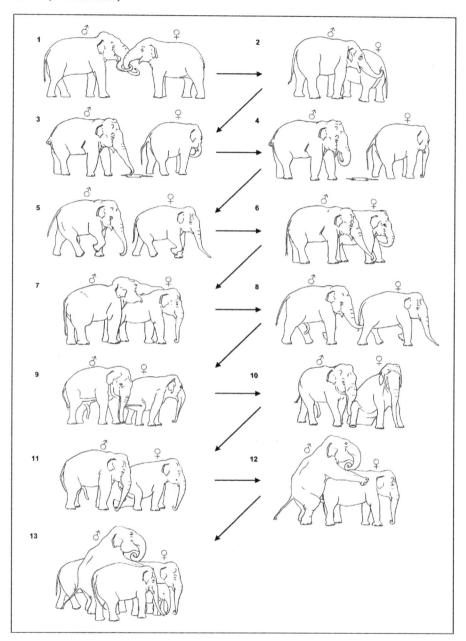

FIGURE 4.1 An ethogram of courtship and mating behaviour in Asian elephants. Source: *Reproduced with permission from Wiley-Blackwell. First published in Rees, P.A., 2015. Studying Captive Animals: A Workbook of Methods in Behaviour, Welfare and Ecology. Chichester, West Sussex. Drawings by Dr A. J. Woodward based on photographs taken by the author.*

(*Continued*)

BOX 4.1 (Continued)

10. The cow lowers her rear to the ground and kneels on one leg. The bull stands behind the cow resting his head on her rump and touching/sniffing her genitals.

11. The bull moves closer to the cow with his penis erect, that is, bent into an 'S' shape.

12. The cow stands up and remains still. The bull rises up on his hind legs and balances with his forelegs astride the cow's back, above or just behind the shoulders. The bull flexes his penis and inserts it. The bull curls his trunk under his lower jaw. Intromission takes just a few seconds. The cow may walk slowly forwards 'carrying' the bull with her as he attempts intromission. Copulation is silent and the bull may be seen making several gentle thrusting movements with his pelvis in order to achieve intromission.

13. Other members of the herd, including juveniles and very young calves, may follow the couple throughout the courting sequence. When mounting and intromission finally take place the couple may be surrounded by the herd. Individuals may vocalise loudly, flap their ears and urinate in a 'mating pandemonium'.

Table 4.2 summarises the behavioural elements described in various accounts of courtship and mating in Asian elephants.

TABLE 4.2 Behavioural elements described in various accounts of courtship and mating in elephants.

Behaviour[a]	Carrington (1958)	Eisenberg et al. (1971)	Eltringham (1982)	Spinage (1994)	Rees (2015)
1 Cow investigates bull's temporal gland		•			•
2 Anal stimulation of cow by bull's trunk					•
3 Bull tests cow's urine and/or faeces with trunk		•			•
4 Bull touches/smells cow's vulva with his trunk		•	•	•	•
5 Bull places trunk in mouth after touching cow's vulva		•	•	•	•
6 Entwining of trunks		•	•	•	•
7 Head butting		•		•	•
8 Bull drives cow forward		•	•		•
9 Cow exhibits coquettishness	•		•	•	•
10 Bull lays trunk and tusks along cow's back/ bull bites cow's back		•	•	•	•
11 Bull holds cow's forefoot/smells her feet?		•			•
12 Cow kneels on one back leg to allow bull to mount		•			•
13 Bull butts cow's rear with his head to position her for mounting					•
14 Bull mounts as cow stands still	•	•	•	•	•
15 Mating pandemonium		•		•	•

[a] List of behaviours is not necessarily sequential.

diverted to a different cow and pursue her instead. Juvenile bulls may interfere with mating by attacking the cow during mounting (Fig. 4.2). Juvenile interference with mating has been recorded in other mammalian species, for example chimpanzees (*Pan troglodytes*) (Goodall, 1971). It is possible that this interference by other elephants may reduce the number of successful mountings.

Mating in elephants often — but not always — concludes with a 'mating pandemonium' in which many members of the herd participate (Fig. 4.3). Such ceremonies have been

FIGURE 4.2 A juvenile bull Asian elephant (*Upali*) attacking an adult female (*Kumara*) while she is being mounted by an adult bull (*Chang*).

FIGURE 4.3 Mating pandemonium within the herd of Asian elephants at Chester Zoo, United Kingdom. Where elephants are kept in social groups, including a breeding bull, courtship and mating may conclude in a social behaviour whereby the other elephants gather around the mating couple in a state of enhanced excitement involving vocalisations, urination and defaecation.

described from the wild in African elephants (Moss, 1988) and consist of the herd gathering around the mating couple in a state of enhanced excitement which involves vocalisations, urination and defaecation.

Within the herd of Asian elephants at Chester Zoo, the calves often followed the courting adults so closely that they had to be pushed out of the way by the adult bull. During attempted or actual intromission, the female calf *Sithami* was observed sniffing and touching the genital areas of the adults involved.

Chemical signals from the adult cow are important in the sexual behaviour of Asian elephants (Rasmussen and Schulte, 1998). Adult bulls test samples of urine and faeces using the Jacobson's organ, a specialised gland located in the roof of the mouth. Chemical communication in elephants is discussed in Section 5.3.4.

4.4 Chemical control of reproduction

4.4.1 Musth

Musth is a condition of adult male elephants and has been likened to the rutting behaviour of ungulates. It is characterised by increased aggressiveness, urine dribbling, temporal-gland enlargement and secretion as well as elevated androgen levels (Fig. 4.4).

Musth has been well documented in wild bull elephants of both genera (e.g. Lincoln and Ratnasooriya, 1996; Poole and Moss, 1981), but has not received a great deal of attention in their captive counterparts. Captive studies have shown that the condition is associated with changes in androgen levels.

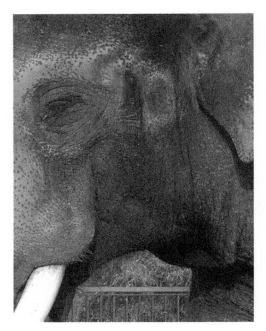

FIGURE 4.4 Temporal gland of an Asian bull elephant. During musth the temporal glands discharge a strong-smelling fluid and the elephant becomes extremely aggressive.

A study of musth in a 37-year-old bull African elephant in Kansas City Zoological Park, United States, recorded significantly higher concentrations of androstenedione (a precursor of testosterone) and luteinising hormone (LH) in his urine when he was in musth than when he was not (Brannian et al., 1989). These increases in urinary androgen levels may have been the result of LH stimulation of testicular steroidogenesis.

Testicular endocrine function in captive African elephants has been studied by Ganswindt et al. (2002). They collected urine and faecal samples over 3 years from five bulls (aged 7–24 years) living in three mixed social groups. In addition, keepers independently recorded indications of musth. Substantial quantities of testosterone and epiandrosterone were found in musth-phase samples of faeces and urine. This was particularly apparent in the two oldest bulls. The authors suggested that measurement of urinary and faecal androgens provides new opportunities to improve bull welfare and management in captivity and to examine physiological correlates of reproductive function in free-ranging bulls.

In a later study, Ganswindt et al. (2005) examined musth in 14 captive bull elephants, making long-term observations of behavioural and physical changes while also collecting physiological data on testicular and adrenal function. They found musth periods to be highly variable in length, but unrelated to age. During musth, androgen levels were elevated (resulting in urine dribbling and temporal-gland secretion) but, usually, glucocorticoid levels decreased, indicating that the phenomenon is not a physiological stress mediated by the hypothalamic-pituitary-adrenal axis. Raised androgen levels were associated with, and presumably the cause of, increased aggression.

Mahouts traditionally bring bulls out of musth more quickly by restricting their food intake (Sukumar, 1994). Some senior elephant keepers in zoos claim to be able to prevent bulls entering musth by their presence and if they take a period of leave the bulls may enter musth while they are absent.

4.4.2 Endocrine monitoring of females

The importance of endocrine monitoring in elephants as a tool to assist captive breeding has been reviewed by Brown (2000). She emphasised the need to identify the cause of reproductive failure and develop mitigating treatments by undertaking endocrine evaluations, ultrasound examinations and behavioural assessments.

The classic mammalian oestrous cycle exhibits a single LH peak during the follicular phase that induces ovulation. Lueders et al. (2011) have shown that, in elephants, two distinct waves of follicles develop during the follicular phase, each terminating in an LH peak. The two peaks are 3 weeks apart. Ovulation and a rise in progesterone only occur after the second peak. They studied this phenomenon by combining ovarian ultrasound examinations with hormone measurements in five female Asian elephants housed at the African Lion Safari in Cambridge, Ontario, Canada.

Niemuller et al. (1997) demonstrated that it is possible to determine pregnancy in Asian elephants by analysis of the circulating 17α hydroxyprogesterone: progesterone (17αONP: P) ratio. During weeks 2–7 of early pregnancy this ratio fell from ≥ 0.7 to <0.7 compared with the same time period in elephants in a nonconceptive cycle.

Analysis of serum prolactin can confirm pregnancy in Asian elephants after around 4 months of gestation, and parturition may be predicted by daily monitoring of serum or urinary progestogens. Brown and Lehnhardt (1995) collected weekly blood and urine samples from a single elephant from 10 months prior to conception until 10 months after parturition, and additional daily samples for 41 days before parturition until 10 days after parturition. Serum progesterone concentrations increased gradually during gestation and, after approximately 13 weeks, were higher than those recorded during the nonpregnant luteal phase, peaking at around 12 months of gestation. They then gradually declined during the last month of gestation and decreased sharply to undetectable levels 2 days prior to parturition. Cyclicity resumed after a 12-week lactational anoestrus. Serum prolactin concentrations increased 100-fold around midterm — from the base level observed during the first 16 weeks of gestation — and remained elevated until after parturition. Serum and urinary cortisol concentrations were unaffected by pregnancy, but both increased markedly after parturition and remained high for several weeks.

As an alternative to analysing urinary samples for progestins as a means of monitoring ovarian cycles, Illera et al. (2014) developed and validated an enzyme immunoassay for these chemicals in saliva using samples taken from five nonpregnant African elephants. A high positive correlation was found between the results of urine and saliva samples.

Hodges et al. (1997) have shown that the 5α-reduced metabolites 5α-pregnane-3,20-dione (5α DHP) and 3α-hydroxy-5α-pregnan-20-one (5α-P-3α-OH) are the principal progestins biosynthesised by the African elephant corpus luteum and found in circulation. Heistermann et al. (1997) analysed samples of urine collected from five female African elephants and established that measurements of 5α-pregnane-3-ol-20-one (5α-P-3-OH) in urine provide a reliable noninvasive method of monitoring luteal function.

Yamamoto et al. (2012) investigated the secretory pattern of inhibin during the oestrous cycle and pregnancy in two African and two Asian elephants. Elephant ovaries have multiple corpora lutea during the oestrous cycle and gestation, and the results suggest that these structures secrete a large amount of progesterone, but not inhibin, during pregnancy.

4.5 Behavioural indicators of oestrus

In addition to hormonal indicators of oestrus, behavioural indicators have also been investigated. A subadult male African elephant housed with four females at Disney's Animal Kingdom, Florida, United States, followed the females more, and performed more genital inspections, flehmen and trunk-to-mouth behaviours towards cycling females during the ovulatory phase of their oestrous cycles than at other times (Ortolani et al., 2005). Two of the females cycled during the study and the other two did not. This study demonstrated that the subadult bull could discriminate between a female in the ovulatory phase of her cycle from one in the mid-luteal phase and between noncycling females and cycling females in the ovulatory and anovulatory follicular phases. The cycling females also showed changes in their behaviour, exhibiting a greater variety of tactile/contact behaviour towards the bull while in the ovulatory phase compared with other phases of the cycle.

Slade-Cain et al. (2008) have shown that the phase of oestrus affects investigative, aggressive and tail flicking behaviour in captive female Asian elephants. Investigative trunk-tip contacts were more prevalent towards female elephants who were in their follicular phase compared with their luteal phase, and the most frequently contacted region was the genital area (which may release chemical signals relating to reproductive state). Aggression rates were elevated when senders were approaching ovulation and receivers were in the luteal phase. Tail flicking may disperse chemical signals in mucus and urine in addition to acting as a tonic signal. This behaviour could indicate the approach of ovulation for elephants and human observers.

The influence of rank and ovarian activity on the social behaviour of 33 African elephant females – 18 cycling and 15 noncycling – was studied in zoos by Freeman et al. (2010a). When they compared the behaviour of individuals based on the ranks determined by keepers, they found that dominant individuals were more likely to approach, displace and push others than were subordinates. Subordinate elephants more frequently presented their hind quarters and held their ears erect than did dominant animals. Behaviours initiated by one individual towards another did not vary between cycling and noncycling elephants, except when the interaction of social rank was tested. The authors concluded that social rank influences the interactions between female African elephants more than does ovarian cyclicity status. Consequently, it is not possible to predict which cycling elephants are most likely to become acyclic by monitoring behavioural interactions.

4.6 Gestation, pregnancy management and birth

Dale (2010) examined gestation in 49 elephant births that occurred in nationally accredited zoos in Canada, Europe, Japan and the United States, except for those at the Ringling Bros. Center for Elephant Conservation and at Riddle's Elephant and Wildlife Sanctuary, both in the United States. He found that the mean gestation period was 657.8 days in Asian elephants and 640.1 days in African elephants. Oerke et al. (2006) found that the equivalent values for 45 elephants in European facilities were 655 and 642 days, respectively. Dale commented that the similarity in these results was compelling, considering that the two studies had just eight individuals in common.

The successful management of six pregnancies in African elephants at Disney's Animal Kingdom, Florida, has been described by Sullivan et al. (2016). The husbandry management programme included staff-directed exercise (e.g. walking) with the goals of maintaining or reducing body weight, increasing fitness and muscle tone in preparation for parturition, and reducing or eliminating ventral oedema. Weight gains were closely monitored during the study; diet was carefully managed and included high fibre levels, vitamin E supplements and low levels of starch and fat. Each pregnancy resulted in the birth of a healthy calf.

Births usually occur at night in African and Asian elephants (Dale, 2010; Kowalski et al., 2010) (Fig. 4.5). They usually produce a single calf, but twins have been reported from the wild and in captivity from over 50 years ago (e.g. Seth-Smith and Parker, 1967; Tun, 1962).

Records available for 19 cows in elephant camps in Tamil Nadu, India, recorded 101 births (including three sets of twins) between 1960 and 1986 (Krishnamurthy, 1995). Twins therefore occurred in 1 in 98 pregnancies. This concurs with the 1% of births

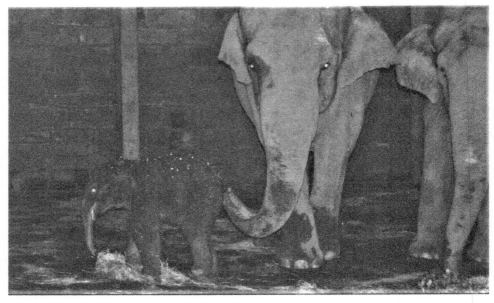

FIGURE 4.5 A 1-hour-old male calf, *Pochin*, born at night to *Jangoli* (left) at Chester Zoo, United Kingdom. His father was *Chang*.

resulting in twins recorded by Laws (1969) from culling operations in Uganda. However, higher and lower incidences have been recorded.

Some individuals and families are known to be genetically predisposed to twinning in humans and some other species (White and Wyshak, 1964) and it appears likely that this is the case in elephants. Foley (2002) observed a high incidence of twin births in Tarangire National Park, Tanzania. Among 291 recorded births, 14 produced twins (5%). One female produced three consecutive sets of twins over a period of 7 years. In contrast, Moss (2001) recorded only one incidence of twinning in 1192 births (0.08%) over a period of 30 years in Amboseli National Park, Kenya.

Infanticide is known in wild populations of a number of mammal species, for example langurs (Cercopithecidae; Borries, 1997), polar bears (*Ursus maritimus*; Derocher and Wiig, 1999) and rats (Rodentia; Brown, 1986). Wanghongsa et al. (2006) reported what they believed to be the first incidence of infanticide in wild Asian elephants. The carcass of a 6- to 9-month-old calf was found on the boundary of Dong Yai Wildlife Sanctuary, Burirum Province, Thailand. After examination of the carcass by police, forest personnel and a veterinary surgeon, and the discovery of the footprints of a large bull at the scene, it was concluded that the calf had been fatally stabbed by a large 25-year old musth bull who had been behaving aggressively towards humans in the vicinity 2 days earlier.

Infanticide occurs in elephants living in zoos. Saragusty et al. (2009) found that stillbirth and infanticide were the major causes of premature deaths in elephants in European and North American zoos. The Asian elephant cow *Thi* at Chester Zoo gave birth to a female calf in September 1993 and killed her shortly thereafter. In December 1995, *Thi* produced a second female calf (*Karha*) but rejected her. Keepers hand-reared *Karha*, but she died in May 1997. *Thi* subsequently produced six more calves between December 1997 and August 2015. One was stillborn but the other five were successfully reared by their mother. This example illustrates that a cow who kills or rejects a calf may go on to be a successful mother (see Fig. 6.16).

4.7 Parenting and calf development

4.7.1 Developmental milestones and birth statistics

Until relatively recently little systematically collected data have been available for newborn elephant calves. Dale (2010) has provided a comprehensive account of birth statistics for elephants under human care. He collected data for 218 newborn calves from 74 institutions over 30 years and found that Asian elephants exceeded African elephants in mean birth height (93.7 and 89.9 cm, respectively) and mean birth weight (116.5 and 103.5 kg, respectively), with sex differences in weight and height observed in African calves, but only in height in Asian calves. An analysis of records of the time of birth of 47 African and 91 Asian elephant births in this population found that they tended to occur at night. Dale also compared the birth weights of 12 (9.3) stillborn Asian elephant calves (all born between 1987 and 2006) with live-born calves and found that the mean weight of the stillborn individuals (136.3 kg, both sexes combined) was significantly higher than that of the live-born individuals (117.9 kg), possibly due to differences in the gestation period.

A second study of the development of 12 African elephant calves found that they were all born at night (between 1900 and 0700) and they showed progression in their behavioural and physical development similar to that observed in the wild (Kowalski et al., 2010). Emphasis was placed on birth-related events, changes in trunk use, first instances of behaviours and interactions with other elephants (usually adults).

A study of developmental milestones among zoo-born African elephant calves on their first day of life found that they developed at comparable rates to those observed in the wild (Bercovitch and Andrews, 2010). Elephant calves are able to stand unaided without falling after about 30–40 minutes, and some have been observed walking within minutes of birth (Spinage, 1994). Bercovitch and Andrews found that zoo-born calves could stand and walk on their own for the first time at the same age as those born in the wild. Calves born in the zoo took slightly longer to begin successful nursing compared with wild calves, but the difference was not statistically significant (Fig. 4.6). Successful nursing requires coordination between mother and calf, so is not entirely dependent on the development of the calf. There was no difference in the rate of development of male and female zoo-born calves, but there was insufficient data available to compare these with the development rates of wild calves.

A study of a single newborn African elephant calf found that suckling occurred for only 2 hours in any 24-hour period, and that significantly more suckling took place during the night than during the day (Andrews et al., 2005). Throughout the first 3 months of life, weight was gained at a rate of 0.385 kg/day. This was slower than that recorded for hand-reared elephant calves, but the suckling patterns observed were almost identical to those seen in wild calves.

FIGURE 4.6 *Pochin*, about 2 weeks old, feeding from his mother *Jangoli* at Chester Zoo, United Kingdom. He could not reach his mother's nipples at birth by standing on all four feet. Shortly thereafter he adopted a posture that allowed his head to point upwards towards the nipple. He continued to suckle in this position when he was much older and would have been perfectly able to reach the nipple while standing upright.

4.7.2 Parenting and allomothering

Elephants have been known to go to extraordinary lengths to protect their calves. Williams (1951) recounts an event while working with elephants in Burma. A mother, *Ma Shwe*, became trapped in a fast-rising river with her 3-month-old calf. The calf was washed downstream and *Ma Shwe* chased it and pinned it against the rocky bank.

> Then with a really gigantic effort, she picked it up in her trunk and reared up until she was half standing on her hind legs, so as to be able to place it on a narrow shelf of rock, five feet above the flood level.

The mother was then washed further down the river, but managed to climb onto one of the banks and was seen reunited with her calf some time later.

Allomothering is the behaviour whereby an animal provides care for a juvenile of the same species (usually) that is not its offspring. The term 'allomother' literally means 'other' or 'different' mother. Elephant allomothers play an important role in early calf development by, for example standing guard over calves while they sleep and providing reassurance to them when they are distressed.

In the wild, related adult cows (aunts, great-aunts, grandmothers) sometimes nurse the calf of a close relative (Lee, 1987; Payne, 2003), and comfort-suckling is often performed on cows who are too young to lactate (Moss, 1988). Lee (1986) has analysed the early social development of African elephants and the phenomenon of allomothering in this species (Lee, 1987). During the first 2 years of life, an African savannah elephant spends as much of its nonnursing time with allomothers as with its mother (Lee, 1989).

Many female elephants living in zoos are past their prime reproductive age and are unlikely to participate directly in captive breeding programmes. However, such females may play an important part in rearing of the calves of others. Allomothers may assist during and after the birth of a calf. In 1997, the first Asian elephants to be born within a large captive (zoo) group were produced at Emmen Zoo, The Netherlands (Anon., 1998). Two separate births took place within a group of seven adult cows and two juveniles (one of either sex). In both cases, an experienced mother protected the newborn calf from its mother and the other elephants for several hours after the birth (Rees, 2001a,b).

Rapaport and Haight (1987) provided a descriptive account of allomothering in Asian elephants at Washington Park Zoo, Portland, Oregon, and Schulte (2000) examined social structure and helping behaviour in captive elephants in North America. Schulte summarised the possible benefits of allomothering within captive elephant herds, but there appears to be no comprehensive account of the contribution allomothers make to the welfare and survival of calves born in zoos.

Studies of wild African elephants suggest that calf mortality in the first 24 months of life is higher among those born to families with no allomothers than in those where allomothers are present (Lee, 1987). Although the risks to calves kept in captivity are lower than would be expected in the wild — at least in relation to death caused by drought, food shortage and predation — dangerous situations may nevertheless arise.

During my research at Chester Zoo, allomothers protected the calf *Sithami* from excessive attention from the juvenile bull. They also protected her from the more serious risk of falling into the moat around the paddock. When she was 20 months old her back legs slipped into the moat and she was pulled to safety by several of the adult animals. Two adult Asian elephants at Grand Park Zoo, Seoul, South Korea, were recorded cooperating to assist a baby elephant who fell into the deep end of a pool in their enclosure. The adults could not enter the deep end so walked around to the shallow end, entered the water and stood side-by-side holding the baby at the surface until they could lead it to the shallow end where it could walk onto dry land (Evans, 2017).

Aspects of the parenting and alloparenting of a female Asian elephant calf at Chester Zoo are discussed in Box 4.2.

Dublin (1983) suggested that kinship plays a role in the cooperative vigilance behaviour of elephants, since groups contain siblings, mothers and daughters. However, the genetic interpretation of alloparenting does not explain why captive elephants care for an unrelated calf. The caretaking behaviour of these elephants appears to be part of a more general concern for the welfare of young animals. Dublin argues that reciprocity may be important in group vigilance behaviour because female elephants are often in the same group for most, if not all, of their lives. These long-lasting relationships should therefore favour the evolution of reciprocal cooperation.

The social facilitation of suckling may be important in the husbandry of captive elephants. Some captive-born calves are reluctant to suckle and may even be rejected shortly after birth. The presence of other suckling calves may stimulate newborn calves to begin feeding. There is some evidence that social facilitation is important in the normal

BOX 4.2 Allomothering in Asian elephants at Chester Zoo, United Kingdom

During a detailed study of the behaviour of the Asian elephants at Chester Zoo, I recorded the times at which a female calf (*Sithami*) suckled from her mother (*Thi*) and any other adult cows for a total of 228 hours over a period of 49 days between January 1999 and October 1999. At the beginning of the study *Sithami* was just over 1 year old. I also recorded her associations with other members of the herd.

Suckling and comfort suckling

Between January 1999 and the end of October 1999 suckling by *Sithami* was observed on 264 occasions: 219 (83%) events involved the mother (*Thi*) and 45 (17%) events involved another adult cow, *Sheba*, who was the oldest cow in the herd and unrelated to the mother. The calf was never observed attempting to suckle from any of the other cows.

During the early stages of the study (January and February 1999), when the calf was between 12 and 14 months old, there was no significant difference in the number of times she was observed suckling from her mother and comfort-suckling from *Sheba* ($\chi_1^2 = 2.96$, $P = .085$) (Table 4.3). This is unusual as, in the wild, elephant allomothers are usually young, nulliparous cows who are unrelated to the mother or cows that are too young to lactate (Moss, 1988). As *Sithami* became older the frequency of comfort-suckling was reduced and *Sheba* was observed kicking the calf away when she attempted to suckle on 7 May 1999. *Sithami* was last observed comfort-suckling from *Sheba* on 28 June 1999, when she was 18 months old.

During the 10 months of this study there was very little variation in the observed frequency of *Sithami* suckling from the mother. On average, suckling occurred approximately once per hour (Table 4.4) and was not affected by comfort-suckling contacts with *Sheba*. During the 6 months over which comfort-suckling with the allomother was recorded, suckling contacts (with either female) occurred at mean intervals of between about 30 and 45 minutes.

TABLE 4.3 Number of suckling (*Sithami/Thi*) and allomother suckling (*Sithami/Sheba*) events.

Period 1999	Calf's age (months)	*Thi* (*n*)	*Sheba* (*n*)	χ^2 (*d.f.* = 1)	*P*
Jan/Feb	12–14	35	22	2.965	>.05 N.S.
Mar/Apr	14–16	35	17	6.231	<.05
May/Jun	16–18	53	6	37.441	≪.005
Jul/Aug	18–20	56	0	–	–
Sep/Oct	20–22	40	0	–	–

Source: *Author's unpublished data.*

TABLE 4.4 Suckling intervals for *Sithami* (bimonthly periods).

Period 1999	Calf's age (months)	Suckling from *Thi* only (mins)			Suckling from *Thi* or comfort-suckling from *Sheba* (mins)[a]		
		Mean	σ	*n*	Mean	σ	*n*
Jan/Feb	12–14	50	37	26	30	33	49
Mar/Apr	14–16	57	42	25	40	37	42
May/Jun	16–18	54	34	43	47	36	49
Jul/Aug	18–20	64	32	44	–	–	–
Sep/Oct	20–22	53	25	32	–	–	–

[a] *No records of suckling from* Sheba *in Jul/Aug or Sep/Oct.*
Source: *Author's unpublished data.*

(*Continued*)

BOX 4.2 (Continued)

When the calf was standing in close proximity to her mother and allomother, she sometimes switched her suckling between the two. Where such 'switches' were recorded within a 5-minute period ($n = 26$) the calf was twice as likely to begin with her mother ($\chi_1^2 = 2.46$, $P = .117$). Sometimes the calf would switch twice. For example on 7 February 1999 she suckled from *Sheba* (at 1421), switched to *Thi* (at 1424) and then immediately back to *Sheba* (at 1424).

Associations

A total of 1850 recordings were made of *Sithami*'s closest associates at 5-minute intervals (154.2 hours), between the end of February and the end of October 1999, when she was between 14 and 22 months old. An associate was defined as any elephant located within one adult elephant's length from the calf. If the calf was within a group of more than three elephants, it was recorded as being within a herd.

In all, 2791 associations were recorded (including 245 observations made alone and 170 within the herd). Sometimes the calf associated with a single elephant, but at other times she was part of a group. Over the 8-month period as a whole, the calf spent 35.5% of her time closely associating with her mother and 35.8% of her time associating with her primary allomother (*Sheba*). These two females were the calf's most frequent associates throughout the study (Fig. 4.7).

When all recordings were considered, regardless of whether or not the calf was part of a group, there was no significant difference between the number of times the calf was observed closely associating with her mother and the number of times she associated with her primary allomother ($\chi_1^2 = 0.037$, $P = .847$). The calf's next most important associates were, respectively, *Maya* (13.9%), *Jangoli* (12.8%), *Upali* (12.2%), *Chang* (9.7%) and *Kumara* (8.7%). These figures are significantly different from a 1:1:1:1:1 ratio ($\chi_4^2 = 31.04$, $P << .001$).

The amount of time the calf spent on her own (more than one adult elephant length from the nearest individual) increased from about 12% when she was 15 months old to around 20% when she was 22 months old. This probably reflected her reduced dependence on her mother and primary allomother as she grew older.

FIGURE 4.7 Allomothering: *Thi* (left) and *Sheba* guarding *Thi*'s calf *Sithami*. *Sheba* was not genetically related to *Sithami*.

development of sexual behaviour in bull Asian elephants (Rees, 2004b). Synchronisation of behaviour may have a protective value. Suckling may occur only when calves and their mothers decide that it is safe. I have suggested elsewhere that the synchronisation of defecation in wild African elephants may reduce the possibility of young calves being left behind in a moving herd, and temporarily exposed to hazards (including predators) (Rees, 1983).

The presence of allomothers may be important in the development of calves in zoos. They may support the mother by providing protection and reassurance to the calf. However, if a calf is rejected by its mother at birth the existence of other adult cows within the herd is unlikely to assure the calf's survival. Hand-rearing is difficult. Baby elephants raised on cow's milk commonly die as a result of chronic diarrhoea. The fat of elephant milk is closer to coconut oil in composition than to the fat of cow's milk. Intolerance to cow's milk is likely to be due to differences in fat composition and droplet size, which could increase the difficulty of their absorption (Garton, 1963; McCullagh and Widdowson, 1970; McCullagh et al., 1969).

In preparation for the possibility that a calf could be rejected by the dam, or unable to suckle, veterinary staff at a zoo may use blood donated by another elephant as a source of antibodies to add to bottles of milk prepared in advance of the birth (Fig. 4.8). The donor is likely to be an acyclic elephant with no breeding potential as the process requires the animal to be anaesthetised and so carries some risk.

Although elephants have a sophisticated social life and perform altruistic acts to assist other members of their group, they do not always act benevolently towards young calves.

FIGURE 4.8 A female Asian elephant (*Maya*) at Chester Zoo donating blood from a vein in her ear while anaesthetised, and under the supervision of veterinary staff from the zoo and the University of Liverpool. The blood was later centrifuged to extract the plasma. This contains antibodies and was used to make up bottles of feed in preparation for the possibility that a pregnant female in the herd (*Jangoli*), who was approaching full term, might not be able to feed her calf. This was subsequently given to the calf (*Pochin*) when he had difficulty reaching his mother's nipples (see Fig. 4.6). Source: *Reproduced with permission from Wiley-Blackwell. First published in Rees, P.A., 2011. Introduction to Zoo Biology and Management. Wiley-Blackwell, Chichester, West Sussex.*

Adoption of calves by unrelated families has been recorded in the wild, but so has the rejection of orphans.

The hunter T. Murray Smith describes a horrific incident in which he discovered an African elephant cow and her calf stuck in a muddy waterhole (Smith, 1963). The cow was stuck fast in the liquid mud in the middle, while the calf lay exhausted near the bank with its forelegs held by the mud. The cow repeatedly raised herself from the mud and lunged for the bank, but each time she sank back further until only the tip of her trunk was visible. The next time the top of her head became visible, Smith shot her dead. After a 3-hour struggle, Smith and his associates freed the calf. The next morning he found spoor showing that three of the herd had come and taken the calf away.

M. Woodford and S. Trevor attempted to foster an abandoned wild African elephant calf into a family unit (Woodford and Trevor, 1970). The calf's first approach to the wild herd was met by a bull who knocked it over. The calf was undeterred, and its second rush brought it to a nursing cow with a calf. The cow suckled immediately and the rest of the herd showed the same interest in it as in a newborn calf. The calf was still seen alive a week later.

Other attempts at fostering African elephant calves in the wild by well-meaning scientists have not ended as well. Iain Douglas-Hamilton was less successful in his attempt at introducing a calf to a new family (Douglas-Hamilton and Douglas-Hamilton, 1975). The calf was accepted but it was not suckled and later died. Keith Eltringham also made a failed attempt. His orphan was forcefully rejected when the matriarch of the group tossed it into the air with its tusks (Eltringham, 1982).

In spite of these failures, one person has achieved spectacular success. In Kenya, Daphne Sheldrick was been extremely successful at hand-rearing orphaned elephant calves using a specially formulated milk at an orphanage that was originally located near Voi at the Headquarters of Tsavo National Park (East), but was later moved to Nairobi National Park (Sheldrick, 1977). By 2019, the Orphan's Project had reared over 32 elephant calves, most of which have been successfully released back into the wild (Anon., 2019b).

Fostering of suckling calves is problematic in small captive groups because of the limited number of potential foster mothers. Hand-rearing a female Asian elephant calf named *Karha* was attempted at Chester Zoo, United Kingdom, but was ultimately unsuccessful as she died at the age of 17 months (Fig. 4.9). Staff at the Noorder Zoo in Emmen, The Netherlands, have reported on the husbandry challenges of hand-rearing an Asian elephant (Rietkerk et al., 1993).

4.7.3 The effect of a calf on social interactions in the herd

In the wild, elephant calves are born within social groups consisting of adult cows and their offspring, where they are raised by their mothers and one or more allomothers or 'aunts'. These allomothers are often relatives of the mother, but in captivity this role may be assumed by unrelated females.

The effects of the birth of a calf on the social interactions between the mother and three other female Asian elephants (all of whom were related), along with changes in their behaviour, were studied by Whilde and Marples (2012) at Dublin Zoo, Ireland. The mother and the aunt of the calf spent significantly less time walking after the birth than before. The aunt also spent less time standing still after the birth. The mother spent most of her time close to

FIGURE 4.9 *Karha* was born at Chester Zoo in January 1995. She was rejected by her mother, *Thi*, and subsequently hand-reared by keepers. She died at the age of 17 months.

the calf, but the proportion of the time she spent with each of the other females was unaffected by the birth. Similarly, the aunt's associations did not change after the birth. The newborn calf's older sister increased the time she spent nearest to her mother after the calf was born and reduced the time she spent close to another young group member.

In captivity, it is common to see young calves receiving pushes, kicks and trunk slaps from other members of the group. They may even be attacked by an adult bull if they interfere with his attempts to court and mount their mother (see Section 3.7 and Box 3.2).

4.8 Early sexual behaviour

4.8.1 Early male sexual behaviour

Chemical communication is important in the social and sexual lives of elephants (see Section 5.3.4; Fig. 5.5). Male elephant calves begin showing an interest in elephant urine and faeces at a very early age and, if the opportunity arises, will closely associate with adults during their courtship and mating (Box 4.3). Although reproductive behaviour may be largely innate, it is possible that there is an element of social learning involved.

4.8.2 Juvenile mounting

In my study of the behaviour of the Asian elephant herd at Chester Zoo I was able to examine the effect of the presence of mating adults on the behaviour of juveniles (Box 4.4). At the time of this study there was no other captive herd in the United Kingdom where a

BOX 4.3 Early sexual behaviour in a male Asian elephant calf

Early manifestation of sexual behaviour, or an interest in such behaviour in others, may take several forms within a herd environment:

- Young bull calves exhibit interest in the urine and faeces of adult cows.
- Penile 'erections' occur in bull calves.
- Calves exhibit courtship behaviour, culminating in mounting.
- Calves may be present during adult courtship and mating, and during a 'mating pandemonium'.

A bull calf, *Pochin*, was born to *Jangoli* at Chester Zoo on 18 July 2000 (Fig. 4.5). He was observed testing a cow's urine at age 40 hours (Fig. 4.10) and at age 15 days was observed standing next to the hind legs of an adult female as she was mounted by an adult bull (Fig. 4.11). *Pochin's* early sexual experiences are summarised in Table 4.5.

FIGURE 4.10   *Pochin*, an Asian elephant bull (aged 40 hours), testing urine produced by an adult female. Source: *From Rees, P.A., 2003. Early experience of sexual behaviour in Asian elephants (*Elephas maximus*). International Zoo News 50, 200–206.*

(Continued)

BOX 4.3 (Continued)

FIGURE 4.11 Very young calves may witness adults mating at very close quarters. *Chang* mounting *Sheba* with *Pochin* (aged 15 days) just visible behind *Sheba*'s rear legs. Source: *From Rees, P.A., 2003 Early experience of sexual behaviour in Asian elephants* (Elephas maximus). *International Zoo News 50, 200–206.*

TABLE 4.5 Early sexual behaviour and experiences in a male neonate Asian elephant.

Age (days)	Behaviour
0	Born within cow herd
1	Observed testing adult cow urine aged 40 h (Fig. 4.10)
13	Observed eating faeces of another elephant
15	Introduced to *Chang* (father)
	Very near adults during courtship and mating (Fig. 4.11)
	Present as juveniles play mount
17	Near adults during courtship and mating
	Present as juveniles play mount
	Observed eating mother's faeces

Source: *Based on Rees, P.A., 2003. Early experience of sexual behaviour in Asian elephants* (Elephas maximus). *International Zoo News, 50, 200–206.*

BOX 4.4 The social facilitation of sexual behaviour in Asian elephant calves

I recorded sexual behaviour within a herd of eight Asian elephants at Chester Zoo for approximately 230 hours on 50 days over a period of 10 months (Rees, 2004b). On each day observations were made continuously so all mountings were recorded. A single adult (*Chang*) and a single juvenile bull (*Upali*) were observed mounting cows more than 160 times.

When *Upali* was between 4 years 2 months and 4 years 8 months old he exhibited mounting behaviour towards the juvenile cow (*Sithami*) only on days when adult mounting had previously occurred. In some cases the sexual behaviour in the youngsters occurred just a few seconds after the adult sexual behaviour (Tables 4.6 and 4.7; Fig. 4.12). Beyond the age of 4 years 8 months the juvenile bull exhibited spontaneous mounting behaviour in the absence of adult mounting (Fig. 4.13; Tables 4.8 and 4.9). The results of this study suggest that mounting behaviour may develop as a result of social facilitation. It certainly appears to occur in wild elephants. During a short visit to Minnerya National Park in Sri Lanka, while watching a herd of some 50 or more elephants, I witnessed a large bull mounting a cow and shortly thereafter a younger bull mounted a nearby female.

TABLE 4.6 Juvenile sexual activity stimulated by adult sexual activity (17 May 1999).

Time of adult sexual activity (*Chang* mounts *Sheba*)	Time of juvenile sexual behaviour (*Upali* mounts *Sithami*)[a]
09:42	
10:52	10:52
	10:58
	11:01
	11:04
	11:25
	11:37
	11:38
	11:46
	11:48
	11:59
13.31	

[a]*Upali* aged 4.5 years, Sithami *aged 1 year 5 months.*
Source: *From Rees, P.A., 2003 Early experience of sexual behaviour in Asian elephants (*Elephas maximus*). International Zoo News 50, 200–206.*

(*Continued*)

BOX 4.4 (Continued)

Further studies are needed to determine the significance of the presence of sexually active adult elephants in the normal development of sexual behaviour in juveniles. If their presence is important, the establishment of larger captive herds containing adults and calves of both sexes should be encouraged.

TABLE 4.7 Total number of mountings observed during the study.

Bull	Recipient of sexual mounting behaviour			Total
	Adult cow	Cow calf	Juvenile bull	
Adult	60	0	12	72
Juvenile	2	102	N/A	104
Total	62	102	12	176

Source: *Reproduced with permission from Taylor & Francis. First published in Rees, P.A., 2004. Some preliminary evidence of the social facilitation of mounting behavior in the Asian elephant* (Elephas maximus). *Journal of Applied Animal Welfare Science 7, 49–58.*

FIGURE 4.12 Asian elephants mating at Chester Zoo United Kingdom. Inset: Juveniles often attempt to mate shortly after observing mating in adults. *Upali* (aged between 4 and 5 years old) attempting to mount *Sithami* (aged between 1 and 2 years old). Source: *Reproduced with permission from Wiley-Blackwell. First published in Rees, P.A., 2011. Introduction to Zoo Biology and Management. Wiley-Blackwell, Chichester, West Sussex.*

(Continued)

BOX 4.4 (Continued)

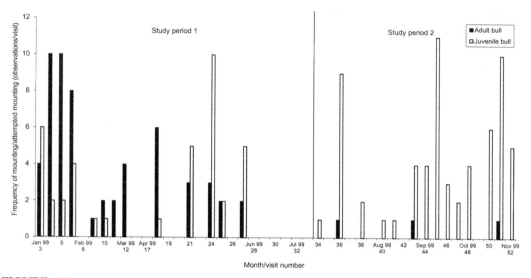

FIGURE 4.13 Frequency of mounting by adult and juvenile bulls. Source: *Reproduced with permission from Taylor & Francis. First published in Rees, P.A., 2004. Some preliminary evidence of the social facilitation of mounting behavior in the Asian elephant* (Elephas maximus). *Journal of Applied Animal Welfare Science 7, 49–58.*

TABLE 4.8 Number of days on which adult and juvenile sexual behaviour was observed during study period 1. ($\chi^2 = 15.76$, $P < .0005$; significant association between adult and juvenile bull mountings.)

		Adult bull		
		Mounting	No mounting	Total
Juvenile bull	Mounting	11	0	11
	No mounting	2	13	15
	Totals	13	13	26

Source: *Reproduced with permission from Taylor & Francis. First published in Rees, P.A., 2004. Some preliminary evidence of the social facilitation of mounting behavior in the Asian elephant* (Elephas maximus). *Journal of Applied Animal Welfare Science 7, 49–58.*

TABLE 4.9 Number of days on which adult and juvenile sexual behaviour was observed during study period 2. ($\chi^2 = 0.171$, $P > .25$; no significant association between adult and juvenile bull mountings.)

		Adult bull		
		Mounting	No mounting	Total
Juvenile bull	Mounting	3	11	14
	No mounting	0	5	5
	Totals	3	16	19

Source: *Reproduced with permission from Taylor & Francis. First published in Rees, P.A., 2004. Some preliminary evidence of the social facilitation of mounting behavior in the Asian elephant* (Elephas maximus). *Journal of Applied Animal Welfare Science 7, 49–58.*

breeding bull was being kept with adult cows, a juvenile bull and a young cow calf, so Chester Zoo provided a unique opportunity to study the effect of adult sexual behaviour on the behaviour of young elephants.

Until relatively recently there has been very little contact between the sexes in zoos prior to being brought together for breeding purposes because so few zoos kept bulls. There is anecdotal evidence that some captive bulls show little interest in females and may not know how to mate. Furthermore, until relatively recently, few zoos have been able to keep breeding adults alongside young animals due to low reproduction rates and poor calf survival. Consequently, there has been little opportunity to study the effect of adult sexual behaviour on that of juveniles.

Lack of social experience may be an important factor causing low libido in some male elephants and there is some evidence that social facilitation may be important in the development of normal sexual behaviour in juvenile bulls.

The future breeding potential of juvenile elephants may be enhanced by early association with other young animals, especially in bulls. There is evidence from other mammalian studies that the role of experience in mating is greater in males than in females (Beach, 1944, 1947; Carr et al., 1965) and particularly important in higher mammals (Beach, 1967). Opportunities for such experience to occur from soon after birth have been reported for some species (Hanby, 1972; Harlow and Harlow, 1965; Zuckerman, 1932) and it may be significant that very young bull calves show an interest in adult urine and faeces when they are exposed to them (Rees, 2003a), considering the important role chemical communication undoubtedly plays in elephant reproduction (Rasmussen and Schulte, 1998) (see Box 4.3).

4.9 Reproductive challenges and solutions

4.9.1 Acyclicity and sperm quality

Captive breeding programmes for elephants have been hampered by the large numbers of acyclic females in the populations. A survey of the reproductive cyclicity status of Asian and African elephants in North America published in 2004 determined that 49% of Asian and 62% of African elephant females were being monitored for ovarian cyclicity using weekly serum or progestogen analyses (Brown et al., 2004a). Of these, many were either not cycling at all or exhibiting irregular cycles (14% of Asian and 29% of African females). Animals over 30 years were more likely to show ovarian inactivity than younger animals in both species, but it was found in all age groups in African elephants. Of the elephants for which hormonal data were available along with the results of transrectal ultrasound examinations, about 75% of cycling females had normal reproductive tracts, but at least 70% of noncycling animals exhibited ovarian or uterine pathology. Brown et al. concluded that if fecundity is to be increased, females should be bred at a young age, before reproductive pathologies have time to develop. They noted that older Asian elephants and, more importantly, many prime reproductive age (10–30 years) African females were not at that time being reproductively monitored, making it difficult to

TABLE 4.10 The rates of acyclicity and irregular cycling in elephants in North America in 2002, 2005 and 2008.

Year	Asian elephants			African elephants		
	2002	2005	2008	2002	2005	2008
Acyclic (%)	13.3	10.9	11.1	22.1	31.2	30.5
Irregular cycling (%)	2.6	7.6	14.3	5.2	11.8	15.8
Acyclic + irregular (%)	15.9	18.5	25.4	27.3	43.0	46.3

Source: Based on Brown, J.L., Olson, D., Keele, M., Freeman, E.W., 2004. Survey of the reproductive cyclicity status of Asian and African elephants in North America. Zoo Biology 23, 309–321; Proctor, C.M., Freeman, E.W., Brown, J.L., 2010. Results of a second survey to assess the reproductive status of female Asian and African elephants in North America. Zoo Biology 29, 127–139; Dow, T.L., Holásková, I., Brown, J.L., 2011. Results of the third reproductive assessment survey of North American Asian (Elephas maximus) and African (Loxodonta africana) female elephants. Zoo Biology 30, 699–711.

determine the factors affecting reproductive health and to develop measures to reinitiate reproductive cyclicity.

A similar survey was conducted in 2005 (Proctor et al., 2010a). Surveys were returned for 100% of Asian and 79% of African elephants. Of these, 79.3% of Asian and 92.1% of African elephants were being monitored weekly to assess ovarian cyclicity using progestogen assays. The acyclicity rate remained similar in Asian elephants in the two surveys. However, it increased in African elephants. The percentage of elephants exhibiting irregular cycles increased between the two surveys in both species (Table 4.10).

The majority of acyclic elephants were not suffering from reproductive tract pathologies. Some elephants who were cycling in 2002 were not cycling in 2005, and the opposite was also true, therefore the condition is not irreversible. In both species, animals over 30 years showed a greater prevalence of acyclicity, but in the African species it also occurred in the reproductive age groups.

The results of a third survey of the reproductive status of female elephants held in Association of Zoos and Aquariums (AZA)/Species Survival Plan (SSP) facilities in North America in 2008 were reported by Dow et al. (2011) and showed further increases in irregular cycling in both species. Ovarian acyclicity rates did not change with time in either species, but were higher overall in the African than in the Asian species. In 2008 the percentage of animals — of both species — with irregular cycles and acyclic and irregular cycles was similar to 2005, but increased compared with 2002. Acyclicity increased with age in both species. Rates of reproductive tract pathologies were higher in acyclic than in cycling females, but pathologies did not account for the majority of acyclicity. In both species, increased cyclicity rates were associated with the presence of a bull and with management under free contact rather than protected contact. In the 2008 survey, 42.2% of Asian and 30.2% of African females were no longer being hormonally monitored, so Dow et al. (2011) concluded that abnormalities in reproductive cycles in North American elephant populations could have been worse than the data available at that time suggested.

The existence of a follicular cyst was identified as the cause of acyclicity in an African elephant at Pittsburgh Zoo in the United States using endocrine and ultrasound evaluation (Brown et al., 1999). Serum progesterone analysis indicated the presence of ovarian cycles and then their cessation. A transrectal ultrasound examination identified a large follicular structure on the right ovary. An intravenous injection of (500 μg) gonadotropin-releasing hormone (GnRH) stimulated a small increase in serum LH, but failed to cause a resumption of ovarian activity. A 5 mg dose of GnRH administered 5 months later had the same result. An ultrasound examination confirmed the cyst was still present. The elephant was treated with human chorionic gonadotropin (10,000 IU, intramuscularly). This induced oestrus and breeding, but not ovulation or luteinisation of the cyst and she remained acyclic. The authors of this study concluded that conventional treatment methods for treating follicular cysts that have been developed for other species may not be effective in elephants.

Freeman et al. (2009) examined the influence of social factors on ovarian acyclicity in African elephants. Three surveys were returned by 46 facilities housing a total of 106 female elephants (64 cycling, 27 noncycling and 15 undetermined). Females were more likely to be acyclic if they had a large body mass index and had resided longer at a facility with the same herd mates. Transferring females between facilities had no major impact on ovarian activity, but in 19 of the 21 facilities that held both cyclic and acyclic animals the dominant female was acyclic. The authors suggested that weight control might be a useful first step to mitigate oestrous cycle problems.

Following work that showed dominant female African elephants were more likely to be acyclic than were subdominant individuals (Freeman et al., 2004), Proctor et al. (2010b) examined the possibility of a relationship among dominance status, serum cortisol concentrations and ovarian acyclicity to test the theory that dominant females may expend more energy maintaining 'peace' within a captive herd than supporting reproduction. They suggested that adrenal glucocorticoid activity would be increased in dominant acyclic elephants — due to their experiencing increased stress — compared with subdominant individuals. Weekly blood samples were collected and analysed over a 2-year period from 81 elephants whose dominance status and cyclicity status were known. Samples from acyclic dominant females did not have a higher mean serum cortisol concentration than those from subdominant animals. Consequently, the authors concluded that alterations in adrenal activity are neither related to dominance status nor do they directly contribute to acyclicity in African elephants.

When bulls are selected for breeding it is clearly important that they are capable of producing high-quality sperm. Semen samples were collected from six bull Asian elephants from the Myanmar Timber Enterprise at Ngalaik Reserved Forest by electroejaculation to evaluate sperm quality (Mar et al., 1995). Although the sample size was too small to make inferences for the population as a whole, incidences of abnormal sperm in formol-saline fixed samples ranged from 24.5% to 36.5% in four ejaculates collected from three bulls.

Thongtip et al. (2008) examined semen collected from 13 Asian elephant bulls aged 10–72 years housed at the National Elephant Institute, Forest Industry Organization, Lampang, Thailand. Age and seasonality affected semen characteristics. Semen quality was highest at age 23–43 years. Percentages of progressive motility and viable sperm were lowest at age 51–70 years, but the lowest sperm concentration was found in animals aged 10–19 years. Bulls aged 23–43 years had the highest percentage of sperm with normal morphology. Seasonal effects were observed, with the percentage of viable sperm and cell concentration both being highest in the rainy season and lowest in the summer months. The percentage of morphologically normal sperm was highest in summer and lowest in the rainy season.

4.9.2 Obstetrics and birthing problems

Obstetrics is complicated in elephants because of their size and complex reproductive anatomy. Veterinary intervention options are limited when caring for elephants during birth or dystocia. Dystocia may be caused by intrauterine infection, uterine inertia and urinogenital tract pathologies. Caesarean section is not an option as it is lethal, and episiotomy and foetotomy are sometimes required to save the dam. These interventions may result in postsurgical complications or even death.

The procedures used by veterinary surgeons when dealing with birthing problems in captive elephants have been reported. Birth and dystocia management in elephants have been reviewed by Hermes et al. (2008). The successful delivery of a dead foetus by vaginal vestibulectomy performed on a pregnant Asian elephant named *Brahmaputri*, aged approximately 45 years, who was used to transport tourists at Kanha Tiger Reserve, Madhya Pradesh, India was described by Chandrapuria et al. (2014). Thitaram et al. (2006) described the successful removal of a dead foetus from a 32-year-old nulliparous female at the Thai Elephant Conservation Center, Lampang, Thailand, who was used daily for tourist trekking trips, following retention of the foetus for 12 months after parturition began. They emphasised the importance of regularly monitoring serum progestogen concentrations in mated elephant cows to verify the establishment of pregnancy and to better estimate the onset of calving.

4.9.3 New techniques in reproductive physiology

In the early 1980s Schmidt (1982) predicted that AI would become a 'practical technique for breeding elephants in captivity'. Although no pregnancy resulting from AI had been achieved at that time, this assertion was based on information gained from studies of Asian elephant reproduction conducted at Washington Park Zoo (now Oregon Zoo) in the United States. These included studies of the oestrous cycle, pregnancy, semen collection, bull management and health care, and pheromones.

The AI of elephants presents a considerable technical challenge, in part because of their size and unusual anatomy. Balke et al. (1988) examined the dead bodies of three mature nulliparous Asian elephants — who had spent most of their lives in captivity — using a fibreoptoscope and by dissection to identify the anatomical obstacles to AI. They determined these to be the length of the urinogenital canal (85–97 cm), constriction at the urinogenital–vaginal junction and tight cervix.

Studies of the reproductive physiology of elephants, including electroejaculation and semen characteristics (Howard et al., 1989; Mar et al., 1995) have culminated in the successful development of AI techniques. By 2012 almost 40 elephant calves (Asian and African) had been born worldwide as a result of AI (Hildebrandt et al., 2012) (Figs 4.14 and 4.15).

The first successful elephant pregnancy resulting from AI was reported by Dickerson Park Zoo, Springfield, Missouri, United States (Schmitt, 1998). An Asian cow was

FIGURE 4.14 Artificial insemination of an Asian elephant. Source: *Rees, P.A., 2011. Introduction to Zoo Biology and Management. Wiley-Blackwell, Chichester, West Sussex.*

FIGURE 4.15 *Ganesh Vijay* was conceived by AI and born to *Noorjahan* on 6 August 2009 at Twycross Zoo, United Kingdom. The semen was donated by *Emmett* from ZSL Whipsnade Zoo, United Kingdom. Source: *Reproduced with permission from Wiley-Blackwell. First published in Rees, P.A., 2011. Introduction to Zoo Biology and Management. Wiley-Blackwell, Chichester, West Sussex.*

inseminated on 25, 26 and 28 January 1998 and successfully gave birth in November 1999. AI of elephants is difficult because of the complex anatomy of the female urinogenital tract. However, a breakthrough in this technology was achieved by the use of an endoscope and ultrasound to guide the semen to the correct location.

Ultrasonography of the urinogenital tract has been used to assess female and male reproductive function. Hildebrandt et al. (2000a) performed ultrasonographic examinations of 280 captive and wild African and captive Asian female elephants. The primary pathological lesions that influenced reproductive rates in these females were uterine tumours, endometrial cysts and ovarian cysts that resulted in acyclicity. These examinations provide a reference for ultrasound specialists involved in breeding programmes and should assist elephant managers in deciding which animals to include in breeding programmes. An ultrasonographic study of male elephants of both species found that observable reproductive tract pathology in adult males was low (14%), even in older bulls (Hildebrandt et al., 2000b). However, apparent infertility of nonorganic cause in these otherwise healthy bulls was high (32%). Semen quality varied markedly in ejaculates from the same bull and in samples taken from different bulls. Hildebrandt et al. suggested that this inconsistency raises the question of the reliability of some individuals as participants in breeding programmes, and that the apparent inhibitory effect of suppressive social interactions on male reproductive potential should be investigated.

The African and Asian elephant European Endangered species Programmes have long recommended ultrasonography of cows in early pregnancy to exclude the possibility of twins, and the examination of every potential breeding bull annually from around the age of 8 years to determine the development of the reproductive organs (Dorresteyn and Terkel, 2000a).

TABLE 4.11 Advantages and disadvantages of using artificial insemination in elephants.

Advantages	Disadvantages
May increase fecundity of zoo population	Not yet widely available
May increase effective population size by allowing mating between animals in geographically separated zoos	May encourage over-dependence on a technological solution to low fecundity
May increase gene pool if semen is imported from nonzoo populations	Reduces experience of normal sexual behaviour in cows
Allows more accurate prediction of date of birth, thereby allowing zoos to make the necessary preparations	Reduces exposure of calves to normal sexual behaviour during critical periods in their development
Allows better control of timing of births (avoiding winter births in temperate climates)	Requires sophisticated technology and technical expertise
Reduces need to move animals between zoos for mating	Does not address the causes of low fecundity
Allows inclusion of sexually incompetent animals in breeding programmes	Expensive
May remove the social influences that affect fecundity	

In 2004, a team of scientists, most of whom were from the National Zoological Park in the United States and the Institute for Zoo Biology and Wildlife Research in Berlin, Germany, reported the advances in AI that had led to 12 elephants conceiving in a 4-year period (1998−2002) (Brown et al., 2004b). These advances resulted in the development of an endoscope-guided catheter and the use of transrectal ultrasound to deliver semen into the anterior vagina or cervix. The new technique also involved using the double LH surge to determine the appropriate time for insemination; the first (anovulatory) LH peak predictably occurs 3 weeks before the second (ovulatory) peak. The female Asian elephant at the National Zoological Park was the first elephant to be inseminated using this approach, but only the fifth to conceive. After six AI trials, beginning in 1995, she conceived in 2000 and subsequently produced a male calf.

Thongtip et al. (2009) reported the successful insemination of Asian elephants using chilled and frozen-thawed semen at the National Elephant Institute, Forest Industry Organization, Lampang, Thailand. This research resulted in the first successful birth using chilled semen and a successful pregnancy using frozen-thawed semen. The authors noted that this was an important stepping stone towards the application of AI technology for the genetic improvement of the elephant population.

AI has the potential to increase the effective population size of populations, but it may also have some negative effects (Table 4.11). In the long term, if zoos come to rely on this technology they may breed generations of sexually inept individuals that have neither witnessed normal adult sexual behaviour, nor experienced normal rearing within a herd. If zoos are serious about maintaining an insurance population of elephants, they need to ensure that the behavioural repertoire of their elephants is not compromised.

CHAPTER 5

Elephant cognition, communication and tool use

He is wise and brave and he learns by experience.

T. Murray Smith (1963)

5.1 Introduction

Elephants are special. There can be absolutely no doubt of this. Much of what we believe to be special about elephants stems from our knowledge of how they think and how they behave towards each other and towards humans. Experiments and observational studies conducted on captive elephants have made important contributions to our understanding of the nature of elephant cognition and the complexity of elephant society.

5.2 Cognition

5.2.1 Historical perspectives

Animal cognition may be defined as 'the study of the mental lives of animals' (Wynne, 2001). It is concerned with the mental process of acquiring knowledge through thought, by experience and through the senses; what animals know and how they know it.

An introductory text on animal cognition published in 2001 mentioned elephants just twice (Wynne, 2001); once to comment in passing on the large size of their brains and once to state that:

> '...other species tested, [using a mirror self-recognition test] *including fish, dogs, cats, elephants and parrots, all react to their reflection (if they react at all) as if it were another animal.*'

Plotnik et al. (2010) noted that, although the field of animal cognition had been growing steadily since the 1960s, it had focussed on animals easily kept in laboratories. Elephants had largely been ignored apart from a number of long-term field studies (e.g. Moss et al., 2011).

Elephants have long been considered to be highly intelligent, based on anecdotal evidence of their behaviour in the wild (Smith, 1963). More recently, scientific attention has turned to their cognitive abilities and, as a result, captive elephants have been the subject of a number of experiments. Elephants are good subjects for cognitive studies because they can be asked to manipulate objects with their trunk, they are amenable to training, and will perform tasks for small food rewards.

There has been very little cognitive research conducted on elephants of any species (Irie and Hasegawa, 2009; Plotnik et al., 2010). Plotnik et al. (2010) discuss the rationale for conducting cognitive research on elephants and Irie and Hasegawa (2009) have emphasised its importance in improving captive elephant welfare. Nissani (2008) has reviewed a number of elephant cognition experiments.

The value of anecdotal evidence in understanding cognition in elephants has been discussed by Bates and Byrne (2007). They draw attention, in particular, to the long-term study of elephants in Amboseli National Park, Kenya, by Moss et al. (2011) who followed the lives of hundreds of individual animals. Byrne and Bates (2011) have drawn together evidence of the cognitive skills of elephants from field studies of wild elephants, experimental studies and anecdotes gathered from trainers of semicaptive, working elephants.

5.2.2 Self-awareness: do elephants know they exist?

When chimpanzees (*Pan troglodytes*), gorillas (*Gorilla gorilla*) or orangutans (*Pongo* sp.) see themselves in a mirror they appear to be able to recognise themselves. However, when monkeys see themselves in a mirror they react to their own image as if it were another monkey.

A number of experiments have used the 'red-spot test' that involves a spot of red dye being placed on the animal's face under anaesthetic or while asleep, and then observing the animal's response to the spot when it looks in a mirror. Apes appeared to show self-awareness by touching the spot more often than a control area. For example they touched an ear marked with a spot more often than the other (unmarked) ear (Rodgers, 1998). Chimpanzees also demonstrated self-awareness by using a mirror to examine parts of the body that could not be seen directly (Povinelli and Preuss, 1995).

The experimental evidence for self-recognition in apes is inconsistent — some of the apes tested in red-spot experiments did not respond to the spots — and is therefore inconclusive.

Gallup suggested that the capacity for self-recognition may not extend below humans and the great apes (Gallup, 1970; Gallup et al., 1995). However, self-examination behaviour has been recorded in dolphins (*Tursiops* sp.) exposed to mirrors and video images. They will also examine marked areas of their bodies in a mirror (Marten and Psarakos, 1995). It has been suggested that the tendency of dolphins and killer whales (*Orcinus orca*) to adorn themselves with seaweed around their fins or flukes, or by carrying dead fish on their snouts, might be evidence of self-awareness (De Waal, 1996).

What happens when elephants are given the opportunity to see themselves in mirrors? Using a mirror, Daniel Povinelli tested two Asian elephants at the National Zoological Park in Washington, DC, United States, and found they paid little attention to their images and concluded they did not show self-recognition (Povinelli, 1989). However, an elephant's eyes are on the side of its head and the experiment has been criticised for using mirrors that were too small to allow the animals to see the entire side of their bodies at one time.

The value of mirror experiments is unclear. Experiments using mirrors have found that pigeons (Columbidae) respond to coloured dots in a similar way to Gallup's chimpanzees. At the time the experimenters concluded that this could not possibly be interpreted as complex behaviour or self-awareness in pigeons and offered a simpler explanation in terms of 'environmental events' (Rodgers, 1998). However, we know that pigeons are capable of some very complex behaviour (Epstein et al., 1981).

Plotnik et al. (2006) studied self-recognition in three adult female Asian elephants at the Bronx Zoo in New York by applying visible marks and invisible sham-marks to their heads and then exposing them to a large mirror to see if an individual would spontaneously use the mirror to touch an otherwise imperceptible mark on its own body (Fig. 5.1). One elephant, *Happy*, passed the test on the first day of the study, and repeatedly touched the visible mark with her trunk while standing at the mirror. The other two elephants, *Maxine* and *Patty*, failed the test. Plotnik et al. noted that these results were not inconsistent with some studies of chimpanzees where fewer than half of the individuals tested passed the test.

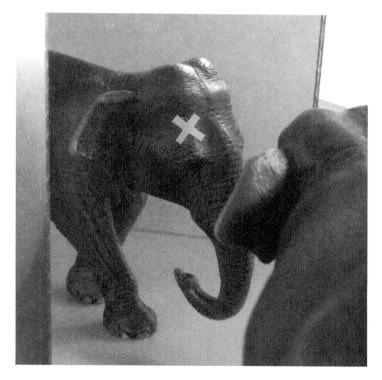

FIGURE 5.1 A representation of the apparatus used in a mirror self-recognition test performed on Asian elephants at the Bronx Zoo, New York, by Plotnik et al. (2006).

5.2.3 Discrimination between objects and between quantities

A comparative study of the ability of African elephants and California sea lions (*Zalophus californianus*) to perform a two-choice object discrimination task found that while both species were able to perform the task, the sea lions took fewer trials than did the elephants to reach a criterion of 10 consecutive correct responses (Savage et al., 1994). The elephants, but not the sea lions, exhibited a gradual learning of the task, which may have been due to differences in the species' visual abilities, or differences in their cognitive functioning.

Many species have demonstrated the ability to discriminate between quantities. When an animal is given a choice between two sets of food it will choose the larger amount. However, the ability to do this accurately decreases as the ratio between the two quantities increases. Irie-Sugimoto et al. (2009) reported that elephants did not exhibit ratio effects based on a study of Asian elephants using animals kept at Ueno Zoological Garden and Ichihara Elephant Kingdom in Japan. However, Perdue et al. (2012) found evidence of ratio effects for visible and nonvisible sequentially presented sets of food in a study of two African elephants at Zoo Atlanta in the United States. They concluded that elephants represent and compare quantities in a similar manner to other species, including humans, when they are prevented from counting.

5.2.4 Insightful behaviour

Insightful behaviour — the sudden ability to solve a problem without a period of trial-and-error learning — is well known in humans and some other species including chimpanzees and birds (Shettleworth, 2012). Three Asian elephants (1.2) housed at the National Zoological Park, Washington, DC, were tested for this ability by presenting them with sticks and other objects to assist them in obtaining food that was placed out-of-reach and overhead (Foerder et al., 2011). The 7-year-old bull moved, and then stood on, a large plastic cube to reach the food. In the absence of the cube, he generalised this technique to other objects and, when presented with smaller objects, stacked them in an attempt to reach the food. The researchers concluded that this elephant had demonstrated insightful behaviour and that previous failures to demonstrate this in other experiments were the result of poor experimental design.

The Asian elephant in Fig. 5.2 is standing on a log that she moved into place to make it easier to reach a suspended feeder in the elephant house at Blackpool Zoo in the United Kingdom (Clara and Elliot Clark, personal communication). Insight learning involving moving and using objects to extend their reach has been widely reported in chimpanzees (Kohler, 2018).

5.2.5 Pointing

Referential gesturing has been identified in a number of nonhuman species including great apes (Hominidae), ravens (*Corvus*) and some fish species. A study of cross-species referential signalling in domestic dogs (*Canis familiaris*) found evidence of 19 referential gestures performed by dogs during everyday communication with humans (Worsley and

FIGURE 5.2 This Asian elephant at Blackpool Zoo, United Kingdom, was observed moving a log from the pile on the right into position under the suspended feeder so that she could extend her reach by standing on it. Source: *Courtesy Clara and Elliot Clark*.

O'Hara, 2018). Recent studies of 11 captive African elephants at Wild Horizons — a tourist facility — in Zimbabwe demonstrated that they were able to use human pointing cues to locate hidden food without prior training (Smet and Byrne, 2013) (Fig. 5.3). This ability to understand and respond to human pointing cues has undoubtedly been important in sustaining elephants' long association with humans.

5.2.6 Memory

Anecdotally, elephants have long memories, and zoos offer an opportunity to test this. An adult Asian elephant cow (*Tuy Hoa*) at Portland Zoo, Oregon, took just 6 minutes to achieve 20 consecutive correct responses — and made only two errors — when retested on a light–dark discrimination problem that she had learned 8 years earlier. In studies with other elephants, no individual acquired the discrimination with only two errors, so the researchers concluded that she had remembered the task (Markowitz et al., 1975). Two other adult cow elephants retested in this study had developed retinal problems in the intertest period, which invalidated their results.

FIGURE 5.3 The experimental apparatus used by Smet and Byrne (2013) to demonstrate that African elephants can interpret human gestures. Source: *Reproduced with permission from Elsevier. First published in Smet, A.F., Byrne, R.W., 2013. African elephants can use human pointing cues to find hidden food. Current Biology 23, 2033−2037.*

When a female Asian elephant, *Shirley*, arrived at The Elephant Sanctuary in Tennessee, United States, in 1999, she and the resident female *Jenny* engaged in an intense greeting ceremony. Staff later discovered that some 23 years earlier, for just a few months, both animals had been part of the travelling Carson & Barnes Circus and had not been in contact since (Ritchie, 2009).

Studies of wild elephants have confirmed the importance of memory to elephant society and survival. During a drought in Tanzania in 1993, the calves of younger mothers were less likely to survive than those of older mothers, and family groups that remained in Tarangire National Park suffered higher calf losses than those that left the park. This may have been the result of older females who survived the previous severe drought in 1958−61 leading their groups to food and water outside the park using their previous knowledge. The oldest female in the clan that remained in the park was 33 years old in 1993 and so had no memory of the earlier drought (Foley et al., 2008). As climate change influences the ecology of arid regions, matriarchs' knowledge of the locations of refugia may be essential to the survival of elephants.

5.3 Communication

5.3.1 Introduction

Elephants exhibit an impressive range of methods of communication including vocal, chemical, tactile and seismic techniques, and much of what we know about elephant communication has been discovered from experiments conducted in captivity (Langbauer, 2000). In addition to their importance in the study of the behaviour, reproduction and social lives of free-ranging and captive elephants, studies of communication have also considered how elephant senses could be used to benefit humans and reduce human–elephant conflict.

5.3.2 Vocal communication

Vocal communication in the African elephant has been studied in the wild and in zoos. In a review of this research, Soltis (2010) claimed that the vocal repertoire of this species consists of 'perhaps nine acoustically distinct call types'. The most common and acoustically variable of these call types are low-frequency 'rumbles', with fundamental frequencies in the infrasonic range. The acoustic structure of rumbles is shaped by the trunk and oral cavity, which appear to facilitate the production of the resonant frequencies detected in rumbles. Elephants respond to low-frequency sounds at distances of up to 2.5 km.

Spatially separated females who are closely bonded engage in exchanges of rumbles that function as contact calls. These calls assist in coordinating movements and reuniting animals. 'Mate attraction' rumbles are produced by both sexes and may advertise reproductive status. Female African elephants do not produce rumbles at random, but are almost twice as likely to rumble after rumbles from other group members (Soltis et al., 2005a). Rumble variation reflects the individual identity and emotional state of callers in African elephants (Soltis et al., 2005b).

One way in which female African elephants advertise changes in their reproductive condition to other elephants is by increasing the rate of production of low-frequency vocalisations. Males respond to these vocalisations, but they may function primarily as a signal to nearby females rather than distant males. Leong et al. (2005) monitored behaviour, vocalisations and ovulatory cycles in six females at Disney's Animal Kingdom in Florida. The frequency of different types of close-range interactions changed during different phases of oestrus and whether or not low-frequency vocalisations were given in association with these interactions appeared to depend on rank. The researchers concluded that changes in low-frequency vocalisations may have been the result of changes in the social dynamics among females.

An understanding of elephant vocalisations may be useful in mitigating elephant–human conflict. O'Connell-Rodwell (2011) has shown that adult musth and subadult non-musth male African elephants may be attracted away from conflict areas by playing calls of females in oestrus.

A system for automatically classifying elephant vocalisations using a hidden Markov model has been devised using vocalisations collected from captive elephants living in a naturalistic environment, namely one adult male and six adult females living at Disney's Animal Kingdom in Florida (Clemins et al., 2005). The system classified vocalisation types with an accuracy of 94.3% and identified speakers with an accuracy of 82.5%.

5.3.3 Human speech imitation

A bull Asian elephant, *Koshik*, at Everland Zoo in South Korea imitated human speech according to a recent report (Stoeger et al., 2012). He matched Korean formats and fundamental frequency so well that Korean native speakers could readily understand and transcribe the imitations (Fig. 5.4). The elephant achieved this by modulating the shape of his vocal tract by inserting his trunk in his mouth. The elephant had been kept alone at the zoo for approximately 7 years and had been exposed to human speech (Korean) from zoo staff and visitors. The authors of the study believed that *Koshik* may have developed human speech as a result of social isolation from conspecifics during an important period in his social development, and humans were his only social contact.

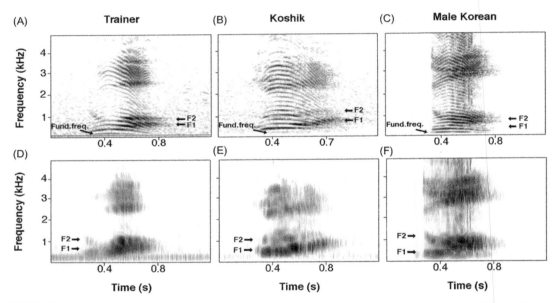

FIGURE 5.4 Spectral comparison of the speech utterance 'nuo' made by a bull Asian elephant (*Koshik*) (B and E), his trainer (A and D), and a male Korean native speaker (C and F). A–C represent narrow-band spectrograms of 'nuo' and D–F give wide-band spectrograms of each 'nuo' utterance, respectively. The fundamental frequency (fund. freq.) and the first and the second formant (F1 and F2) are indicated. Formants are the resonant frequencies of the vocal tract. Source: *Reproduced with permission from Elsevier. First published in Stoeger, A.S., Mietchen, D., Oh, S., Silva, S., Herbst, C.T., Kwon, S., Fitch, W.T., 2012. An Asian elephant imitates human speech. Current Biology 22, 2144–2148.*

5.3.4 Chemical communication

Studies of captive Asian elephants have been invaluable in the study of chemical communication in this species (Rasmussen and Krishnamurthy, 2000). Elephants raise their trunks to sniff the air, and use the tips of their trunks to explore the ground — especially urine and faecal material — and to sniff the mouths, temporal glands, feet, and genitals of other elephants (Fig. 5.5).

FIGURE 5.5 A bull elephant (*Chang*) testing the urine of a cow (*Sheba*) at Chester Zoo, United Kingdom: (A) *Sheba* urinates and defaecates; (B) *Chang* collects a sample of urine; (C) *Chang* tests the urine with the Jacobson's (vomeronasal) organ in the roof of his mouth.

Chemical senses are used by elephants to recognise relationships, determine metabolic state and social status and to exchange information about reproductive condition within and between the sexes. They are also important in long- and short-range navigation.

Elephants release complex mixtures of chemicals in their breath, urine, temporal glands, interdigital glands and ears. Some of these chemicals originate in the blood and may be metabolites, while others are secretory products that are sometimes under hormonal control. Elephants emit chemicals including proteins, hormones and volatile compounds. Certain volatile ketones function as chemical signals and an acetate

[(Z)-7-dodecenyl acetate] functions as a pheromone. Some of these chemicals identified from Asian elephants living in the United States have been found in exudates from elephants in India. The similarity is demonstrable for three metabolic conditions, namely pregnancy in females, and males in pre- and postmusth conditions (Rasmussen and Krishnamurthy, 2000).

Imagine if elephants could identify criminals who had killed their relatives and elephant friends and help to bring them to justice by providing evidence to the courts. This sounds fanciful, but research funded by the US Army has found that African elephants can be trained to match human scents to corresponding samples. In theory, this ability could be used to trace scents left by poachers on mobile phones or clothing to samples of the scents of suspects (Von Drückheim et al., 2018). Protocols similar to those used in the training of forensic canine units in Europe were used to train — using operant conditioning — three elephants from the tourist attraction Adventures with Elephants in Bela Bela, South Africa, to match human body scents to corresponding samples. Elephant handlers were blind to the experiment details and conducted 470 trials on 26 humans of differing ethic groups, sexes and ages. In each trial an elephant was required to match a target scent with scents from any of nine glass jars by exhibiting an alert behaviour (Fig. 5.6). The elephants demonstrated an ability to contextualise and classify humans into categories and to discriminate accurately between individual human scents and among closely related individuals.

In addition to human scents, elephants are also able to detect explosives. Miller et al. (2015) used three African elephants under controlled conditions to test whether they were able to detect and reliably indicate the presence of trinitrotoluene (TNT) — a common component of landmines — using olfaction. Using operant conditioning, they were able to show that the elephants detected and indicated TNT at levels greater than chance even in the presence of volatile distractor odours. The elephants' ability surpassed that of TNT-detection dogs under similar conditions.

FIGURE 5.6 A study of human scent matching in African elephants. Alert behaviour exhibited by three African elephants during scent matching to sample performance: (A) *Chova*; (B) *Chishuru*; (C) *Mussina*. Source: *Reproduced with permission from Elsevier. First published in Von Drückheim, K.E.M., Hoffman, L.C., Leslie, A., Hensman, M.C., Hensman, S., Schultz, K., Lee, S., 2018. African elephants* (Loxodonta africana) *display remarkable olfactory acuity in human scent matching to sample performance. Applied Animal Behaviour Science 200, 123–129.*

Use of the secretion of the temporal gland of the African elephant as an elephant repellent in Kruger National Park, South Africa, was examined by Gorman (1986) using wild elephants and seven young animals kept in captivity.

5.3.5 Tactile and seismic communication

The role of touch in the social interactions of six female Asian elephants at Busch Gardens Tampa Bay, Florida, was studied by Makecha et al. (2012). The body part most frequently used to initiate and receive tactile behaviour was, unsurprisingly, the trunk.

Seismic communication is a method of conveying information using mechanical vibrations of the substrate. It is used by a number of animal taxa including insects, arachnids, crustaceans, mammals, birds, reptiles and amphibians. Elephants communicate seismically by stamping their feet (O'Connell-Rodwell, 2007).

Bouley et al. (2007) studied the distribution, density and three-dimensional histomorphology of mechanoreceptors known as Pacinian corpuscles in the foot of the Asian elephant to examine their potential role in seismic communication. They studied samples from preserved elephant feet provided by the Leibniz Institute for Zoo and Wildlife Research, Berlin, presumably from zoo elephants, but the paper does not specify their provenance. The authors concluded that Pacinian corpuscles are one possible anatomical mechanism elephants use to detect seismic waves.

5.4 Visual acuity and visual discrimination

High visual acuity is important to elephants as it facilitates the eye–trunk coordination required for feeding, drinking and social interactions. An African elephant cow tested on a discrimination task using Landolt-C stimuli demonstrated a visual acuity of 48′ of arc, representing the ability to discriminate a gap of 2.75 cm in a stimulus 196 cm from her eye (Shyan-Norwalt et al., 2010). This standard acuity test requires the subject to determine where the gap in a C-shaped ring is located when the symbol is presented in various orientations (Fig. 5.7).

After studying simultaneous visual discrimination in Asian elephants conducted with 20 animals in a Burmese logging camp, Nissani et al. (2005) concluded that their results could not be reconciled with the widely held view that elephants are exceptionally

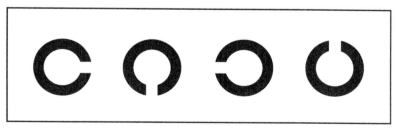

FIGURE 5.7 Typical images used in a Landolt-C test.

intelligent. In their first experiment, seven elephants acquired a white—black discrimination, reaching criterion in a mean of 2.6 sessions and 117 discrete trials. However, four elephants acquired discrimination in a mean of 5.3 sessions and 293 trials. One elephant failed to reach the criterion in the task in 9 sessions and 549 trials, and two other elephants failed to reach the criterion in 9 sessions and 452 trials.

In a second experiment, three elephants learned a large—small transposition problem, reaching the criterion within a mean of 1.7 sessions and 58 trials, but four other elephants failed to reach the criterion in a mean of 4.8 sessions and 193 trials.

Both discrimination tests showed an age effect, and Nissani et al. suggest that elephants beyond the age of 20—30 years either may be unable to acquire these visual discriminations or may require a very large number of trials to do so.

5.5 Tool use

Very few animals are known to use tools. Tool use and tool making are considered by zoologists to be among the highest level skills an animal may possess, although some simple tool use is seen in species from groups that we do not normally consider to be particularly mentally advanced, such as some birds (e.g. Hunt, 1996) and invertebrates [ants (Formicidae): Fellers and Fellers (1976); octopuses (Octopoda): Finn et al. (2009)].

Tool use is commonplace in elephants and its frequency and diversity appear to be similar to those found in tool-using primates (Beck, 1980). An elephant will use a small stick held in its trunk to scratch its skin (Fig. 5.8). It may even break off a long branch from a tree, snap off an appropriate length using its trunk and foot, and use it to reach parts of its

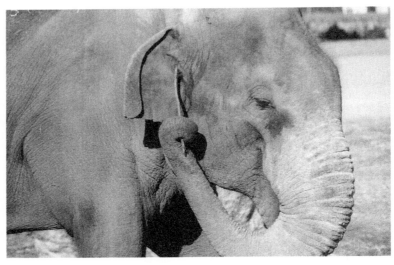

FIGURE 5.8 An Asian bull elephant (*Chang*) at Chester Zoo, United Kingdom, scratching his ear with a twig.

body that are otherwise inaccessible or to repel insects — an example of tool fabrication? Elephants have also been seen using clumps of grass held in the trunk to wipe cuts on their bodies (Douglas-Hamilton and Douglas-Hamilton, 1975).

Tool use has been studied in Asian and African elephants in zoos in order to develop a taxonomy of tool use and to compare their behaviours with those of wild African elephants (Chevalier-Skolnikoff and Liska, 1993). The authors used the very broad definition of tool use employed by Beck (1980):

> ...the external employment of an unattached environmental object to alter more efficiently the form, position, or condition of another object, another organism, or the user itself when the user holds or carries the tool during or just prior to use and is responsible for the proper and effective orientation of the tool.

They studied three juvenile African and six adult Asian elephants at Marine World/Africa USA, California, none of which had been trained to perform tasks requiring tools. These captive elephants performed tool use at frequencies of 22.8 acts/h (African) and 11.6 acts/h (Asian), whereas tool use in wild African elephants living in Samburu-Buffalo Springs Game Reserve, Kenya, was measured at a frequency of just 1.45 acts/h.

Nine different types of tool use were observed during the study of wild African elephants. These were grouped into four categories:

1. rubbing or swatting the body with vegetation held in the trunk,

2. collecting dirt or vegetation with the trunk and throwing it onto the body,

3. siphoning water and dust into the trunk and blowing it onto the body, and

4. throwing dirt at others with the trunk or front foot.

Most of these behaviours occurred in the functional contexts of body care, aggression (intraspecific and interspecific) and feeding another elephant. The captive elephants exhibited 551 acts of tool use, which were grouped into 21 types. In addition to those recorded in the wild elephants, captive animals engaged, among other things, in the following behaviours:

1. using vegetation to sweep an area of ground for placement of food;

2. twisting hay into special shapes (ropes, rings made of these ropes, balls) and placing them on the body, especially on the head and trunk and around the neck (ropes) (function unknown);

3. scrubbing a wall with vegetation (function unknown);

4. blowing away birds that try to eat food;

5. regurgitating liquid into the trunk and squirting it on the body (for cooling);

6. blowing air on body with trunk (for cooling and to remove flies); and

7. throwing and rolling objects at birds trying to eat food and flailing an object held in the trunk at them.

FIGURE 5.9 A female Asian elephant (*Maya*) using a wall as a support while breaking a branch with her foot.

Chevalier-Skolnikoff and Liska concluded that tool use in elephants occurs mainly in relation to parasite control, body care and body cooling. They suggest that it may have an adaptive function in compensating for a lack of biological equipment in species that, although they are land mammals, are largely hairless (Alcock, 1972).

Hart and Hart (1994) studied the use of branches as fly switches by adult Asian elephants used to give rides to visitors in a national park in Nepal. At the time of day when fly activity was most intense (1100), they compared fly counts around the elephants for 10 minutes before and 10 minutes after they were presented with branches similar to those they spontaneously collected during the day and found that the use of these branches reduced fly counts by 43%. Branches were sometimes modified by the elephants by removing side stems or shortening them. Hart and Hart concluded that the animals used fly switches to reduce fly harassment and their use was not related to diurnal changes in temperature and was not a stereotypic behaviour.

Occasionally individuals may use an object as a fulcrum in order to apply appropriate forces to break a branch. I observed Asian elephants at Chester Zoo standing branches against walls and then applying pressure with a forefoot to snap them in half (Fig. 5.9). They were also observed placing one end of a branch against a fallen tree trunk, the other end on the ground, and then stepping on the centre of the branch to break it into two (Fig. 5.10).

Elephants will pick up loose soil with the trunk and rub it into the skin, presumably to protect their skin from the sun or insect bites, and use water to cool their bodies. They use branches and stones as weapons, throwing them, often with great accuracy, at other animals, including people. This particular skill can be a problem in zoos and safari parks. In

FIGURE 5.10 A female Asian elephant (*Jangoli*) using a log as a fulcrum to break a piece of tree trunk.

May 2000 a 9-year-old girl suffered a broken jaw and lost ten teeth when she was struck in the face by a large stone thrown by an elephant at Sorocaba Zoo in Brazil. In 2016, a 7-year-old girl was killed by a stone thrown by an elephant in Rabat Zoo, Morocco (Anon., 2016a).

Although elephants do not possess the opposable digits that primates use to manipulate objects, nevertheless, the possession of a trunk that terminates in a highly sensitive and mobile tip makes it possible for them to perform quite complex, manipulative operations. They are capable of picking up extremely small objects, such as peanuts and small coins, and captive elephants will even attempt to unscrew the D-rings that hold their chains together, presumably after watching keepers secure them by screwing the parts together. Some elephant trainers have even taught them to manipulate and play musical instruments including harmonicas and a xylophone (with a rock) (Anon., 1976; Croke, 1997). Others have been taught to collect litter from their zoo enclosures and place it in litter bins (Fig. 7.29A).

Elephants working in logging camps are taught to respond to a series of verbal and tactile commands and others are taught to play games to amuse tourists (Table 5.1). Asian elephants have been trained to play football, and elephant polo is an established sport in some Asian countries.

5.6 Knowing when to cooperate

Elephants appear to know when they need assistance from another elephant to achieve a particular goal. Plotnik et al. (2011) performed experiments that required two elephants to pull simultaneously opposite ends of the same rope to obtain a reward. The elephants

TABLE 5.1 The main commands used to direct elephants used in logging operations in Sri Lanka.

Command	Behaviour
Hathderi	Bite
Udderi	Lift
Puru	Push
Dana Puru	Kneel down
Hide	Lie down
Theth theth	Go back
Pitchet	Release
Daha	Proceed
Deri Daha	Proceed with load
Dana Daha	Kneel and push
Diga	Spread down
Bila	Lift log

Source: Based on FAO, 1999. Forest Harvesting Case-Study 5. Elephants in logging operations in Sri Lanka. Publications Division, Food and Agriculture Organization of the United Nations, Viale delle Terme di Caracalla, 00100 Rome, Italy. http://www.fao.org/3/v9570E/v9570e00.htm#TopOfPage (accessed 08.03.19.).

successfully cooperated in this task, and furthermore, if one elephant's arrival was delayed, the other elephant 'inhibited the pulling response' (waited) for up to 45 seconds. If one elephant was present, but had no access to the rope, the other individual understood that there was no point in pulling the rope. Tests of this type were originally designed to be used with chimpanzees and Plotnik et al. suggest that their results with elephants suggest that the two taxa have evolved similar cooperative skill levels by convergent evolution.

CHAPTER 6

Elephant ecology and genetics

On the High and Far-Off Times the Elephant, O Best Beloved, had no trunk. He had only a blackish, bulgy nose, as big as a boot, that he could wriggle about from side to side; but he couldn't pick up things with it.

The Elephant's Child, Rudyard Kipling (1902)

6.1 Introduction

Although elephants living in zoos, and other captive elephants, clearly do not have the same ecological relationships that they would have in their natural habitats, they nevertheless have *an* ecology and this is worthy of investigation. Some ecological investigations are only — or at least principally — relevant to a captive context, for example enclosure use or studies of population dynamics in zoos or forest camps (Leighty et al., 2010; Sukumar et al., 1997; Wiese, 2000). However, other captive studies may assist in the understanding of the ecology of wild elephant populations (e.g. Rees, 1982). Captive elephants can make a contribution to our knowledge of elephant genetics and evolution. Additionally, our understanding of their genetics may assist in the management of elephant conservation breeding programmes. Although this chapter considers the population ecology of captive elephants living in zoos and logging camps, the biology of reproduction and birth statistics (e.g. length of gestation, birth weight) are discussed in Chapter 4, Elephant reproductive biology.

6.2 Ecophysiology

6.2.1 Introduction

Physiologists have found elephants an attractive subject of study at least in part due to their status as endothermic gigantotherms: animals that generate heat physiologically, but are so large that their low surface area-to-volume ratio makes it difficult for them to lose heat over their body surface. Elephants are also physiologically interesting because they must extract large quantities of nutrients from poor-quality plant material that is difficult to digest and yet, unlike ruminants, they have a simple stomach.

Captive elephants are often trained to perform routine tasks and understand commands from their handlers, either because they perform work (e.g. in logging camps) or as part of their husbandry. This makes it possible to train them to cooperate in experiments, some of which require them to wear respiratory apparatus (e.g. Langman et al., 2012), stand on weigh bridges, wear inertial sensors (Rothwell et al., 2011), walk in front of video cameras and across force platforms (Panagiotopoulou et al., 2016).

6.2.2 Acclimatisation to new environments

In the wild, mammals exposed to seasonal climatic changes respond by changes in coat insulation. This is more economical than increasing heat production to protect against the cold due to the high energy costs. Langman et al. (2015) used a noninvasive method to measure mammal coat insulation in a number of captive species, including African elephants. Elephants have very little body hair and, although some individuals may be subjected to quite a wide range of daily temperatures in the wild, they do not experience substantial seasonal changes in temperature. Consequently, elephants have no ability to acclimatise to the seasonal changes in ambient temperatures that occur in the northern hemisphere by increasing the insulation afforded by body hair [e.g. unlike the Amur tigers (*Panthera tigris altaica*) and mountain goats (*Oreamnos americanus*) that were included in this study]. This has important implications for elephant husbandry. When the ambient temperature is below 21°C, elephants will use body energy to maintain heat balance so they require housing that is capable of maintaining the ambient temperature above this value.

Jachowski et al. (2013) examined the stress responses of African elephants translocated to five fenced areas in South Africa and found that they needed an extended period of time to adapt physiologically to their new surroundings (see Section 7.2.6).

6.2.3 Thermoregulation

A number of studies of thermoregulation in elephants have been conducted in zoos. The role of the pinnae (ears) in temperature regulation has been studied by Phillips and Heath (1992). They measured the surface temperatures of the pinnae of four female African elephants at ambient temperatures between 14°C and 32°C using infrared thermography and calculated instantaneous heat loss to be from 10.67 to 76.2 W. They estimated that these heat losses accounted for 0.65%–4.64% of the elephants' standard metabolic rates (considering one side of one ear only) and concluded that movement of the pinnae and vasodilation could meet 100% of an African elephant's heat loss needs.

Heat transfer in an adult African and an immature Asian elephant at the San Diego Zoo has been studied by Williams (1990). Using infrared thermography, she measured skin temperatures and determined that heat transfer by free convection and radiation accounted for 86% of total heat loss at an ambient temperature of 12.6°C. At this temperature, heat transfer across the ears represented less than 8% of total heat loss. The thermal conductance of the elephants was three to five times higher than that predicted from an allometric relationship for smaller mammals. Williams concluded that the high thermal conductance of elephants could be attributed to the absence of fur and that it appeared to counteract reduced heat transfer associated with a low surface area-to-volume ratio.

Weissenböck et al. (2010) examined thermal windows on the body surfaces of six African elephants using infrared thermography. They found independent thermal windows — highly vascularised skin areas — on the whole body and also sharply delimited hot sections on the elephants' pinnae. The frequency of these thermal windows increased with ambient temperature and with body weight. They believed that the restriction of an enhanced cutaneous blood flow to thermal windows might enable an elephant to react flexibly to its heat loss needs.

Rowe et al. (2013) studied heat storage in two adult female Asian elephants during periods of submaximal exercise at Audubon Zoo in New Orleans, United States. They measured the elephants' core body temperature and skin temperature pre- and postexercise using thermography. In ambient temperatures in the range of 8°C–34.5°C, when exercised in full sunlight, the elephants stored 56%–100% of active metabolic heat production in core body tissues. The authors estimated that during nocturnal activity — in the absence of solar radiation — core tissues would store 5%–64% of metabolic heat production. They concluded that potentially lethal rates of heat storage in active elephants could be regulated behaviourally by choosing to engage in nocturnal activity.

The influence of season and roofing material on thermoregulation by captive Asian elephants and its implications for captive elephant welfare have been studied by Vanitha and Baskaran (2010). Their subjects were six elephants managed by six Hindu temples in Tamil Nadu, India (Fig. 6.1). The frequency of ear-flapping increased with increase in ambient temperature in these elephants. The authors compared ear-flapping rates of elephants kept in houses with roofs made of asbestos, reinforced cement concrete (RCC) and

FIGURE 6.1 An elephant at Kateel Shri Durgaparameshwari Temple, Karnataka, India.

coconut fronds, as well as the ambient temperatures in these buildings, and found that buildings with asbestos roofs had the highest maximum and lowest minimum temperatures. They suggested that asbestos roofs should be replaced with coconut fronds because these provide a more suitable thermal environment for elephants as well as concerns about the negative health implications associated with the use of asbestos. Vanitha and Baskaran also suggested that false roofs made of fronds should be constructed on top of RCC roofs or they should be shaded by trees. They advised that those working with captive elephants should consider the diurnal and seasonal patterns in temperature variation when scheduling the work of their animals.

I studied dusting behaviour in a captive herd of eight Asian elephants at Chester Zoo and found that they exhibited this behaviour when the maximum daily temperature exceeded approximately 13°C (55.4°F), and dusting frequency increased as the environmental temperature increased (Rees, 2002a) (Fig. 6.2). Individual animals showed variation in dusting frequency but this was not related to body mass, suggesting that the function of dusting is not primarily thermoregulatory. The function of dusting behaviour could not be determined from the data obtained, but it may be involved in skin care, protection from insects or other parasites, temperature control, protection from radiation or some combination of these. Synchronisation in the timing of dusting behaviour within the herd suggests that it may have a function in the maintenance of social cohesion.

Elzanowski and Sergiel (2006) were unable to find any increase in dusting frequency with increase in temperature in their study of a single female Asian elephant in Wroclaw Zoo, Poland. Dust bathing in wild African elephants at waterholes in Etosha National Park, Namibia, was studied by Barandongo et al. (2018) over a period of 27 days in a range of temperatures from approximately 25°C to 34°C. They found no relationship

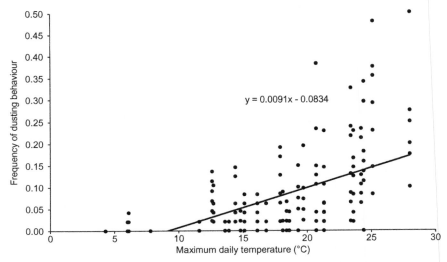

FIGURE 6.2 The effect of environmental temperature on the frequency of dusting behaviour in five adult Asian elephants. Source: *Reproduced with permission from Elsevier. Originally published in Rees, P.A., 2002. Asian elephants* (Elephas maximus) *dust bathe in response to an increase in environmental temperature. Journal of Thermal Biology 27, 353—358.*

between temperature and dust-bathing behaviours. However, this study was not directly comparable with the previous two studies as the authors measured dust bathing in terms of intensity (number of throws of soil) and the proportion of elephants dust bathing within a group rather than dusting frequency. Also, the range of environmental temperatures in the African study (approximately 9°C) was much smaller than that experienced by the Asian elephants at Chester Zoo in the United Kingdom [approximately 24°C (4°C−28°C)].

6.3 Feeding ecology and energetics

6.3.1 Food preferences

Elephants living in zoos and circuses must eat what they are given apart from any plant material that they can gather from fields and enclosures. There is a limit to what studying their diet can tell us about the ecology of wild elephants. However, this is not the case for elephants living in logging camps that gather food from the forests where they live and work.

In the wild, elephants eat a wide range of plant species and exhibit preferences for particular foods. For example elephants in the Kuldiha Wildlife Sanctuary in Odisha, India, ate only 71 (24%) of the 290 floral species found in the area, belonging to 37 families (Mohapatra et al., 2013). Of these, 56% were tree species. In Sumatra, 14 free-foraging elephants at the Seblat Elephant Conservation Center were recorded eating at least 273 species of plants belonging to 69 families (Sitompul et al., 2013). These elephants spent 56.3% of their feeding time browsing and 43.1% grazing. In the Sengwa area of Zimbabwe, Guy (1976) recorded elephants eating 133 plant species from 41 families. Bull elephants spent on average 18.9% of their time grazing and 28.5% browsing. The corresponding figures for cows were 12.4% and 27.0%, respectively. However, this varied with season and grazing increased to 44.6% in bulls and 32.8% in cows in the wet season.

A single elephant observed for a single day in Tsavo National Park (East), Kenya, consumed 64 plant species from 28 families (Dougall and Sheldrick, 1964). In Lake Manyara National Park, Tanzania, Douglas-Hamilton (1972) reported that elephants fed on at least 134 (20.6%) of the 650 plant species available.

Some keepers may be unreliable sources of information on the food preferences of captive elephants. Gaalema et al. (2011) found that the actual food preferences of African elephants did not always coincide with food preference ratings indicated by their caregivers. They suggested that the food preferences of one particular animal may be falsely generalised to all animals of that species. In contrast, Campos-Arceiz et al. (2008) found mahouts at an elephant camp in central Myanmar (Burma) to be an important source of information on the feeding ecology of wild elephants. They interviewed mahouts and veterinarians to determine the diet of Asian elephants in mixed-deciduous forest. They consumed 103 plant species from 42 families and mostly browse (94% of plant species). They were highly selective with respect to the plant parts taken. Many trees were eaten exclusively for their bark (22%) or their fruits (14%). The elephants ate the fruits of 29 plants species, several of which were found in their dung, suggesting a role in seed dispersal.

Vanitha et al. (2008) evaluated the food and feeding of captive elephants in Tamil Nadu, southern India, with regards to the elephants' health. They found that privately kept elephants and temple elephants were stall-fed with significantly less food, and poorer-quality food (often just one or two plant species), than was received by Forest Department elephants who had access to natural food as well as receiving a supplementary diet. They attributed poor health — including deficiency diseases — among the privately owned and temple elephants to these dietary inadequacies.

6.3.2 Feeding methods

During a study of the feeding ecology of seven juvenile African elephants (aged approximately 11 years) at Knowsley Safari Park, United Kingdom, I studied the relationship between feeding method and the quantity of food collected when feeding on various grasses (Rees, 1977). The collection of food was usually achieved by gripping the grasses near the base of the shoots and tearing them from the ground using a twisting action of the trunk. The trunk was twisted around the grass near the bases of the leaves and rotated so as to bring the trunk tip up from the ground, thereby breaking the blades of grass near their bases. The proboscideal processes were then used to lift the food to the mouth. If the grass was pulled up by its roots and these had soil attached, elephants would usually remove the soil by either shaking the grass or striking it against their bodies.

Three of the elephants in this group used their feet to assist the trunk in ripping up grass. *Ndovu* simply placed one of her forefeet against the trunk after it had been wrapped around the grass and then used it to assist the trunk in pulling up the grass. *Adega* used a different technique which involved swinging one of her forelegs away from her body and then kicking the tip of the trunk and/or the bases of the blades of grass. 'Foot-assisted' grabs in the three elephants who used these methods were 2.9%, 40.0% and 40.9% of all grabs collected.

In addition to grazing, captive elephants will browse if suitable material is available. They will use their trunks, teeth, tusks and feet to break bark off tree trunks and strip leaves and twigs from branches. The use of browsing as an enrichment is discussed in Section 7.3.2.

6.3.3 Calculating food consumption

How much food do elephants eat? Estimates of the amount of vegetation consumed by wild elephants are of interest when examining energy flow in ecosystems and estimating carrying capacities. They are particularly important where elephants constitute a high proportion of the herbivore biomass. Food consumption can be measured directly by recording feeding rate and the amount of food collected in each trunkful, or indirectly by weighing stomach contents or using gross assimilation efficiency (GAE).

I attempted to measure food consumption directly by recording the feeding behaviour of the seven elephants at Knowsley Safari Park on Ilford Motion Picture film Mk.5 at 18 frames per second using a Bolex H16 reflex cine camera fitted with an Angenieux zoom type $20 \times 12B$ lens (focal length 12–240 mm) (Rees, 1977). Thirty-six 20 second sequences were filmed and elephants were recorded collecting food (grasses) from the ground on 175 occasions. The rate at which food was collected averaged 13.9 grabs (trunkfuls)/min in November (calculated from the film) and 9.5 trunkfuls/min in February (recorded with a stop-watch) when the grass was in poorer condition. This is similar to the rates observed by Wyatt and Eltringham (1974) of 14 trunkfuls/min measured in a single wild adult African bull in Uganda feeding on bush vegetation; however, the mean rate in females ($n = 6$) feeding on grassland was lower at 3.5 trunkfuls/min in the morning and 8.5 trunkfuls/min in the evening.

I collected samples of vegetation similar in size to those gathered by the elephants at Knowsley and used the weights of these samples (mean = 76.7 g) along with feeding rates, to calculate food consumption, subtracting the quantities of food they collected and then dropped while feeding. The mean wet weight of food eaten by the elephants was calculated as 712 g/min in November (calculated from the film) and 488 g/min in February, when the grass was shorter (Rees, 1977).

Guy (1975) measured feeding rates of between 1.5 and 4.2 trunkfuls/min in wild elephants browsing in Zimbabwe, and between 2.0 and 3.7 trunkfuls/min when they were grazing. He used a mean weight per trunkful of 75 g, which agrees well with my study. Guy observed the highest feeding rates when food was most abundant, which supports my observations on captive elephants at Knowsley.

Phillipson (1975) drew attention to the fact that the computation of food consumption by elephants in the southern section of Tsavo National Park (East), Kenya, from data available in the literature at that time (Benedict, 1936; Coe, 1972; Petrides and Swank, 1966; Wing and Buss, 1970) resulted in very different estimates, depending on the assumptions made in the calculations. Some of the data used in these calculations had been derived from studies of captive elephants.

Two indirect methods have been used to calculate food consumption by wild elephants. The first requires an estimate of defaecation rate and GAE [i.e. apparent dry matter digestibility (DMD)] (Coe, 1972; Petrides and Swank, 1966). The second depends on a knowledge of the weight of the stomach contents of dead elephants and the food passage time (Laws, 1970a; Laws and Parker, 1968; Laws et al., 1975).

Using the first method, if the quantity of dung an elephant produces is known, it is possible to use information about its assimilation efficiency to calculate the amount of food it took to produce this amount of dung. The GAE of an elephant measures the efficiency with which it converts food into its own body mass. This is a crude measure and is calculated as:

$$GAE = \frac{I - E}{I} \times 100$$

where I is the dry weight of food consumed and E is the dry weight of dung produced over the same period (Fig. 6.3). If we know that the GAE is 50%, then for every 1 kg of dry dung produced 2 kg of dry food must have been eaten.

To calculate the GAE of an elephant — or the digestibility of different foods — it must be confined so that its food intake and dung production can be carefully controlled and measured.

Benedict calculated the GAE of an Asian circus elephant called *Jap* in New Jersey, United States, over 9 days and determined it to be 43.8% (Benedict, 1936). This figure was subsequently used by Petrides and Swank (1966) to calculate the food consumption of wild African elephants in the Queen Elizabeth National Park, Uganda, as 42.3 kg/day

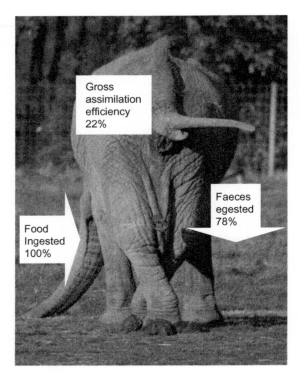

FIGURE 6.3 Gross assimilation efficiency in the African elephant. Source: *Reproduced with permission from Wiley-Blackwell. Originally published in Rees, P.A., 2011. Introduction to Zoo Biology and Management. Wiley-Blackwell, Chichester, West Sussex, based on data in Rees, P.A., 1982. Gross assimilation efficiency and food passage time in the African elephant. African Journal of Ecology 20, 193–198.*

(dry weight) using data collected on dung production from a single wild bull over a period of 12 hours. At that time, no GAE value was available for African elephants.

A GAE value that is too high will result in an overestimate of food consumption. One that is too low will result in an underestimate of food consumption. In an attempt to improve food consumption estimates for wild African elephants, I calculated a GAE value for two juvenile (11-year-old) female African elephants kept at Knowsley Safari Park (Rees, 1982). The elephants were confined in the elephant house for 7 days during which the dry weights of all the food consumed and dung voided were determined. The mean GAE of the two elephants was calculated to be 22.4%. This was a little more than half of that calculated by Benedict (1936), and if this lower GAE is applied to the data of Petrides and Swank (1966) it results in a food consumption estimate of 30.7 kg/day, approximately 72.5% of that calculated using Benedict's figure.

Coe (1972) calculated that the Tsavo elephant population consumed 60.34 kg/km^2 per day using data he collected on dung production from four captive African elephants housed at an elephant orphanage in Tsavo National Park (East), the population data of Laws (1969) and Benedict's assimilation figure. This estimate is reduced to 43.75 kg/km^2 per day using an assimilation figure of 22.4% (Rees, 1982). Applying this lower assimilation efficiency figure could theoretically increase calculations of elephant carrying capacities — the numbers that could be supported by a particular area at a given time — based on food availability alone by 38%.

6.3.4 Digestibility

The elephant digestive system is extremely inefficient. If a carrot is placed on an elephant's tongue and it swallows it, without chewing, the carrot may later appear whole in its dung. Elephants sometimes eat undigested food from their own dung.

Digestibility is a measure of the nutritive value of food. It varies between different foods when consumed by the same animal and may vary between individuals of the same species feeding on identical diets. In addition, the digestibility of a particular plant food may vary at different times of the year due to changes in its chemistry. Knowledge of the digestibility of different foods is important in the management of elephants in captivity and a number of studies have considered this.

Although the total faecal collection method of determining DMD in large animals is the standard method of analysis, it is time-consuming and labour-intensive. The acid-insoluble ash (AIA) marker technique has been successfully used as an alternative in ruminants and monogastric mammals. Pendlebury et al. (2005) studied three bull African elephants at Disney's Animal Kingdom, Florida, to compare DMD values determined using total faecal output and an AIA marker technique. The elephants were feeding on a Bermuda grass hay-based ration. Total dry matter intake and total faecal output were measured over a 7-day period, and the feed ingredients and faeces were analysed for AIA. This study found no statistically significant difference between the DMD values calculated using the two methods and the authors concluded that AIA can be used to estimate DMD accurately in captive African elephants.

A study of the apparent digestibility of timothy hay (*Phleum pratense*) by two juvenile African elephants found it to be 39% in summer and 35% in winter, for dry matter (Roehrs et al., 1989). This difference was not statistically significant, but the digestibility of acid detergent fibre present — the percentage of the material that is difficult to digest — was significantly higher in summer (36%) than in winter (24%) (Table 6.1). Dry matter intake was 1.5%–1.6% of body weight (BW) and provided an average of 144 kcal of digestible energy/kg $BW^{0.75}$. Other studies of feed digestibilities in elephants in captivity have been reviewed by Clauss et al. (2003).

6.3.5 Food passage time

The time taken for food to pass through the gut — food passage time — along with the weight of the stomach contents of a dead elephant may be used to estimate food consumption.

TABLE 6.1 The composition of timothy hay in summer and winter.

Variable	Summer (%)	Winter (%)
Dry matter	39	35
Gross energy	43	32
Crude protein	45	30
Acid detergent fibre[a]	36	24

[a]*Significant difference* (P < .05).
Source: *Based on Roehrs, J.M., Brockway, C.R., Ross, D.V., Reichard, T.A., Ullrey, D.E., 1989. Digestibility of timothy hay by African elephants. Zoo Biology 8, 331–337.*

If the food passage time is 24 hours, the stomach contents represent the quantity of food consumed in a day. If the food passage time is 12 hours the stomach contents represent the amount of food consumed in 12 hours, and so on.

To estimate food passage time, I fed 6.35 kg of beetroot to an 11-year-old African elephant at Knowsley Safari Park, United Kingdom. The time taken for this to pass through the elephant's gut was estimated to be between 21.4 and 46 hours (based on the appearance of colour in the faeces) (Rees, 1982).

The first estimate of food passage time in an elephant was made by Benedict (1936), using an Asian elephant, by measuring the amount of time taken for pieces of rubber that had been inserted in loaves of bread to appear in the faeces. He concluded that the significant fact resulting from four tests was that in each test the first piece of rubber emerged after between 21 and 30 hours. He attributed this fairly rapid passage time to the fact that the food consumed was hay and this may have affected the peristaltic action of the gut. Bax and Sheldrick (1963) reported that oranges fed to a tame African elephant at the orphanage in Tsavo National Park (East) were voided between 11 and 14 hours after feeding, but continued to appear in the dung up to 19 hours later.

The food consumption estimates made for African elephants by Laws and Parker (1968) and Laws (1970a) used stomach weights of culled animals and a food passage time of 12 hours (based on the results of Bax and Sheldrick, 1963). They do not appear to have taken any account of the quantity of water drunk per day, which could increase the weight of a stomach by 100 kg (Guy, 1975), thereby leading to gross overestimates of food consumption. A further error is produced by not considering the weight of food present in the rest of the gut. The study conducted by Guy in the Sengwa Wildlife Research Area in Zimbabwe used a food passage time of 19 hours, that is the mean of figures given by Benedict (1936) and Bax and Sheldrick (1963).

My estimate of a (minimum) food passage time of 21.4 hours in an African elephant compares well with Benedict's figure of 21 hours for an Asian elephant (Rees, 1982). If around 24 hours is a more realistic estimate of food passage time in the African species than is the 12 hours used by Laws and Parker (1968) and Laws (1970a) their food consumption estimates would be reduced by half.

The determination of food passage time is not as simple as I have suggested here. Accurate estimates can only be made using quantitative markers and recording the percentage passed with time. It should also be noted that food passage time may be affected by the quality of the diet and by feeding rate.

Wild elephants feed for long periods and have a high food intake. In herbivores, increased food intake usually reduces food passage time through the gut, lowering digestive efficiency. Clauss et al. (2007) used the data collected by Hackenberger (1987) on 50 captive Asian and African elephants to show that increased food intake in these species leads to only a very moderate increase of ingesta passage, thus theoretically allowing the optimisation of energy gain by eating more. They recounted the observation of McKay (1973) that wild Asian elephants have a propensity to develop 'extremely bulging bellies' after just 5 or 6 hours of intensive feeding, drawing the conclusion that:

> ...if the rate at which an elephant can ingest food greatly exceeds the rate at which the intestines and caecum can process it, there would be a distinct advantage to the possession of a very elastic stomach.

6.3.6 Defaecation

The quantity of dung produced by elephants, and diurnal variations in this, are of interest to ecologists because dropping counts can been used to investigate population size and movements in the wild (Laws et al., 1975; Wing and Buss, 1970).

Coe (1972) studied patterns of defaecation in four orphaned elephants aged 1–10 years at the Voi headquarters of Tsavo National Park (East), Kenya, for 4 days and three nights. He found that the quantity of dung produced with each defaecation reflected the growth curve of the species. The defaecation rate did not vary significantly with age and the temporal pattern of defaecations exhibited a mid-morning low peak and a mid-afternoon high peak.

I conducted a similar study using two juvenile African elephants at Knowsley Safari Park (Rees, 1983). These elephants exhibited a synchrony in the temporal pattern of defaecation as a result of social facilitation (Table 6.2). This behaviour may confer a selective advantage on wild individuals. Elephants generally defaecate standing still with their rear legs spread. Any individual, especially a calf, who did not defaecate at the same time as others in the herd could be left behind if it stopped while the rest of the herd was moving, increasing its vulnerability to predators such as lions (*Panthera leo*).

TABLE 6.2 Synchronisation of defaecation in two female African elephants kept at Knowsley Safari Park, United Kingdom.

Cefrose	*Tissa*		
	Elimination	No elimination	Total
Elimination	8	18	26
No elimination	27	339	366
Total	35	357	392

Values are the number of 7.5-min periods in which elimination occurred independently and synchronously in the two elephants.
Source: *From Rees, P.A., 1983. Synchronization of defaecation in the African elephant* (Loxodonta africana). *Journal of Zoology 201, 581–585.*

6.3.7 Food supplementation

Many elephants living in zoos rely entirely on keepers to provide their food (e.g. hay, potatoes, carrots, bread, etc.) because they have no, or very little, access to natural vegetation in their enclosure. The substratum of many enclosures is essentially bare soil or sand and any ornamental shrubs or trees present are likely to be protected by electric fences. Other elephants may be kept in fields and have access to grasses and other vegetation for perhaps up to 8 h/day. In temperate latitudes, where plant growth occurs in the spring and summer and day length is short in winter, access to wild plants in the enclosure is seasonal, with more natural plant material being available in the warmer months and less in the colder months. To compensate for this, the quantity of supplementary food provided by the zoo is likely to vary seasonally.

Fig. 6.4 shows the variation in the amount of food provided for seven juvenile African elephants kept at Knowsley Safari Park in 1976–77 including, hay, straw, vegetables and fruits (Rees, 1977). During the day these animals were able to feed in large 'reserves' (fields) containing natural vegetation, predominantly grasses. During winter the elephants

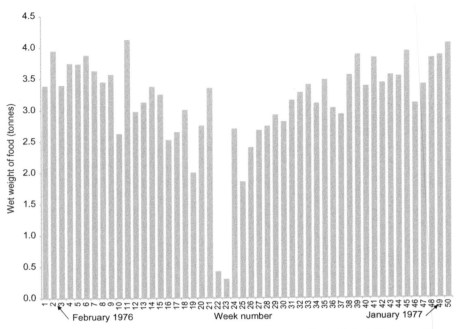

FIGURE 6.4 Supplementary food fed to elephants at Knowsley Safari Park, United Kingdom. Source: *Based on records kept by Mr Ivor Rosaire, Knowsley Safari Park.*

spent less time feeding in these reserves than they did in the summer, due to seasonal differences in their management. The seasonal variation in the requirement for supplementary feeding is clearly apparent. Weeks 22 and 23 fell in July when the highest maximum temperature recorded in the region was 33.7°C (HMSO, 1977a) and the animals spent around 8 h/day grazing outside. In January of 1977 the highest maximum temperature recorded was 12.0°C (HMSO, 1977b) and the animals were fed inside and on hard standing (concrete) areas on very cold days.

6.4 Energetics

It is interesting to study the energetics of elephants because of their status as the largest living land animals and because they can help us to understand how energy requirements alter with size.

Langman et al. (1995) studied the energetics of walking in the African elephant using three young males at Zoo Atlanta, Georgia, United States, who were trained to wear loose-fitting face masks enclosing the mouth and trunk. Air was pumped to the face mask via a tube; the pump was carried by a motorised golf cart that the elephant followed at walking speed. Langman et al. concluded that, contrary to what might be expected based on its size, the energy cost of locomotion in the African elephant was the lowest of any living land mammal.

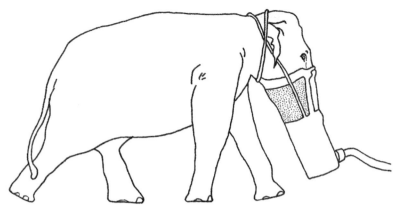

FIGURE 6.5 An elephant wearing the mask used by Langman et al. (1995) to study the energetics of locomotion. The tube on the right is connected to a pump carried on a motorised golf cart as the elephant moves forwards. Note that the original published photograph was of an Asian elephant from a different study. Source: *Drawing by Dr A. J. Woodward based on a photograph in Langman, V.A., Roberts, T.J., Black, J., Maloiy, G.M.O., Heglund, N.C., Weber, J-M., Kram, R., Taylor, C.R., 1995. Moving cheaply: energetics of walking in the African elephant. Journal of Experimental Biology 198, 629–632.*

At optimum walking speed, large animals have lower mass-specific energy requirements for locomotion than small animals. In other words, the quantity of energy required per unit mass reduces with size. Langman et al. (2012) measured the minimum energetic cost of walking in two adult female Asian elephants at Audubon Zoo, New Orleans, and combined these measurements with those obtained from smaller African elephants (Langman et al., 1995) in an attempt to answer the question 'Do we really need a bigger elephant?' in order to reduce these costs (Fig. 6.5). *Panya* was 31 years old and had a mass of 3545 kg; *Jean* was 23 years old and had a mass of 2682 kg. The study determined that the mass-specific minimum cost of transport (COT_{min}) (J/kg per m) occurred in elephants at walking speeds between 1.3 and approximately 1.5 m/s.

The authors suggested that it is unlikely that the COT_{min} in a larger elephant (with a body mass of 7500 kg, equivalent to a large bull African elephant) would be significantly lower than the values they reported. So, it seems that making elephants bigger would not reduce the mass-specific energy costs of walking.

Elephants are still used as draft animals in the forestry industry of several countries in Asia. Gajaseni (1993) calculated the energy value of elephant labour and determined that a working elephant expended approximately 13,000 kcal/h (54,392 kJ/h).

6.5 Exhibit design and enclosure use

6.5.1 Introduction

In the wild, some elephants range over large areas. In northern Kenya some elephants have been recorded migrating 140 km twice a year during the two rainy seasons, between small dry-season ranges and larger wet-season ranges (Thouless, 1995). In northern Kenya, 20 radio-collared female elephants had home ranges of between 102 and 5527 km^2 (Thouless, 1996). In private nature reserves in South Africa the home range size of female elephants ranged from 115 to 465 km^2 and in males it ranged from 157 to 342 km^2 (De Villiers and Kok, 1997). Sukumar (1989) measured home ranges to be between 105 and 320 km^2 in southern India. In southern Sri Lanka the mean home range size of 10 elephants studied by VHF radio telemetry was approximately 115 km^2 (Fernando et al., 2008).

The elephants studied by Wyatt and Eltringham (1974) in Uganda did not walk very far in 24 hours and often moved in a circle. The furthest distance covered was 38.6 km in 67 hours. The average walking speed was 0.5 km/h, but this did not take into account the amount of time when they stopped to feed at thickets (sometimes for up to 15 minutes at a time), so the actual speed was much greater. Eltringham (1982) quotes an average speed of 0.129 km/h (with a range from 0.01 to 1.5 km/h) calculated from data collected by McKay (1973).

Clearly, zoos cannot provide space on the scale of that available in the wild. The smallest wild ranges identified here are around 100 km^2, or 100,000,000 m^2. *Elephant Eden* at Noah's Ark Zoo in England is one of the largest elephant enclosures in the world and has an area of some 80,000 m^2; around 0.08% of the smallest home ranges found in the wild.

In zoos the opportunity to walk long distances does not arise and, of course, there is little biological need for them to do this because they have access to the resources they need for survival at close quarters. The provision of enclosures of appropriate size and the extent to which elephants exhibit naturalistic behaviour with respect to walking are important husbandry concerns for zoos.

6.5.2 Enclosure use

We know something about how wild elephants use their home ranges, but very little about how elephants living in zoos use their enclosures. I studied enclosure use by the Asian elephant herd at Chester Zoo, United Kingdom, and found that this varied with time of day, with a concentration of activity around the door to the elephant house at the end of the afternoon (Box 6.1).

BOX 6.1 Enclosure use by the Asian elephant herd at Chester Zoo

In a study of eight Asian elephants at Chester Zoo, I found that the pattern of enclosure use varied depending on the time of day. The location of each elephant was recorded every 15 minutes between 0930 and 1515 on each of 35 days between January and November 1999, while recording data on activity budgets (Rees, 2009b). The outdoor enclosure was divided into 16 zones that could easily be identified from landmarks, but which varied in area. Zones O and N were two halves (front and back) of a bull pen. Zone X was a fenced corral that was only accessible to the juvenile bull, *Upali*, because he could step over the lower rail and under the upper rail. The percentage of all recordings of all elephants in each zone is shown in Table 6.3.

When the elephants were released into the enclosure at 0930−1000 each day, they immediately began eating the food provided by the keepers. Most of this was in zone G, resulting in their spending over 30% of their time here between 0930 and 1015 (Table 6.3).

(Continued)

BOX 6.1 (Continued)

TABLE 6.3 Use of enclosure zones at Chester Zoo.

Time 0930–1015 (n=344)

	B	A	X	O	
B	0.29	0.00	0.00	2.33	O
E	1.16	2.33	3.20	13.37	N
H	7.85	31.69	0.87	14.83	L
K	7.85	8.43	1.45	4.36	M
	K	J	I	M	

Time 1030–1115 (n=512)

	B	A	X	O	
B	6.45	3.52	1.37	7.81	O
E	2.34	1.37	3.71	10.94	N
H	7.23	16.41	5.27	14.06	L
K	8.98	3.91	2.34	4.30	M
	K	J	I	M	

Time 1130–1215 (n=536)

	B	A	X	O	
B	11.01	7.65	1.68	10.07	O
E	4.48	2.43	6.34	9.14	N
H	4.29	5.97	3.92	10.26	L
K	8.77	4.10	5.04	4.85	M
	K	J	I	M	

Time 1230–1315 (n=488)

	B	A	X	O	
B	13.73	14.14	2.05	9.63	O
E	3.89	1.64	12.30	7.38	N
H	3.28	4.51	4.10	10.04	L
K	4.51	1.23	4.10	3.48	M
	K	J	I	M	

Time 1330–1415 (n=440)

	B	A	X	O	
B	19.09	24.32	1.82	11.14	O
E	1.82	2.50	9.32	3.18	N
H	2.73	1.14	5.00	12.27	L
K	1.59	0.00	1.36	2.73	M
	K	J	I	M	

Time 1430–1515 (n=160)

	B	A	X	O	
B	11.25	35.00	1.25	10.00	O
E	5.63	0.00	9.38	0.63	N
H	0.63	1.25	3.75	2.50	L
K	0.00	8.75	5.00	5.00	M
	K	J	I	M	

Total 0930–1515 (n=2480)

	B	A	X	O	
B	10.58	12.06	1.44	8.57	O
E	3.04	1.88	7.37	8.09	N
H	4.65	10.18	3.97	11.42	L
K	5.97	3.65	3.13	4.01	M
	K	J	I	M	

Key

%
>= 30
>= 20
>= 10
>= 5

House

B	A	X	O
E	D	C	N
H	G	F	L
K	J	I	M

Locations of all 8 elephants every 15 min from 0930 to 1515 (grouped into 45-min periods). Data is percentage of all recordings made in each 45-min period (based on 2480 recordings). The thick black line indicates the position of the door to the elephant house. Zones were not of equal area. Zones A and B were very small. Zone A was immediately in front of the door to the elephant house. Zone X was a corral that could only be accessed by *Upali*. Zones O and N were a bull pen. Source: *Author's unpublished data.*

(Continued)

BOX 6.1 (Continued)

Zone A was a small area adjacent to the door to the elephant house. This door was closed during the day. In the first 45 minutes of the day the elephants spent no time in this area. Between 1430 and 1515 they spent 35% of their time immediately outside the door. This was undoubtedly in anticipation of the door being opened so that they could gain access to their evening feed. Throughout the day the combined use of zone A and the three adjacent zones (B, E and D) increased with time until the elephants spent approximately 52% of their time in these zones at the end of the afternoon immediately before returning to the elephant house (Figs 6.6 and 6.7), at which time the frequency of stereotypic behaviour increased in the five adult females (see Fig. 2.6A).

FIGURE 6.6 Elephants at Chester Zoo congregating at the entrance to the elephant house in the late afternoon. This area was designated as zone A in the study of enclosure use (see Table 6.3).

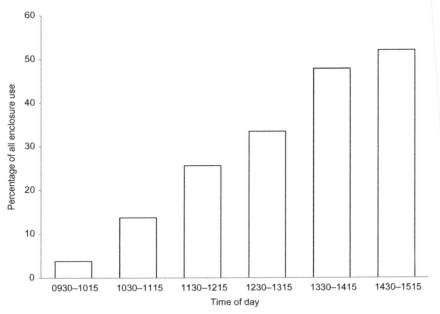

FIGURE 6.7 Percentage of recordings of all elephants made in zones A, B, D and E of the outdoor enclosure (i.e. near the entrance to the elephant house) at different times throughout the day.

6.5 Exhibit design and enclosure use

Leighty et al. (2010) studied the use of exhibit space and resources by a herd of five adult female African elephants at Disney's Animal Kingdom, Florida, United States, using collar-mounted global positioning system (GPS) recording devices. They found preliminary evidence that position in the dominance hierarchy affected the percentage of space occupied and contributed to access to resources. Dominant animals used a higher percentage of the available space than did subordinates, and the latter avoided narrow or enclosed regions of the enclosure. Dominant females occupied the watering hole more than subordinates, but the mud wallow was used equally by all animals.

To determine the distances African elephants walk in a captive environment Leighty et al. (2009) fitted seven adult females at Disney's Animal Kingdom with collar-mounted GPS systems. These recorded their movement rates while in 'outdoor guest viewing habitats'. They found that the elephants moved at an average rate of 0.409 ± 0.007 km/h during data collection periods of 9 hours, which equates to 3.68 km travelled. The authors claimed that this is comparable to that recorded in the wild. The movements of four of the elephants demonstrated a significant positive correlation with temperature and females in the largest social group showed increased movement rates when in larger enclosures. Leighty et al. found some evidence that increased social complexity may be associated with increased walking rates (including the presence of infants) and that factors such as reproductive and social status may constrain movements.

Rothwell et al. (2011) used an accelerometer, a pedometer and a GPS unit to measure walking distances in African elephants at the San Diego Wild Animal Park, California. They found that the accelerometer overestimated step count and hence walking distance because it included false counts of steps. Actual distance walked was measured by the GPS unit and did not match the distance calculated by extrapolating from stride length. However, the GPS walking distance correlated well with the activity level output measured by the accelerometer. The authors concluded that the elephants in this study exhibited a walking rate similar to that reported in other zoos. To assist zoos using accelerometers with their elephants Rothwell at al. provided a linear regression equation to convert accelerometer outputs to walking distances.

It is important to note that the studies of Leighty et al. (2009) and Rothwell et al. (2011) were conducted in institutions that kept elephants in naturalistic social groups within relatively large enclosures. It would be interesting to measure walking rates in solitary elephants and those kept in pairs and small groups, and also walking rates in animals kept in smaller enclosures than those available at Disney's Animal Kingdom and San Diego Wild Animal Park.

Holdgate et al. (2016a) used GPS data loggers attached to anklets to measure the daily distances walked by 56 adult elephants (33 African and 23 Asian) housed in 30 North American zoos and found that, on average, they walked 5.3 km/day with no significant difference between species. Elephants fed on a temporarily unpredictable feeding schedule walked 1.29 km/day more than those on a predictable schedule. Those in larger social groups walked more than those in smaller groups, and older animals tended to walk less than younger individuals. Exhibit size did not appear to affect distance walked. However, there was a small, but significant, negative correlation between distance walked and night time space experience. This was defined as 'the average weighted (by per cent time) size of all environments in which an elephant spent time', and allowed the authors to account for complex housing conditions, in which elephants may be shifted between environments of different sizes for varying amounts of time. Space experience effectively took into account the time a particular animal spent in each different available environment. The authors suggest that the fact that increased space experience at night was negatively correlated with distance walked challenges the idea that if elephants are provided with more space they will voluntarily walk greater distances, although the effect was small. Holdgate et al. found no relationship between distance walked and physical health (e.g. foot or joint health, body condition) or the performance of stereotypic behaviour and so could not determine how differences in walking may have an impact on welfare.

Occasionally, zoo elephant enclosures are extended and this provides an opportunity to study the response of elephants to being given access to a novel area. When this occurs, some individuals may exhibit neophilia (attraction to novel things) and other neophobia (fear of novel things). When a new bull pen was added to the elephant enclosure at Chester Zoo, I studied the introduction of the herd to this novel area (Rees, 2000). Of the eight animals in the group, only two — one adult cow (*Kumara*) and a juvenile bull (*Upali*) — entered the new area on the day it was first made accessible via a gate connecting it to the main enclosure. Although the herd was not observed on every day, it was 113 days after the opening of the bull pen before all eight herd members were recorded using it (Figs 6.8 and 6.9).

It is possible that personality differences between the animals may have affected the time taken for each animal to begin using the novel area. The last animals to be recorded using the area were an adult cow (*Thi*) and her calf (*Sithami*).

6.5.3 Substrate and indoor versus outdoor preferences

Floors of elephant houses have traditionally been made of concrete, while many modern facilities have sand or rubberised floors. The substrate preferences of six Asian elephants at the Oregon Zoo were studied by Meller et al. (2007) when the elephant house was being renovated from having concrete floors to poured rubber flooring. The elephants had access to seven rooms but the study found that flooring substrate did not affect room-use choice. However, there was a reduction in 'discomfort behaviours' on the rubber floor. Normal locomotion and stereotypic locomotion increased on the rubber floor and resting behaviour changed so that it more closely resembled that observed in the wild where elephants typically sleep in a standing position and spend very little time lying down. The authors concluded that rubber flooring may provide a more comfortable surface for locomotion and standing—resting behaviour.

In a study of nocturnal behaviour in Asian elephants, Williams et al. (2015) found that only two of the eleven individuals that had access to rubber flooring engaged in lying rest on it. In a second study of nocturnal behaviour, Walsh (2017) found that all eight of the Asian elephants at Dublin Zoo, Ireland, preferred to sleep on sand rather than concrete when given a choice.

Powell and Vitale (2016) examined the preferences of three female Asian elephants at the Bronx Zoo in New York for spending time indoors or outdoors at night. Two of the elephants spent most of their time outdoors: *Maxine* (26.4%—93.7% outside);

6.5 Exhibit design and enclosure use

			Name								
	Days since previous visit	Days since first opened	Sithami	Upali	Chang	Thi	Kumara	Maya	Jangoli	Sheba	
			F	M	M	F	F	F	F	F	Sex
Visit	Date		1	4	17	17	32	30	30	43	Age (years)
1	29.4.99	0	0								
2	1.5.99	2	2								L
3	7.5.99	6	8								C
4	17.5.99	10	18								
5	20.5.99	3	21								
6	25.5.99	5	26								
7	31.5.99	6	32								
8	8.6.99	8	40								
9	14.6.99	6	46								
10	24.6.99	10	56								
11	28.6.99	4	60								
12	1.7.99	3	63								
13	5.7.99	4	67								
14	8.7.99	3	70								
15	12.7.99	4	74								
16	19.7.99	5	79								
17	23.7.99	4	83								
18	26.7.99	3	86								
19	29.7.99	3	89								
20	2.8.99	4	93								C
21	11.8.99	9	102								
22	22.8.99	11	113								
23	26.8.99	4	117								C
24	1.9.99	6	123								
25	8.9.99	7	130								
26	16.9.99	8	138								
27	21.9.99	5	143								
28	1.10.99	10	153								
29	6.10.99	5	158								
30	13.10.99	7	165								L
31	22.10.99	9	174								
32	4.11.99	13	187								

FIGURE 6.8 Elephants recorded present in the new bull pen at Chester Zoo, United Kingdom, at different lengths of time after it was opened. Based on Rees (2000). Key: ▓, Observed in bull pen; C, Bull pen closed all day; L, Bull pen opened late and inaccessible during the early part of the day.

Patti (27.1%−88.9% outside). The third elephant, *Happy*, preferred to be inside (11.1%−58.3% outside). This study appeared to show that the choice to go outside was more important to these elephants than the presence of any stimuli or additional behavioural opportunities provided by the outside enclosure (see Section 2.4).

The housing environments of 255 African and Asian elephants living in 68 North American zoos have been studied by Meehan et al. (2016a). They found that elephants spent on average 55.1% of their time outdoors, 28.9% indoors and 16.0% in areas with a choice between being indoors or outdoors. The time spent on hard flooring ranged from 0% to 66.7%, with night values significantly higher than day.

FIGURE 6.9 Elephants in the bull pen at Chester Zoo, United Kingdom.

6.5.4 Multispecies exhibits

Some zoos keep Asian and African elephants and, at least in the past, it was common to see both species in the same enclosure as they both require a lot of space and the same handling facilities. Keeping related species together was a common practice in the past. Polar bears (*Ursus maritimus*) have been housed with brown bears (*U. arctos*), and lions (*Panthera leo*) were kept with tigers (*P. tigris*). The result was the production of hybrids. These animals were considered rare novelties by zoos and their visitors half a century ago. Nowadays, of course, hybrids of this kind are of no interest to zoos that breed animals for conservation purposes.

It is currently generally accepted that elephants of different species should not be housed together because of the possibility of disease transmission between them and differences in their biology (AZA, 2019a; Walter, 2010).

In 1994 Safari Beekse Bergen in the Netherlands began housing five female African elephants with a group of over 30 hamadryas baboons (*Papio hamadryas hamadryas*) in a 1.3 hectare outdoor enclosure during the daytime (Deleu et al., 2003). This provided considerable enrichment for both species. There were occasional agonistic encounters between the two species, but baboons were able to escape to rock formations within the enclosure. A single incident of cross contamination with salmonella occurred, but close monitoring of the health of both species appeared to assure that this did not become a problem. The interspecific interactions observed included baboons searching for undigested food items in elephant dung and in holes dug by elephants. Remarkably, baboons (particularly juveniles and subadults) enjoyed riding on the backs of the elephants. The elephants invited the baboons to do this by stretching their trunks out to individuals sitting on rocks.

The elephants may have gained some benefit from this association as the baboons were observed removing seeds and insects from their skin. Stereotypic behaviour was reported to be almost absent in both species; however, no comparison was made with a control situation.

6.5.5 Rotational exhibits

Some zoos are developing 'rotational exhibits' to make more — and more varied — space available to some of their larger animals. Essentially these exhibits consist of several spaces interconnected in such a manner that different animals — possibly different individuals or species — may use each space at different times. For example a rotational exhibit accommodating two species could have species A in enclosure 1 and species B in enclosure 2 this week; next week species A could be in enclosure 2 and species B in enclosure 1. In effect, both species time-share a much larger space than if each were confined within just one of the enclosures. Apart from the benefit of access to additional space, the animals also receive the additional stimulation provided by the odour of the urine and faeces of the species with which they share the space.

The *Trail of the Elephants* exhibit at Melbourne Zoo, Australia, is a rotational exhibit consisting of three interconnected enclosures so that a bull and cows with calves can use all three areas alternately (Fifield, 2014).

Rotational exhibits are not a new idea. In the 1970s Knowsley Safari Park, United Kingdom, moved its African elephants (one bull and six cows) between three enclosures through which visitors could drive in their own vehicles, sometimes creating temporary multispecies exhibits. On some days they were kept alone, but the adjacent enclosure contained lions they could see and smell. On other days they were kept with a herd of Cape buffalo (*Syncerus caffer*). Sometimes they were kept with hippopotamuses (*Hippopotamus amphibius*), giraffe (*Giraffa camelopardalis*), wildebeest (*Connochaetes taurinus*) and other African ungulates. At least one keeper in a Land Rover accompanied the elephants continuously throughout the day to prevent agonistic interactions with other species. At the end of the day the elephants were walked through the park back to the elephant house. Interspecies agonistic interactions were rare, but did occasionally occur between the elephants and a large bull buffalo.

While I was working as an elephant keeper at Knowsley Safari Park in the late 1970s, I was left in charge of the elephants in a large enclosure that also contained a pair of hippopotamuses. Prompted by an unusual incident reported in *New Scientist* magazine — a wild hippopotamus attempting to mate with a domestic cow — I felt compelled to write a letter to the editor about an equally unusual occurrence:

Bottom biting

Your article ... which describes an observation of a hippo attempting to mount a cow brought to mind an amusing incident [I] witnessed ... while engaged on a research project in an English Safari Park. I had occasion to separate (with a Land Rover) a bull African elephant and a cow elephant he was trying to mount, in order to move them. The cow elephant allowed me quietly to escort her back to the rest of the herd. When I returned for the bull I found him attempting to mount an uncooperative hippo he had found. To the best of my knowledge he had never seen a hippo before and did not quite know how to approach the problem. Having discovered that the conventional position resulted in his trunk hanging just in front of the hippo's open jaws he resorted to reversing onto her from the rear. It took two Land Rovers to separate them!

Rees (1979)

Although this attempted mating incident had no possibility of success, a hybrid elephant was produced at Chester Zoo in 1978 as a result of keeping an African elephant bull in the same enclosure as an Asian elephant cow (see Section 6.7.2).

6.5.6 Elephants as agents of landscape change in zoos

Wild elephants are keystone species and have a dramatic effect on their habitats, especially where they occur at high densities. In Africa they can cause a rapid change from woodland to grassland with a consequent effect on other animal species (Birkett and Stevens-Wood, 2005; Laws, 1970b; Valeix et al., 2011).

There appears to be no published work on the effects of elephants living in zoos on the ecology of their habitats. It is important to prevent trees from being pushed over or destruction by ringing, that is the removal of the bark that includes the vascular tissue. In Zimbabwe, Guy (1976) recorded that out of 115 trees that were pushed over by wild elephants, 92 (80%) were destroyed by bulls and 23 (20%) by cows (Fig. 6.10A and B). Where their enclosures include trees, it is common to find that browsing by elephants is prevented by electric fences and other barriers (Fig. 6.11).

Where elephants are kept in grassland enclosures they may damage vegetation and increase soil erosion by ripping up the turf and exposing the soil underneath (Fig. 6.12). African elephants kept on grassland in Knowsley Safari Park had two substantial effects

FIGURE 6.10 Damage caused by wild African elephants: (A) Elephant damage to a yellow fever tree (*Vachellia xanthophloea*) in the Lerai Forest in Ngorongoro Crater, Tanzania; (B) elephant damage to the bark of a baobab tree (*Adansonia* sp.) in Tarangire National Park, Tanzania.

FIGURE 6.11 Tree protected from damage by Asian elephants by wooden boards, Berlin Zoo, Germany.

FIGURE 6.12 African elephants grazing in their outdoor enclosure at Knowsley Safari Park in 2007. Note the exposed soil where the sward has been damaged.

on the soil and vegetation (Rees, 1977). In the 1970s a herd of African elephants was kept in a large 'reserve' (essentially a field) during the day. As the animals walked on areas of soft, waterlogged ground their feet sank into the soil. This created a landscape temporarily 'potholed' with cylindrical depressions whose sides were colonised by grasses and other plants (Fig. 6.13). Potholes present in the summer of 1976 had completely disappeared 5 months later and presumably were either trampled by elephants in the intervening period or the walls between them collapsed of their own accord. Such structures are also formed in soft ground by wild elephants and are hazardous to some other species, especially those with a running habit and long, thin legs, such as antelopes and painted wolves (*Lycaon pictus*).

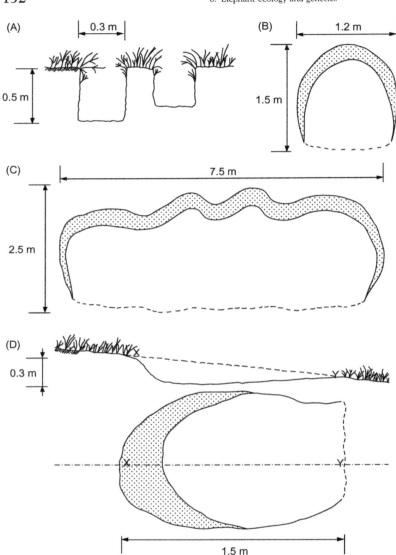

FIGURE 6.13 Damage caused to a grassland landscape by African elephants at Knowsley Safari Park, United Kingdom. (A) The vertical section of a 'pothole' created in soft ground by elephants' feet. These holes were partially filled with water during wet weather. (B) A plan view of a small dust bowl created in grassland after an elephant has pulled up the turf and used the exposed soil for dusting. The stippled arc indicates where the elephant has used its feet to dig into sloping ground. The clear area extending to the broken line is a relatively flat area of exposed soil. (C) A plan view of a larger dust bowl created by several elephants. (D) A vertical section through a small dust bowl (top) and a plan view of the same structure (bottom). The labels X and Y indicate corresponding positions on the two diagrams. Source: *Redrawn from Rees, P.A., 1977. Some Aspects of the Feeding Ecology of the African Elephant* (Loxodonta africana africana, *Blumenbach 1797) in Captivity (Unpublished BSc dissertation). University of Liverpool.*

In Knowsley Safari Park, where no bare soil was available, the elephants used their toenails to pierce the sward (often facing uphill) and then used their feet and trunks to peel the vegetation back, like a piece of carpet, thereby uncovering loose soil that they could use for dusting. This activity exposed large areas of bare soil resulting in soil erosion during dry conditions. The herd was periodically rotated between different reserves to prevent excessive damage to any particular area.

Where outdoor enclosures for zoo elephants have a sand substratum, it is possible for zoo staff to landscape this to create small-scale physiographical features. Such features may provide an interesting source of enrichment, especially for calves. However, elephant

FIGURE 6.14 *Victor*, an Asian bull, digging at Berlin Zoo, Germany.

FIGURE 6.15 Landscaping enclosures can add interest and exercise for elephants: (A) Bulldozer used to landscape the surface of the outdoor enclosure at Chester Zoo, United Kingdom; (B) Asian elephant calves playing in a sand mound at Chester Zoo.

activity may soon flatten or otherwise alter these features requiring staff to invest time in restoring them (Figs 6.14 and 6.15).

6.6 Population ecology

6.6.1 Introduction

Since the middle of the 20th century field biologists have been studying the population biology of elephants. The opportunity to study reproduction and population ecology in wild African elephants from anatomical evidence arose when the national park authorities initiated cropping operations in North Bunyoro, Uganda, to protect the habitat from the increasing destruction caused by high elephant densities in the mid-1960s. This work was undertaken by Richard Laws, Ian Parker and Ronald Johnstone when Laws was Director of the Tsavo Research Project in Kenya. Among many other measurements made, placental scars were counted (to estimate calving interval) and embryos and foetuses were measured. In addition, sperm smears, ovaries and female tracts were examined in the field to determine reproductive status (Laws et al., 1975).

Information and material were obtained from approximately 3200 elephants: 2000 cropped in the Murchison Falls National Park between 1965 and 1967; 300 shot during control operations in the vicinity of Budongo Forest; and 900 from three populations in Kenya and Tanzania. It is inconceivable that scientists would undertake a study of this nature today unless the animals were to be killed in any event. In this case, the contracts for the culling of the elephants in Uganda included a specific requirement that the game management company concerned should cooperate in the research to be carried out by the Nuffield Unit of Tropical Animal Ecology.

Although captive elephants do not have a population ecology analogous to that of wild elephants, nevertheless those kept in logging camps in Asia have been studied at a population level and elephants kept by zoos, although they are generally kept in small groups, can be treated collectively as a metapopulation.

6.6.2 Age determination from teeth

Data on the age at which molars are lost in elephants are useful in determining the age of dead elephants found in the wild. To support their study of the age structure of elephants cropped in Uganda using evidence from their teeth and a method devised by Laws (1966), Laws et al. (1975) referred to studies of two zoo elephants who died at known ages. Laws examined the lower jaw of a bull (*Peter Pan*) who died at Brookfield Zoo, Chicago, in 1971, and Short (1969) examined the dentition of a female (*Dicksie*) who died in London Zoo.

Lang (1980) studied growth and molar change in five African elephants in Basel Zoological Gardens, Switzerland, and circus elephants of the same age. Growth in height decreased at 14–15 years in bulls and levelled off between 15 and 16 years in cows. Molar I was lost between 1 and 2 years; molar II between 3 and 4 years of age; molar III between 9 and 10 years of age; and molar IV between 19 and 25 years of age. The age at which molars V and VI were lost could not be determined. Although this study is interesting, it provides very limited information compared with subsequent studies in the wild. For example Stansfield and Chavatte-Palmer (2015) devised a novel method of estimating age and sex from mandibles using material collected from 231 elephants killed as a result of professional hunting and culling in Zimbabwe and two that died of natural causes.

6.6.3 Longevity and life expectancy in zoos

Longevity, or lifespan, is the time between birth and death. Life expectancy, however, is a statistical measure of the time until death from a specified age. Thus longevity is an observed value, but life expectancy is an expected value based on statistical analysis. Dead elephants have had a lifespan; living elephants have a life expectancy.

Schmid (1998) calculated the mean life expectancy of 81 captive-born elephants in Europe with known dates of death to be just 6.1 years. This low mean resulted from a very high mortality rate in the first 5 years of life. Mean life expectancies of calves born to working elephants in Burma (Myanmar), however, reach nearly 30 years (Schmidt and Mar, 1996).

In an examination of longevity and life expectancy in captive elephants Wiese and Willis (2004) noted that one Asian elephant at Taipei Zoo lived to 86 years old, comparable to the maximum longevity estimated in wild populations. At the time of their study, longevity records had not been set for elephants in North America because the oldest individuals were still alive at 77 years (Asian) and 53 years (African). Using a life table, they estimated the median life expectancy for female Asian elephants ($l_x = 0.50$) to be 35.9 years in North America and 41.9 years in Europe. Average life expectancy estimates calculated from survival analysis for Asian elephants were 44.8 years in North America and 47.6 years in Europe. At the time of the study, the average life expectancy of African elephants in North America was estimated to be just 33.0 years, but this was based on relatively little data and was expected to increase as more data became available.

After analysing data from 4500 animals, Clubb et al. (2008) concluded that elephants in zoos had half the median lifespan of conspecifics in protected populations in range states. However, the survivorship curves produced in this analysis reflect the experiences of elephants in the past and cannot take into account the effect of the improved conditions that now exist in many modern zoos.

More recently, Tidière et al. (2016) examined data on more than 50 species of mammals to test the commonly held belief that animals live longer in zoos than in the wild. They found that this was true for 84% of the species considered, but long-lived species with low reproductive rate and low mortality in the wild benefitted less than short-lived species with high reproductive rate and high mortality, partly because there are fewer opportunities to reduce mortality in long-lived species. This study supported the conclusions of previous work that Asian and African female elephants live longer in the wild than in zoos, although these species were not included in the detailed analysis. The authors conclude that zoos need to improve conditions for 'slow-living species', such as elephants, that are particularly susceptible to extinction in the wild. They acknowledge that many zoos have already improved the living conditions of their elephants and that their analysis — like that of Clubb et al. (2008) — is historical in nature and so reflects past husbandry practices; the effects of improvements in elephant husbandry will not be known for many years.

6.6.4 Birth rates and calf survival

Historically, elephant birth rates have been low in zoos and calf mortality has been high (Kurt, 1994; Taylor and Poole, 1998). There were 141 births in European zoos and circuses between 1902 and August 1996 (Anon., 1996; Haufellner et al., 1993) and 37% of these animals died within their first year; 48% were stillborn and 27% were killed by their mothers (Schmid, 1998).

In 1998 the Asian elephant European Endangered species Programme (EEP) — now the European Association of Zoos and Aquaria Ex-situ Programme — reported just 11 births (7 males, 4 females) from a population of approximately 250 animals (Anon., 2000). Of these, only three bulls and one cow survived. This situation has improved recently. On 1 January 2018 the Asian elephant EEP contained 307 elephants (90.217) and the mean number of births per year was 15 (between 2013 and 2017) (Schmidt and Kappelhof, 2019). Fecundity in the Asian elephant population in North America has also been reported to be low and an analysis from the end of the 20th century showed that first-year mortality was over 30% (Wiese, 2000).

Mortality rates among zoo elephant calves are high, but it would be naïve to imagine that calves always survive well in the wild. It is extremely difficult to make comparisons between zoos and the situation in the wild. Zoo mortality rates can be calculated with a high degree of accuracy because all births and deaths are recorded. The situation in the wild is completely different. Deaths can only be recorded if calves are seen alive and then later found dead or disappear. Alternatively, a pregnant cow may be observed and a calf death may be inferred if no calf is subsequently recorded. Collecting such data is difficult in a savannah habitat and probably impossible in forests. Death rates in most wild populations are therefore likely to be significant underestimates or 'guesses' inferred from mathematical population models.

Of the 1551 calves born in Amboseli, Kenya, between 1972 and the end of 2003, 229 (14.8%) died before reaching 1 year and a further 64 died in their second year (i.e. 18.9% in the first 2 years). Over 50% of these calves died in the first 6 months of life (Lee et al., 2011). Those who experienced drought or a prolonged dry season in their first year were 70% more likely to die than those who did not.

Comparing calf mortality in zoos with that in the wild is problematic. Mortality rates in the wild depend upon prevailing ecological conditions. Very high calf mortality rates have been recorded in some wild elephant populations. In Uganda mortality rates of 36% have been reported in the first year of life (Spinage, 1994). In drought conditions they may be even higher. In Amboseli, 29 calves were born in the drought year of 1976. Of these, 14 (48.3%) died in the first year of life. Only six survived to 1988, three of which were born after the drought had ended. Moss (1988) concluded that adequate nutrition in the first year of life was crucial to the health and strength of an elephant for the rest of its life.

Laws et al. (1975) suggested a postweaning mortality rate of around 5% in their study of elephant populations in Uganda. Over a 12-year nondrought period, average annual calf mortality in Tarangire National Park, Tanzania, was 2%, but during the 1993 drought, 20% of 81 monitored calves died within a 9-month period (Foley et al., 2008). In one clan, 11 (41%) of 27 calves died. Mortality among twins in the first 5 years of life in Tarangire was higher (43%) than that recorded for single births (13%), possibly due to the mother being unable to satisfy their nutritional requirements and the increased chance of separation from the mother (Foley, 2002).

Sukumar (1989) made crude annual mortality rate estimates for elephants in his study area in southern India and found these to be 1.8% for females and 6.4% for males. Considering only individuals more than 5 years old, the rates were 1.7% for females and 11.8% for males. These rates were based on the number of carcasses found and estimates of population size. Sukumar acknowledges that the true rates may have been higher especially for young calves as the carcasses tend to go unnoticed. Sukumar (2003) also reported estimated mortality rates for the Biligirirangans in south-eastern Karnataka, India. In female elephants, he found annual mortality rates to be in the range of 5%—15% for calves under the age of 1 year, and approximately 3% for individuals aged 5—40 years. From birth to the age of 5 years, males died at about twice the rate of females.

Although male and female calves are born in approximately equal numbers in the wild, Saragusty et al. (2009) found that excess males are born in captivity and there is substantial perinatal mortality. They used studbook data from North American and European zoos for both genera and compared the data with that for Asian elephants in timber camps in Myanmar. A significant excess of male births was found in Asian elephants in Europe (ratio 0.61, $P = .044$) and in artificial insemination (AI) births (ratio 0.83, $P = .003$). In North American African elephants there was a nonsignificant 'numerical inclination' (ratio 0.6). In African elephants in Europe and Asian elephants in Myanmar, juvenile mortality was 21%—23%, but in the other populations it was 40%—45%. In European Asian and Myanmar populations premature death (under 5 years) was biased towards males. This skewed sex ratio at birth and high juvenile mortality in breeding programmes both hinder efforts to create self-sustaining populations and contribute to the bull accommodation challenges discussed by Schmidt and Kappelhof (2019) (see Section 11.12).

6.6.5 Sexual maturity and mean calving interval

Cow elephants in European zoos may reach sexual maturity earlier than wild elephants; for example a cow at Vincennes Zoo, Paris, gave birth at 6 years old (Schmid, 1998). However, cow elephants kept in zoos appear to reach reproductive senescence earlier than animals kept in more natural conditions, so overall their reproductive lives are shorter. The oldest reproductive female Asian elephant in North America reported by Wiese (2000) was 32 years old, and the oldest reported by Schmid (1998) was 31, but births to elephants in their 60s have been reported from Asian elephant centres (Taylor and Poole, 1998). In May 2018 *Thi* gave birth to a calf at the age of 35 years at Chester Zoo and had a mean calving interval of 2.75 years (calculated from a total of nine births) (Fig. 6.16).

In his analysis of birth statistics for captive elephants in zoos in several developed countries and two circus facilities in the United States, Dale (2010) found that elephant cows of both species gave birth to their first calves aged 14—18 years at mean ages comparable to, but slightly older than, those observed in primiparous elephant cows in the wild [*Loxodonta* 13.7 years (Moss, 2001); both species, mean of ≤12.5 years in 13/22 studies (Sukumar, 2003)].

In a captive herd kept in Jaldapara Wildlife Sanctuary, West Bengal, Baskaran et al. (2009) found that the mean age at first calving was 19.5 years. Females caught or rescued from the wild calved at a mean of 21.7 years, but captive-born females calved at the younger age of 17.2 years. The mean calving interval was 4.7 years.

Sukumar (1994) reported a mean calving interval of 4.8 years for wild elephants in southern India. However, mean calving interval in the wild is affected by environmental conditions. From 1966 to 1967, 34 calves were born to 98 cows studied in Lake Manyara National Park, Tanzania. Only eight had been born in the previous year.

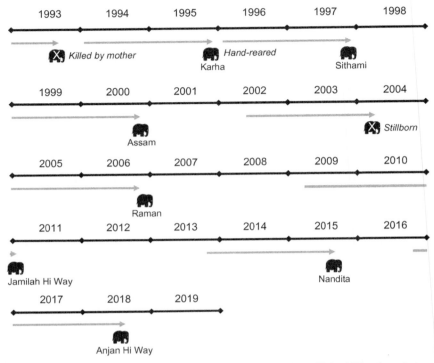

FIGURE 6.16 Calves born to the Asian elephant cow *Thi* at Chester Zoo, United Kingdom, between 1993 and 2018. Arrows indicate approximate duration of pregnancy. *Thi* killed her first calf and rejected her second (*Karha*), but went on to be a good mother. *Karha* died aged 17 months.

Douglas-Hamilton (1972) concluded (like Laws in Uganda) that 'the better the rains the more cows would conceive'. Laws (1969) has recorded a calving interval in wild African elephants of 2.75 years and has also shown from placental scars that some cows may go 13 years between calves (Laws, 1967).

Birth weights, gestation and other birth statistics are considered in Sections 4.6 and 4.7.

6.6.6 Reproductive performance of Asian camp elephants

The records of forest departments in Southeast Asia are an important source of data on the population biology of captive Asian elephants (Fig. 6.17). Studies of the demography of Asian elephants living in timber camps provide a useful comparison with that of this species when kept in zoos. Well-managed captive populations of Asian elephants, such as those found in forest logging camps, may have a breeding performance comparable to wild populations and much better than zoo animals.

Richmond (1932) listed 78 elephants of over 20 years owned by the Madras Forest Department in India. Elephants under 20 were generally not used in logging camps.

FIGURE 6.17 An elephant camp on the Cauvery River in Karnataka, India, in 2001. The camp was maintained by the Forest Department, but at this time there was no work for the 15 resident elephants.

Of these 78 animals, 33 were bulls, giving a sex ratio of 1 bull to 1.4 cows. Four of these had been born in captivity. The oldest animal was a 67-year-old cow.

From 1888 to 1992, elephant camps in Tamil Nadu, India, recorded the birth of 209 calves (i.e. approximately two births per year). Detailed records available for 19 cows recorded 101 births (including three sets of twins) between 1960 and 1986 (Krishnamurthy, 1995). The mean calving interval was approximately 5 years, and the number of calves born to individual cows ranged from three to 12. Cows reached sexual maturity at the age of 15 ± 2 years and some were reproductively active for over 40 years. Bridges (1937) reported that the Forest Department caught 400 animals in one area of Assam, India, in the 1930s. This may have been the last mass capture of its kind.

One timber organisation, the Bombay Burma Corporation, operating in what is now Myanmar, was able to balance deaths from births within its population of 2000 elephants, making wild capture a rare event (Spinage, 1994). Chadwick (1992) claimed that the company's elephant corps numbered more than 3000 at the beginning of the 20th century. When the company scaled down its logging operations some of its elephants were given to the government and kept at Mudumalai Sanctuary Elephant Camp in the Nilgiri Biosphere Reserve. In 1973 the government in Karnataka, India, assumed ownership of the maharajah's elephants. Mahouts were relocated to the national forests to work for the Forest Department (Hart and Sundar, 2000).

Sukumar et al. (1997) analysed data from elephant camps in southern India for the previous century and found the average fecundity to be 0.095/adult female per year, although it was higher during 1969–89 at 0.155, comparing favourably with wild populations (Fig. 6.18). Births in timber camps peaked in January. The sex ratio of calves born did not differ significantly from 1:1 (129 males: 109 females), although mothers 20–40 years old were more likely to have a male calf. Mortality rates were higher in females than in males up to the age of 10 years, but thereafter the situation was reversed. The population growth rate was 0.5% per year based on the average fecundity over the century, but increased to 1.8% per year based on the higher fecundity recorded during 1969—89. This study showed that the timber camps in southern India were potentially capable of maintaining a stationary or increasing elephant population without capturing animals from the wild and using a traditional management system.

A demographic analysis of timber elephants in Myanmar has been undertaken by Mar (2001) using studbook data collected from the elephant log books and the annual reports of the Extraction Department, Myanmar Timber Enterprise of the Union of Myanmar. The data available included records of approximately 9600 elephants captured or born after 1875 including 3070 calving records. Life table analysis indicated that 32.5% of calves born failed to reach the age of 5 years. Survivorship analysis of animals over 5 years of age showed that wild-caught elephants had higher survival rates than those that were captive born, and females had higher survival rates than males. When elephants under 5 years were analysed, it was found that calves born to wild-caught dams had higher survival rates than those born to mothers who had been born in captivity.

Taylor and Poole (1998) used a questionnaire to assess the effect of reproductive behaviour and husbandry factors on breeding success in 20 western zoos and three eastern

FIGURE 6.18 A calf born in an elephant camp operated by the Forest Department in Karnataka, India (see Fig. 5.16).

elephant centres They showed that breeding success in most of the zoos was notably lower and the percentages of stillbirths and infant mortality were relatively higher than recorded in the institutions in Asia. Although female elephants in zoos appeared to reach sexual maturity and reproduce earlier than those in the Asian facilities, the zoo elephants produced fewer young per female.

6.6.7 Reproductive cessation and the 'mother hypothesis'

The 'mother hypothesis' suggests that it makes more sense, from an evolutionary point of view, for older females to concentrate on rearing their existing calves and ensuring their survival than to risk age-related death during calving in an attempt to leave more offspring (Williams, 1957). If this hypothesis is true, natural selection should result in the mother living long enough after reproductive cessation to rear those offspring still dependent upon her.

The effect of mother death on calf mortality in Asian elephants has been studied using longitudinal data available for elephants in Myanmar timber camps (Lahdenperä et al., 2016). Older females showed no increased immediate mortality after calving, but calves had a 10-fold higher mortality risk in their first year if they lost their mother. By their fifth year this risk decreased to only a 1.1-fold higher risk. The study detected delayed effects of maternal death, including an increased risk of death in calves who lost their mother at an early stage at ages 3—4 years and during adolescence, although these effects were weaker in magnitude. Lahdenperä et al. concluded that the mother hypothesis could account for the first 5 years of postreproductive survival, but that there were no costs of continued reproduction on the immediate mortality risk of mothers.

In an earlier study of reproductive cessation and postreproductive lifespan in Asian elephants in timber camps in Myanmar, Lahdenperä et al. (2014) found that although some female elephants had long postreproductive lifespans (11—17 years), relatively few animals reached this phase and the decline in fertility with age (after age 50 years) paralleled the decline in survivorship. Longer-living elephants continued to reproduce until older ages, some beyond the age of 65 years.

6.6.8 Sustainability of zoo populations

Historically, elephants have not bred well in zoos. Zoo elephant populations in North America have been described as being in 'serious demographic decline' (Hutchins and Keele, 2006). A demographic analysis of Asian elephants in North America conducted in 2000 concluded that the population was not self-sustaining, with a first-year mortality of almost 30% and extremely low fecundity throughout the prime reproductive years (Wiese, 2000). He predicted that in 50 years (i.e. by around 2050) the population would drop to approximately 10 individuals and be demographically extinct unless birth rates substantially increased or additional elephants were imported. Modelling of the population suggested that fecundity would need to increase four to eight times the historical rates to prevent extinction. While Wiese recognised that AI could help the population move towards sustainability, he also acknowledged that the increased production of male calves would create a challenge for zoos because of the demands they would put on accommodation. Wiese suggested that the importation of young females from self-sustaining populations overseas would alleviate the need to create a self-sustaining Asian elephant population in North America.

Olson and Wiese (2000) noted that, historically, captive African elephant populations have received less attention than those of the Asian species. They had reproduced poorly — only 25 calves having been born in captivity in North America up to 1 January 1999 — and juvenile survival had been low. The population at the time of their analysis was relatively young compared with that of the captive Asian elephant, and Olson and Wiese concluded that the African population had the potential to produce a self-sustaining population if reproduction and juvenile survivorship could be increased over the next 10 years (i.e. by 2010), and would not require further importations.

Six years after the studies of Wiese (2000) and Olson and Wiese (2000), Faust et al. (2006) re-examined the decline in the Asian elephant population in North American zoos using an individual-based model rather than the age-based matrix model utilised by Wiese. Their simulations predicted that if the demographic trends prevailing at the time were to continue, the population would continue to decline at 2% per year. It was possible to simulate sustainable populations with a large increase in annual births. Faust et al. examined the effect of reducing infant mortality, altering birth sex ratio and recruiting new individuals from outside the population. These strategies had the greatest impact when combined with increased birth rates. Increases in the male population and decreases in the female population resulted from almost all simulations. This study confirmed Wiese's earlier conclusion that it would be difficult to either increase the size of the Asian elephant population in North America, or sustain it at its current (2006) size.

In a recent study of the demographics of the North American zoo elephant population, Prado-Oviedo et al. (2016) concluded that:

Taken together, the decrease in frequency of importation as well as low rates of captive births have led to a female population with an age structure that is heavily skewed towards older animals. Older animals are more likely to experience health issues such as foot problems and female reproductive problems, including ovarian acyclicity and hyperprolactinemia.

This last condition is associated with infertility and may contribute to the nonsustainability of the North American population (Morfeld et al., 2014a).

The future of the Asian elephant EEP is discussed in Section 11.11.

6.6.9 Importation of elephants from range states

Prior to the establishment of international trade restrictions to protect wildlife from overexploitation, large numbers of elephants were imported to Europe and North America for use in circuses and display in zoos. In the 1970s the circus owner Jimmy Chipperfield imported over 30 young elephants into England from Uganda for the newly created 'safari parks' that he began establishing in the mid-1960s. In 1995 there were 39 Asian elephants in the UK zoo population (including two in Dublin, Ireland); 28 (70%) of these were wild caught, seven (17.5%) were from Burmese logging camps, three (7.5%) were born in European zoos (including one born in Chester Zoo,

United Kingdom) and two (5%) were of unknown origin (Spooner and Whitear, 1995). Therefore at least 87.5% of the elephants were taken from wild or captive populations in Asia, some from as long ago as the 1950s.

Importations of elephants from their range states to western zoos for captive breeding programmes are now rare because the movements of elephants are strictly controlled by the Convention on International Trade in Endangered Species of Wild Fauna and Flora 1973 (CITES). The practical and legal difficulties associated with obtaining further animals from Asia have been discussed by Minion (2001).

In August 2003, under the Association of Zoos and Aquariums' (AZA's) Elephant Species Survival Plan (SSP) and a permit issued by the US Fish and Wildlife Service, three male and eight female African elephants (estimated to be 13 years old) who had been taken from the wild were imported into the United States from Swaziland by the San Diego Zoo's Wild Animal Park and the Lowry Park Zoo in Florida (now known as Zoo Tampa at Lowry Park). Unknown to the zoos, one of the cows was 10 months pregnant at the time of importation and subsequently successfully gave birth (Andrews et al., 2005). Three years later Hutchins and Keele (2006) urged AZA zoos to prepare to position themselves to justify the future importation of more elephants into North America to support breeding programmes. They emphasised the need for zoos to address the deficiencies in elephant management practices and care, and urged zoos to prepare for the ethical, legal, practical and public relations challenges that they would face in attempting to import elephants from range states in the face of increasing public and political scrutiny.

Proposals to loan elephants from Sumatran elephant camps to western zoos in exchange for financial aid to support these inadequately funded camps or for the conservation of wild elephants have been discussed by Hedges et al. (2006). In theory, such an arrangement could have benefitted the zoos and the camps, but Hedges et al. argued that there was little demand from zoos for Asian elephants (especially *Elephas maximus sumatranus*) so any financial resources that could be generated by a loan scheme would have been very limited. In addition, it would have required a permanent moratorium on captures from the wild to prevent illicit captures for sale or loan. Without this, they argued, a loan scheme would directly contribute to the reduction of wild elephant populations.

6.7 Genetics

6.7.1 Introduction

The careful genetic management of captive elephant populations is essential if they are to play a future role in the conservation of in situ populations. The long-lived nature of these animals has resulted in hybrids between subspecies entering the breeding programmes that are not useful for breeding if the established evolutionarily significant units (ESUs) are to be maintained. Developments in molecular genetics now mean that it is possible to produce genetic profiles of both wild and captive individuals, allowing us better to protect populations of these animals from genetic contamination, thereby increasing their conservation value.

6.7.2 Interspecific hybridisation

It is not uncommon for closely related, and sometimes not so closely related, species to exhibit sexual behaviour towards each other when kept together in captivity. Occasionally this may result in hybridisation between species producing, for example a liger (male lion × female tiger) or tigon (male tiger × female lion) (Rees, 2011). Such animals have no conservation value and rarely occur in captivity unless intentionally bred by private owners as curiosities.

The natural ranges of Asian and African elephants do not overlap so there is no possibility of interspecific hybridisation occurring in the wild. However, when zoos keep individuals of each species together there may be unintended consequences. On 11 July 1978 a hybrid elephant was born at Chester Zoo in the United Kingdom: an African elephant × Asian elephant cross. The calf was a male. He was 6 weeks premature and 27 kg underweight. His father was an African bull called *Nobby* and his mother was an Asian cow called *Sheba*. The calf was named *Motty* after the zoo's founder, George Mottershead. Unfortunately, *Motty* died of peritonitis after only 10 days: his colon and umbilical area were infected with the bacterium *Escherichia coli* (Rees, 2001a).

Motty exhibited an interesting combination of African and Asian features. His head, ears and trunk were African, except for the tip which was characteristic of the Asian species, having only one finger. His vertebral column was convex like the Asian species, but shaped like the African species above the shoulder. His forelegs possessed five toenails, but his hindlegs had only four. This is typical of the Asian species but considerable variation is found in African elephants (personal communication, Chester Zoo).

6.7.3 Intraspecific hybridisation

Within the species *E. maximus*, three subspecies are recognised by the International Union for the Conservation of Nature and Natural Resources (IUCN) (see Section 1.2.1). The Asian elephant EEP population consists mainly of *E. m. indicus* and *E. m. maximus*. These subspecies have undergone considerable hybridisation and, since molecular genetic studies do not support the suggestion that the Sri Lankan population of *maximus* should be treated as a distinct species, this is not a problem from the point of view of possible future reintroductions from the EEP (Fernando and Lande, 2000). However, in 2018 the EEP also contained 20 other individuals that were problematic (Schmidt and Kappelhof, 2019) (Table 6.4). Some of these animals were considered to belong to ESUs and therefore were distinct for conservation purposes.

6.7.4 Genetic diversity

Zoo populations tend to be relatively small and therefore prone to inbreeding, that is the breeding of close relatives such that the sire and dam of an individual share common ancestors. This may lead to inbreeding depression in which reduced fitness occurs in the offspring of closely related parents. Outbreeding depression may also occur in zoo populations, that is reduced fitness in the offspring of very distantly related parents. A study of mitochondrial and nuclear DNA loci in 78 Asian elephants from Thai elephant camps showed that they probably suffer from outbreeding depression due to cryptic speciation,

TABLE 6.4 'Problem' animals within the Asian elephant EEP in 2018.

Vernacular name	Subspecies of *Elephas maximus*	Sex	Reason for exclusion from breeding programme
Pure Bornean pygmy	*borneensis*	0.2.0	Taxonomy unclear. Possibly an ESU (Fernando et al., 2003) and should not breed with other subspecies
Pure Sumatran	*sumatranus*	1.4.0	Considered an ESU by the IUCN and should be genetically managed separately from other subspecies
Hybrids	*borneensis* × *indicus* or *maximus*	0.0.11	Hybridisation
Hybrids	*sumatranus* × *indicus* or *maximus*	0.0.2	Hybridisation

Source: Based on data in Schmidt, H., Kappelhof, J., 2019. Review of the management of the Asian elephant Elephas maximus EEP: current challenges and future solutions. International Zoo Yearbook 53, 1–14.

that is the creation of two or more different species that appear morphologically similar (Fickel et al., 2007).

Paternity testing is an important tool in preventing inbreeding in captive populations. Using 18 Asian elephants from three zoos in North America, Bischof and Duffield (1994) demonstrated that DNA fingerprinting with the hypervariable probe M13 can provide a reasonable first estimator of relatedness for first-degree relatives — family members that share approximately 50% of their genes (parent and offspring, siblings) — and for unrelated animals, but would be unreliable for intermediate degrees of relatedness. This methodology is only reliable when all potential sires can be tested.

Zoo populations of large animals often consist of nonreproductive individuals, related animals, unequal proportions of males and females, and individuals whose genes are overrepresented in the population. Such factors mean that the actual population is much larger than the effective population size. It is the latter which determines the future breeding potential of the population and it must be carefully managed if inbreeding is to be reduced to a minimum.

Effective population size (N_e) may be calculated using the equation:

$$N_e = \frac{4 \times N_f \times N_m}{N_f + N_m}$$

where

N_f = the number of breeding females

N_m = the number of breeding males.

It should be noted that, strictly speaking, this formula only applies to a stable population where mating is random and generations are nonoverlapping (Ballou and Foose, 1996).

A population of 50 male and 50 female Asian elephants has an effective population size of 100, assuming all individuals are capable of reproducing. If any are infertile or if the sex ratio is not 1:1, the effective population size decreases. So, a population of 100 elephants consisting of 20 fertile adult males and 80 fertile adult females has an effective population size of 64. A population of 100 elephants consisting of 20 fertile bulls but only 40 fertile cows (the remaining 40 cows being infertile) has an effective population size of 53. If only half of these bulls (10) are fertile the effective population size is just 32, even though the actual population size is still 100.

There is experimental evidence that genetic diversity is lost more quickly with time in small populations compared with larger populations (Schou et al., 2017). There is evidence that genetic diversity decreases with each generation in captive breeding programmes with small effective population sizes, and in some salmonid populations a loss of over 8% of allelic diversity per generation has been estimated (Fraser, 2008). Loss of genetic diversity has been demonstrated to reduce disease resistance in animal populations (Spielman et al., 2004).

An analysis of the Asian elephant EEP population in 2018 suggested that it was in good genetic health. The population had retained 98.44% of the gene diversity of the wild population and an effective population size of 114 (with an actual population of 307 individuals) (Schmidt and Kappelhof, 2019). The inbreeding rate per generation was 0.44% and, therefore, below the maximum of 1% advised by the Food and Agriculture Organization (FAO, 1998). This makes it theoretically possible – under prevailing conditions – for the EEP to reach its goal of preserving 90% gene diversity in the next 100 years.

Hartl et al. (1995) investigated genetic diversity in 26 Asian elephants from zoos in Germany (Berlin, Hamburg and Leipzig) and Switzerland (Zürich). They found that the proportion of polymorphic loci ($P = 4.6\%$) and the average heterozygosity ($H = 1.4\%$) were very low at 44 presumptive blood protein and enzyme coding loci in these animals. These results were in accordance with a previous study of wild populations and consequently the authors concluded that low genetic variation was not only associated with captive breeding, but reflects low effective sizes in wild populations – most of the specimens used in the study were wild born – and species-specific breeding tactics.

Elephants have a long generation time. In theory, this may have a beneficial effect on small captive populations in contributing to a reduction in the rate of genetic diversity loss (Lippé et al., 2006). However, some studies of small populations have not confirmed this. Kekkonen and Brommer (2015) reviewed 13 studies that used a small number of founders to establish isolated populations of animals in the wild. In many cases, they detected a loss of heterozygosity and a loss of allelic richness with time. They concluded that a high population growth rate could help retain genetic diversity, because it would quickly turn small populations into much larger populations. However, those species with longer generation times did not appear to retain more genetic diversity than those with shorter generation times.

Zoos have a dilemma in that they want to increase their elephant populations quickly (because this helps to secure the future of these populations in captivity), but if they do this by allowing animals to breed at a young age, they reduce the generation time and potentially increase the rate of loss of genetic diversity. In addition, early reproductive senescence may occur in females if birth spacing is not reduced in older animals. Using data collected from elephants kept by timber industry camps in Myanmar, Robinson et al. (2012) examined senescence and age-specific trade-offs between reproduction and survival in female Asian elephants. They found that associations between reproduction and survival were positive in early life, but negative in later life. Up to 71% of later-life survival declines were associated with investing in the production of offspring. The authors suggested that working females of 30 years and older should have their workloads reduced — to balance the energetic demands of daily activities with the energy resources available — and that captive breeding programmes should take into account birth spacing for females of this age to prevent early reproductive senescence.

CHAPTER 7

Elephant welfare

> ...overall zoo elephants generally experience poor welfare, stemming from stress and/or poor physical health.
>
> Live Hard, Die Young (RSPCA, 2006)

7.1 Historical perspectives

Although elephants have always been popular with zoo visitors, nevertheless Victorian attitudes to captive animals were reflected in the manner in which they were treated by some visitors and keepers in the 1800s. London Zoo allowed visitors to feed its famous African elephant, *Jumbo*, until they discovered that some people were inserting pins and other objects into the food. The diet provided by the zoo was far from ideal and included 'beer and hard liquor, oysters, cakes and candy' (Anon., 2016b). When *Jumbo* misbehaved, he was flogged by his keepers out of sight of the public (Chambers, 2007).

In 1944 H. S. Wilson, while writing about the baby elephant at New York's Bronx Zoo, claimed that visiting a baby elephant was a childhood privilege that many American children were being denied because there were too few zoos in the United States, and even fewer elephant houses (Wilson, 1944). The pleasures that American children were being deprived of — amounting to a threat to democracy according to Wilson — included feeding a baby elephant with peanuts, tickling it with a stick and playfully squirting it in the eye with a water pistol. Thankfully, attitudes to elephants have changed and, at least in the best zoos, the right of elephants not to be molested by visitors takes priority over concerns for the 'deprivation' of children unable to pet, feed or assault them.

Serious concerns about the welfare of animals began with the passing of the world's first animal cruelty legislation in the United Kingdom in 1822, Martin's Act, an *Act to Prevent the Cruel and Improper Treatment of Cattle*. This was focused on domestic species and at a time when the needs of animals were poorly understood. Everything changed in the mid-1960s. The *Report of the technical committee to enquire into the welfare of animals kept under intensive livestock husbandry systems* (the Brambell Report) was published by the Farm Animal Welfare Council in 1965 (Brambell, 1965). It established the 'five freedoms' that have become the basis of modern animal welfare legislation in many parts of the world. These are (1) freedom from hunger and thirst, (2) freedom from discomfort, (3) freedom from pain, injury and disease, (4) freedom from fear and distress, and (5) freedom to express normal behaviour.

It was another 37 years before the first detailed welfare guidelines in the world were published for elephants in zoos by the Federation of Zoos of Britain and Ireland [now the British and Irish Associations of Zoos and Aquariums (BIAZA)] (Stevenson and Walter, 2002) when zoos in the United Kingdom and Ireland were home to 46 African and 44 Asian elephants. The guidelines were subsequently adopted by the European Association of Zoos and Aquaria (EAZA) in 2003. In North America, the Association of Zoos and Aquariums (AZA) produced a document entitled *Minimum Husbandry Guidelines for Mammals: Elephants* in 1997, but it ran to just two and a half pages (Tuttle et al., 1997).

Elephant husbandry guidelines in zoos have been much improved in recent years as a result of changing attitudes towards keeping elephants in captivity and an increase in the publication of research on captive elephant welfare and behaviour. The current AZA's *Standards for Elephant Management and Care* run to 28 pages, while BIAZA's *Management Guidelines for the Welfare of Zoo Animals: Elephants* (3rd edition) covers 298 pages, although the latter contains a great deal of superfluous general information about elephants (AZA, 2019a; Walter, 2010).

Although the welfare of captive elephants has received a considerable amount of recent attention, such concern has a long history. G. H. Evans published *Elephants and Their Diseases* early in the 20th century (Evans, 1910) and A. J. Ferrier published *The Care and Management of Elephants in Burma* in 1947 (Ferrier, 1947). The ancient Sanskrit text *Hastya Ayurveda* or *Gaja Ayurveda*, written by Palakapya, dealt with elephant anatomy, surgery, diseases, diet, hygiene and training. The history of traditional veterinary medicine in India dates back to the era of Mahābhārata (5000 BCE) (Sharma, 2006).

Academic interest in elephant welfare is relatively recent. Although the major work by Fowler and Mikota (2006) entitled *Biology, Medicine, and Surgery of Elephants* discusses many aspects of the health of elephants, the terms 'welfare' and 'well-being' do not occur in either the chapter headings or the index.

A number of studies have highlighted concerns about the welfare of elephants living in zoos and circuses over the past two decades. However, they have often been conducted by scientists with little or no previous experience of working with or studying captive elephants and have relied heavily on general animal welfare principles in drawing conclusions, rather than surveying the available literature on elephants in detail or collecting a large amount of original data. Where data was collected, it was often insufficient to allow a rigorous statistical analysis.

In 2002 in the United Kingdom, the Royal Society for the Prevention of Cruelty to Animals (RSPCA) funded a *Review of the Welfare of Zoo Elephants in Europe* by scientists at Oxford University (Clubb and Mason, 2002). The RSPCA subsequently published a booklet based on this review entitled *Live Hard, Die Young – How Elephants Suffer in Zoos*, which called for the phasing out of elephants from European zoos (RSPCA, 2006). Most zoos in the United Kingdom refused to cooperate with the study and consequently the review was based largely on secondary sources, no new behavioural data were collected and the authors admitted that:

> In most cases, there were insufficient data to statistically determine what aspects of the captive environment correlate with welfare indicators, as was the remit of this report... Therefore, most data presented here are purely descriptive [sic].
>
> Clubb and Mason (2002)

This review was supported by a list of 757 references, but only about 20% of these were directly relevant to elephant welfare in zoos (Rees, 2002b). The work was widely criticised by the zoo community and the European Elephant Group expressed concern about the accuracy of the data upon which its conclusions were based (Endres et al., 2004).

In 2003 the Federation of Zoological Gardens of Great Britain and Ireland responded to the criticism of the RSPCA with *In Safe Hands – A Response to the RSPCA Report on the Welfare of Elephants in Captivity* (Zoo Federation, 2003).

Three years later, Harris et al. (2006) conducted *A review of the welfare of wild animals in circuses* commissioned by the RSPCA and, more recently, the Welsh Government commissioned a similar study: *The Welfare of Wild Animals in Travelling Circuses* (Dorning et al., 2016). Both reports included an examination of the welfare of elephants. In 2008, researchers at Bristol University reported on a project commissioned by the Department for the Environment, Food and Rural Affairs (Defra) of the UK Government entitled *The Welfare, Housing and Husbandry of Elephants in UK Zoos* (Harris et al., 2008).

Harris et al. (2008) attempted to establish a relationship between the frequency of stereotypic behaviour and a number of environmental variables such as age, sex, the amount of space available, handling period and cortisol levels. They collected data during relatively short sampling periods (30 minutes and 1 hour) spread over a small number of days (a maximum of 9) and amounting to a mean of approximately 21 hours of study per elephant. The authors admitted that:

> ...for some of the analysis, the database was not amenable to the conventional statistical approach...

[and] *that due to repeated analysis on a single dataset, there was an increased probability of a Type I error* [i.e. false positive error].

Their report contained 63 references and only 36 of these (57.1%) were concerned with elephants and only 27 (42.9%) of the references were articles in peer-reviewed journals.

The implications of the Harris et al. study were subsequently considered by the Zoos Forum [now the Zoos Expert Committee (ZEC)] which advises the UK Government on zoo matters, and it recommended the formation of an Elephant Advisory Group (Zoos Forum, 2010). The Zoos Forum's report on the Harris et al. study devoted just over 1200 words to the role of elephants in UK zoos. The account was largely anecdotal and did not quote any directly relevant academic studies, even though such studies existed at the time. Interestingly, neither the study conducted by Clubb and Mason (2002) nor that of Harris et al. (2008) appears to have resulted in any peer-reviewed publications.

In 2010 BIAZA convened an Elephant Welfare Group (EWG) in response to concerns about the welfare of captive elephants in the United Kingdom. Lord Henley, the UK Government's Animal Welfare Minister at the time, tasked the group to:

> ...drive forward a programme of improvements, encourage co-ordination, develop and share husbandry advice and good practice, and monitor progress.

He added that:

> ... [he had] not ruled out the option of looking at the scope for phasing out the keeping of elephants in the UK in the future if there is little or no evidence of improved welfare.
>
> Anon. (2016c)

The EWG published its findings in a number of academic papers (Chadwick et al., 2017; Harvey et al., 2018; Williams et al., 2015, 2018).

In England, the *Secretary of State's Standards of Modern Zoo Practice* (SSSMZP) were updated in June 2017 adding Appendix 8 — *Specialist Exhibits, Elephants* (Defra, 2017). This makes a specific reference to the existence and responsibilities of Defra-nominated elephant inspectors:

> 8.8.1 All elephant facilities will be inspected by at least one nominated Defra-appointed elephant inspector. If only one elephant inspector is present, the elephant inspectors may confer on elephant-related aspects of individual zoo inspection reports, including any recommendations or conditions, to ensure consistency of inspections. In addition to the Standards in this Section (8.8), the inspectors will refer to the current Management Guidelines for the Welfare of Zoo Animals: Elephants (BIAZA) and current recommendations of Defra/ZEC-endorsed elephant management groups such as the UK Elephant Welfare Group, and will use these to inform any recommendation or suggested conditions. Institutions will be expected to provide evidence in support of achieving these standards. Where facilities do not meet these standards, the inspectors will assess the justifications and mitigations.

> 8.8.2 All elephant-holding institutions must produce a documented Long Term Management Plan (LTMP) for the elephant collection and an Individual Welfare Plan for each elephant (IWP).

Around the same time that the EWG was working in the United Kingdom, a large-scale epidemiological study of elephant welfare was undertaken in North American zoos (*Using Science to Understand Zoo Elephant Welfare*), funded by the Institute of Museum and Library Services in the United States, with a view to improving elephant husbandry (Carlstead et al., 2013; Posta et al., 2013b). Its goals were to document the prevalence of positive and negative welfare states in 291 elephants exhibited in 72 zoos and identify the environmental, management and husbandry factors that affect their welfare. The project has resulted in papers that, although primarily concerned with welfare, have also shed light on the demography of the North American elephant population (Brown et al., 2016; Prado-Oviedo et al., 2016).

This project found that daily practices such as social management, enrichment and exercise play a critical role in improving the welfare of elephants living in zoos, and its work led to the establishment of the Elephant Welfare Initiative (EWI). This is a project supported by a community of member zoos with the goal of 'advancing evidence-based elephant-care practices that enhance welfare'. The EWI is supported by a web-based system of software tools and resources allowing real-time analysis and providing metrics for key welfare indicators for elephants (Meehan et al., 2019).

7.2 Measuring elephant welfare

7.2.1 What is welfare?

The welfare of an individual has been defined by Broom (1986) as 'its state as regards its attempts to cope with its environment'. He emphasises that, 'Welfare is a characteristic of an animal, not something given to it, and can be measured using an array of indicators' (Broom, 1991). The physical and psychological well-being of elephants must involve a consideration of the effects of all aspects of their experience, including their housing, handling, feeding, breeding and transportation. Welfare — or well-being — is about feelings such as 'suffering' or 'contentment' and cannot be measured directly, but may be inferred (Mason and Veasey, 2010a).

Most published studies of captive elephants are based on a small number of subjects; sometimes a single elephant. Many are concerned with elephants held in a single zoo, or occasionally, those held by a small number of zoos. In recent years a more ambitious approach has been taken to studying elephants in captivity using the approach taken in epidemiology. Epidemiology is the study of the incidence and causes of disease within a discrete population and is concerned with developing strategies to prevent disease and manage those who already have it. In the current context, this approach has been used to study the welfare of elephants living in zoos.

7.2.2 How can welfare be measured?

Physical welfare is relatively easy to measure. We can assess the body condition of an individual and compare it with others, and we can measure physical functions by, for example, examining an animal's gait and looking for abnormalities. Psychological welfare is more problematic. Keeping individuals in a social group may be good for some individuals, but stressful for others. Chronic stress may be experienced by animals without exhibiting indicative behaviours.

We can assess welfare indirectly by examining the conditions under which animals are kept. If they have a good (appropriate) diet, access to clean water and shade, an appropriate substrate on which to walk and lie down, and are members of a cohesive social group they are more likely to be experiencing good welfare than if they have a poor diet, inadequate water, no shade, inappropriate substrate and are kept alone or within a group where relationships are manifestly antagonistic. If elephants are managed using a system of commands and rewards for appropriate behaviours, they are less likely to experience stress than if they are dominated by aggressive keepers using ankuses. However, using indirect measures of welfare may be misleading. In theory, an enriched environment is likely to provide opportunities for better welfare than would a barren environment. But the mere presence of enrichment devices does not necessarily mean that animals use them. For example the presence of a puzzle feeder does not mean that the elephants use it, or that the keepers fill it with food.

Mason and Veasey (2010b) have reviewed the various welfare indices used to measure the well-being of animals — behavioural and cognitive responses, physiological responses and the potential negative effects of chronic stress on reproduction and health — and have discussed the extent to which they are understood for elephants. They divided these indices into three groups: (1) those that have been best validated; (2) those that are valid or partially validated, and/or validated but not (at the time of the study) used in zoos; and (3) potentially useful indices (Table 7.1).

Individuals who work with elephants possess a significant quantity of expertise relating to their husbandry and welfare, but there have only been a few attempts to capture this knowledge systematically and use it to improve elephant welfare until relatively recently.

In their report on the welfare of animals — including elephants — in travelling circuses, Dorning et al. (2016) asked experts (including animal trainers and circus owners, veterinarians, lawyers, scientists, NGOs and zookeepers) to complete a questionnaire in which they were asked to identify indicators of good and poor welfare. The most common indicators were aspects of health and behaviour. The top indicator selected was physical condition (e.g. body weight, muscle mass, fitness and the condition of coat or skin, eyes, hooves, etc.). The second-most important indicator of good welfare was the ability to perform natural behaviours. The experts considered that the scarcity of natural behaviours and the presence of stereotypies, aggression and fearfulness (indicators of stress) indicated poor welfare. Alertness, responsiveness and interest in the surroundings were considered to be good indicators of psychological health, and some aspects of physiology

TABLE 7.1 Validated, unvalidated and potentially useful indices for measuring elephant welfare.

Status	Welfare index
Best validated	Corticosteroid outputs
	Stereotypic behaviour
Valid, partially validated and/or validated, but not yet (2010) applied in zoos (unvalidated)	Measures of preference/avoidance
	Displacement movements
	Vocal/postural signals of affective (emotional) state
	Startle/vigilance
	Apathy
	Salivary and urinary adrenaline (epinephrine)
	Female acyclicity
	Infant mortality rates
	Skin/foot infections
	Cardiovascular disease
	Premature adult death
Potentially useful indices, but no validation work undertaken (at 2010)	Operant responding and place preference tests
	Intention and vacuum movements
	Fear/stress hormone release
	Cognitive biases
	Heart rate
	Pupil dilation and blood pressure
	Corticosteroid assay from hair (especially tail hairs)
	Adrenal hypertrophy
	Male infertility
	Prolactinaemia
	Immunological changes

Source: Based on Mason, G.J., Veasey, J.S., 2010. How should the psychological well-being of zoo elephants be objectively investigated? Zoo Biology 29, 237−255; Mason, G.J., Veasey, J.S., 2010. What do population-level welfare indices suggest about the well-being of zoo elephants? Zoo Biology 29, 256−273.

(digestion, reproduction and endocrinology) were also considered to be reliable indicators of welfare.

As part of a research project on elephant welfare commissioned by Defra in the United Kingdom, and undertaken by the EWG, Chadwick et al. (2017) used open response focus groups and workshop discussions to gather information about elephant welfare from representatives from 15 elephant-holding facilities in the United Kingdom and other experts in elephant welfare and behaviour. Three broad categories of welfare indicators were described by participants: behavioural, physical and physiological. The resources that the participants perceived to be important to elephants included feeding opportunities, substrate and other aspects of the physical environment, along with social factors such as group size and the relatedness of group members. The purpose of the study was to develop an elephant welfare assessment strategy informed by stakeholders with relevant knowledge and expertise.

As part of the same project, Williams et al. (2018) reviewed the indicators used to assess welfare in captive elephants. They extracted 37 unique indicators of welfare from 30 peer-reviewed papers containing potential welfare indicators for captive elephants. Behavioural measures of welfare ($n = 21$) were found to be more common than either physical ($n = 11$) or physiological ($n = 5$) measures. Unsurprisingly, the presence or absence of stereotypic behaviour was the most frequently used behavioural measure. The most frequently used physiological measure was glucocorticoid level, and body condition scores (BCS) were the most frequently used physical measure. The literature examined supported the notion that improved welfare state was best indicated by reduced stereotypic behaviour, reduced glucocorticoids and improved BCS. The authors also suggested that increased lying rest and positive social interactions were likely to be important indicators. Williams et al. asserted that the information they had gathered forms a 'crucial part of the knowledge required to efficiently monitor and improve the welfare of elephants in captivity'.

7.2.3 Population-level welfare indices

Following an examination of objective population-level welfare indices for elephants living in zoos, Mason and Veasey (2010b) concluded that their management has been 'suboptimal' and that welfare has been 'widely compromised' compared with elephants in other 'protected populations' ('benchmarks'). They examined fecundity, potential fertility, stillbirths, infant mortality, adult survivorship and stereotypic behaviour. Mason and Veasey acknowledged that most of these elements can be affected by factors unrelated to well-being and considered the potential role of these other factors. They found population-level comparisons generally indicated poor reproduction and poor survivorship — in both adults and infants — compared with benchmark populations. Approximately 60% of the zoo elephants studied exhibited stereotypic behaviour. The authors considered this measure to be the strongest evidence of compromised welfare in zoo elephants as they believed it to be the index least open to alternative interpretations. Mason and Veasey noted that there were regional and temporal variations in the data they analysed, as well as differences between species. They also acknowledged that poor well-being was a simplistic explanation for the diverse range of population-level effects that they observed and highlighted the need for more data in order to identify confounding factors.

7.2.4 Body weight and condition scoring

Accurate estimates of body weight are important in determining drug dose levels and in evaluating feeding programmes, nutritional status and general health. Hile et al. (1997) made measurements of 75 Asian elephants and established that heart girth was the best predictor of weight, and that adding body length and foot pad circumference to heart girth slightly improved predictions.

The evaluation of body condition has been used to study the health of wild elephants in Asia (Ramesh et al., 2011) and in Africa (Pinter-Wollman et al., 2009). A number of different methods of condition scoring have been developed using captive elephants.

A body condition score (BCS) is a numerical value that represents the physical condition of an animal's body. It is species-specific, independent of weight and should take into account age, time of year and individual differences (e.g. natural variation in size). Methods vary between species, but generally measure muscle and fat reserves by scoring different parts of the body (Fig. 7.1). Sudden changes in an animal's BCS may indicate a health problem.

Mikota (2006) presented a single body condition index for Asian elephants based on summing scores from 0 to 2 for the appearance of each of nine parts of the body: the head, scapulae, thorax, flanks, lumbar vertebrae, pelvis, axillary fat, brisket fat and tail. This was developed by Dr V. Krishnamurthy, Dr C. Wemmer, and J. Lehnhardt. Subsequently, a number of other indices have appeared in the literature.

Wemmer et al. (2006) reported a method of assessing body condition in Asian elephants using visual assessment to assign numerical scores to six regions of the body which were added up to produce an index ranging from 0 to 11. They examined the relationship between this index and morphometric variables for 119 juvenile and young adult elephants from southern India, Nepal and Myanmar. The mean index value was 7.3 (s.e. = ± 0.2). There was no significant correlation between the index and age in either sex. There was significant sexual dimorphism in the breadth of the zygomatic arch and three measures of subcutaneous fat: neck girth, thickness of the cervical fold and anal flap thickness.

Fernando et al. (2009) described a method for assessing (Asian) elephant body condition that is simpler than that suggested by Wemmer et al. (2006). This system requires the user to compare the subject animal with five reference photographs (Table 7.2) of free-ranging adult Asian elephants taken in Sri Lanka. Intermediate scores are assigned when the appearance of an individual falls between those represented by two photographs (one showing a better condition, the other showing a worse condition), and the range extends from 0 to 10 to allow for animals in poor and better condition than the five examples.

Evans et al. (2013) used a simple five-point scoring system to monitor body condition in a captive-raised adult cow African elephant released into the wild in Botswana, and to compare her condition with that of wild conspecifics (Table 7.3). Half scores were used where a condition was considered to fall between two of the descriptions.

Morfeld et al. (2014b) developed a visual BCS index for female African elephants — using photographs of 50 animals from AZA-accredited zoos and 57 free-ranging

FIGURE 7.1 Elephants kept by the Forestry Department in Karnataka State, India: (A) A female in good condition; (B) An elderly female (perhaps 70 years old) in poor condition; note the visible ribs and pelvis.

TABLE 7.2 The diagnostic characters pertaining to scores in the photographic scale described by Fernando et al. (2009).

Score	Characters
1	All ribs (shoulder to pelvis) visible, some ribs prominent (spaces in between sunken in)
3	Some ribs visible (spaces in between not sunken in), shoulder and pelvic girdles prominent
5	Ribs not visible, shoulder and pelvic girdles visible
7	Backbone visible as a ridge, shoulder and pelvic girdles not visible
9	Back rounded, thick rolls of fat under neck

TABLE 7.3 The body condition scoring system used by Evans et al. (2013).

Score	Description
1	Emaciated
2	Very thin – protruding shoulder blades, pelvic bones and backbone
3	Normal – shoulder blades, pelvic bones and backbone noticeable
4	Good – slight depressions in front of pelvic bones and shoulder blades; backbone protruding slightly
5	Fat – fat hanging on the body; no sign of shoulder blades, pelvic bones or backbone protruding

Half scores were used between the five conditions described.

TABLE 7.4 Simplified version of the body condition scoring system for female African elephants devised by Morfeld et al. (2014b).

Body condition scoring index	Description
1	Individual ribs visible or outline of some ribs visible
2	Pelvic bone clearly visible Gradual sunken area in front and flattened area behind pelvic bone
3	Pelvic bone visible as ridge Slight sunken or flattened area in front and/or behind pelvic bone
4	Backbone visible as a ridge from tail head to mid-back No obvious depression alongside the backbone in lumbar region; fat beginning to accumulate in this region
5	Backbone difficult to differentiate, may be visible from tail head to pelvic bone region and appears to be covered with thin layer of fat In the lumbar region, area alongside backbone is filled in

individuals from Kruger National Park, South Africa – and validated this using ultrasound measurements of subcutaneous fat (Table 7.4). They showed that as the BCS increased, so too did the ultrasound measures of body fat thickness, thereby validating a BCS index for African elephants for the first time. When the two elephant populations were compared with this system the median BCS in the free-ranging elephants (BCS = 3, range 1–5) was found to be lower than that in the zoo elephants (BCS = 4, range 2–5).

More recently, Wijeyamohan et al. (2015) described another visually based BCS system for Asian elephants. Their scale correlated with morphometric indices such as weight, girth and skin fold measures and they claimed that it was quick, inexpensive, noninvasive and user-friendly in the field. The authors emphasised that, apart from information about their health, body condition scores for elephants may provide information on their production potential and the suitability of their environment.

Clearly, great care must be taken when comparing the results of studies that use different condition scoring scales as the values obtained with different scoring systems are not directly comparable.

It is possible to infer the welfare status of elephants indirectly from the conditions in which they are kept. Chatkupt et al. (1999) developed a survey method and a visual assessment to describe and evaluate the health of working elephants and their living conditions in Thailand. They considered a number of different work and living situations, and found these situations to be significantly associated with whether or not an elephant was in good body condition or received appropriate husbandry. They suggested that the working and living conditions of elephants could be used to predict their welfare.

7.2.5 The welfare of elephants working in tourism

Elephants are widely used in tourism in Southeast Asia and, to a lesser extent, in Africa. In recent years attention has been turned to the role of these elephants and their welfare. The reduction in the use of elephants in the logging industry — particularly in Thailand — has meant that many animals are surplus to requirements. Tourism offers alternative work for these elephants and it is generally considered that their welfare depends upon their economic value, utility and the financial position of their owners (Magda et al., 2015).

Duffy and Moore (2010) have argued that elephant-back safaris in Thailand and Botswana have extended neoliberalism — a model of social studies and economics in which the control of economic factors is transferred from the public sector to the private sector — by opening up new frontiers in nature. They argue that the effects are not completely negative, and neoliberalism creates new ways of valuing and conserving elephants.

The global regulation of the use of elephants in tourism has been examined by Duffy and Moore (2011). They considered the use of elephants for trekking and safaris in Thailand and Botswana and concluded that applying universal principles in diverse locations is inherently problematic in relation to working animals. They noted the highly variable working practices between countries and determined that any attempt at developing global standards needs to engage with local practices if they are to be workable and acceptable to animal welfare organisations, tour operators, elephant camp owners and tourists.

Elephants working in the tourism industry frequently have cutaneous skin lesions due to friction from saddles (Fig. 7.2) The prevalence of such lesions in parts of the body in contact with saddles and related equipment was studied by Magda et al. (2015) in 194 elephants living in 18 tourism camps in Thailand. Working hours at these camps ranged from 2 hours in some camps to 10 hours in others; some elephants were given breaks while others were not. Active lesions were found on 64.4% of the elephants examined, most often on the back, but also on the neck, tail and around the girth (Table 7.5).

The following risk factors were identified for lesions: increasing age, use of rice sacks as padding in contact with the skin and the provision of a break for the elephants. The probability of an elephant possessing an active lesion initially increased with the number of days an elephant worked, but then declined, possibly because only elephants without lesions could work longer hours. Magda et al. recommended increased monitoring of older elephants and the back region of all elephants, restricting the working day to less than 6 hours and avoiding the use of rice sacks as padding against the skin.

FIGURE 7.2 An elephant working in the tourism industry in Sri Lanka.

TABLE 7.5 Anatomical distribution of cutaneous lesions in 194 tourist camp elephants in Thailand.

Anatomical region	Percentage of all active lesions
Neck	1.68
Girth	30.73
Back	50.28
Tail	17.32

Source: Based on Magda, S., Spohn, O., Angkawanish, T., Smith, D.A., Pearl, D.L., 2015. Risk factors for saddle-related skin lesions on elephants used in the tourism industry in Thailand. BMC Veterinary Research 11, 117. https://doi.org/10.1186/s12917-015-0438-1.

7.2.6 Stress and distress

An animal may experience both stress and distress. In the United States, the National Research Council (2008) makes this distinction between these two concepts:

> Stress responses are normal reactions to environmental or internal perturbations and can be considered adaptive in nature. Distress occurs when stress is severe, prolonged, or both.

> ...stress denotes a real or perceived perturbation to an organism's physiological homeostasis or psychological well-being. In its stress response the body uses a constellation of behavioral or physiological mechanisms to counter the perturbation and return to normalcy.

> ...Distress as an aversive, negative state in which coping and adaptation processes fail to return an organism to physiological and/or psychological homeostasis.

...Stress and distress are dissociable concepts, distinguished by an animal's ability or inability to cope or adapt to changes in its immediate environment and experience.

In essence, experiencing stress of various types is part of normal life and the body usually copes with this by adapting and returning to its normal state. When it is unable to do this, the animal becomes distressed. Stressors include infection, the threat of physical harm, restraint, hunger and thirst, but they may not be unpleasant and include exercise and sexual activity. The coping mechanisms elicited by stressors include behavioural reactions, the activation of the sympathetic nervous system and adrenal medulla, the secretion of hormones (such as glucocorticoids and prolactin) and the mobilisation of the immune system. However, while a stress response may involve one or several of these responses, none by itself is sufficient to denote stress, and their absence does not necessarily indicate the absence of stress (Moberg, 2000). Knowles et al. (2014) have noted that many of the biochemical variables commonly used as measures of animal welfare reach extreme levels during mating and play, but most people would not consider animals to be experiencing impaired welfare during these activities.

The physiological and behavioural responses to stressors are stressor-specific so the processes used by the body to restore well-being also differ. Furthermore, responses to stressors are species-specific and may vary between individuals of the same species, so an event or situation that causes stress in one individual may not be stressful — or as stressful — for another.

In African elephants kept in North American zoos, low serum cortisol (i.e. reduced stress) was associated with positive elephant behaviours (friendly, affiliative) and interactions with the public (Carlstead et al., 2018). Older elephants and those living in zoos at higher latitudes tended to have higher cortisol levels. This may have been related to the colder climate as it has been shown that elephants exhibit more stereotypic behaviour in lower temperatures (Rees, 2004a; see Section 7.2.8). The risk factors for low cortisol in Asian elephants were attitudes indicating social inclusion of the keepers in elephant groups and elephant interaction with the public. In this species the latitude of the zoo was also a predictor of higher cortisol.

7.2.7 Behaviour as a welfare indicator

Since we cannot ask an elephant how it feels, once we have examined its body condition to assess its physical state, we may turn to an assessment of its behaviour to assess its psychological state.

Vicino and Marcacci (2015) developed an integrated play index for measuring elephant welfare that quantifies the integrated intensity of play bouts after studying African elephants kept at San Diego Zoo Safari Park, California, United States. They assumed that play behaviour is an indicator of positive welfare and that comparing values of the index at different times could detect changes in welfare (Fig. 7.3). However, higher rates of play behaviour may be observed in animals in a zoo environment compared with those recorded in the wild, because play is what Hill and Broom (2009) have called a 'luxurious' behaviour. Such behaviour is made possible in zoos because other behaviours linked to survival — such as antipredator behaviour — are reduced.

FIGURE 7.3 Asian elephant calves playing at Chester Zoo, United Kingdom.

A study of the possible effect of construction noise on several species — including two Asian elephants — at Auckland Zoo, New Zealand, was conducted by Jakob-Hoff et al. (2019) by analysing their behaviour from video recordings made while they were exposed to ambient noise (the control) and an amplified broadcast of prerecorded continuous and intermittent construction noise. In addition, an acoustic map was made of the enclosure. The study found that when the elephants were exposed to the construction noise they appeared to show an increase in behaviours that could indicate stress or agitation. These included increased vigilance and locomotion, indicating that elephants may prefer quieter regions of their enclosure when exposed to noise. The purpose of the study was to inform the zoo management of possible animal responses to noise in anticipation of a major construction project at the zoo.

Captive conditions often constrain the ability of animals to perform natural behaviours, thereby affecting their psychological welfare by preventing motivations from being satisfied. Specific motivations may be frustrated relating to behavioural needs — activities that animals perform instinctively even when they are not necessary for fitness — and also when environmental deficits or external environmental cues elicit strong motivations to perform certain behaviours (such as digging in concrete substratum). As many animals actively seek stimulation, behavioural restriction may harm welfare by causing boredom if they are kept in homogeneous and monotonous conditions (Mason and Burn, 2011). However, Hutchins (2006) has cautioned against assuming that captive environments must necessarily provide conditions similar to those found in nature.

Kiley-Worthington (1990a) assessed the extent to which elephants living in zoos and circuses are 'distressed' and presented her findings to the Conference of the Society of Veterinary Ethology in Bristol, United Kingdom, in 1989. She analysed 59 behaviours during 226 hours of continuous observation on 35 Indian (Asian) elephants and 36 hours on six African elephants and concluded that there was some evidence of distress in these animals under the husbandry conditions at the time.

Kiley-Worthington and Randle (2005) subsequently studied the effect of the behavioural restrictions imposed on elephants by different types of environments on their welfare. They compared the experiences of animals living in five different environments, namely wild/feral, safari park/extensive pasture, a group in a small enclosure, elephants shackled for 12 hours and elephants shacked for 24 hours per day. The study considered the amount of restriction imposed in terms of factors such as movement, gaits possible, self-grooming, social contacts, sexual behaviour, maternal behaviour, feeding, food choice, stimulation from the environment and the acquisition of ecological knowledge. Scores were assigned to each category for the five environments from 0 (unrestricted) to 5 (extremely restricted) and then added together. Low-overall scores indicated better quality of life. Wild elephants achieved the lowest score (0), followed by those in safari parks (3–17), small enclosures (11–25) and those shackled for 12 hours (25–38). Elephants shackled for 24 hours per day, not surprisingly, achieved the highest score (38–47), indicating that these animals had the lowest quality of life as the maximum possible score is 50 (worst welfare). When other factors were added the rankings of the five environments remained the same (18–22, 4–33, 13–40, 28–51 and 43–60, respectively) and the maximum possible score is 75 (worst welfare). These factors were the availability of adequate food, water, shelter, veterinary treatment, emotional stimulation from humans and stimulation from learning new things. Importantly, the addition of these factors recognises that wild elephants may have a lower quality of life than some in captivity because the availability of food, water and shelter may be much lower in the wild. Furthermore, wild elephants do not generally have access to veterinary care.

This approach is misleading because a simple scoring system cannot adequately weigh the benefits of the availability of veterinary care to elephants shackled for 24 hours per day against the welfare losses they incur by virtue of their extreme confinement. Nevertheless, it is important to recognise that living wild is not stress-free for elephants and the price they pay for their freedom may be to die prematurely from a disease that could be treated in captivity, or die of starvation and dehydration during a drought.

7.2.8 Stereotypic behaviours

Captive animals often develop stereotypic behaviours (stereotypies) that are rarely observed in wild or free-ranging animals (Boorer, 1972). Stereotypic behaviour may be defined as any movement pattern that is performed repeatedly, is relatively invariant in form and has no apparent function or goal (Odberg, 1978). In elephants this typically consists of rhythmic head movements (rocking, head-bobbing), swaying from side to side (weaving) (Fig. 7.4), repetitive pacing or route-tracing. Other abnormal behaviours have been observed including trunk-sucking, object-biting, nipple-pulling, trunk-tossing, manipulating and throwing faeces, and – in a former circus elephant – pirouetting.

When Clubb and Mason (2002) reviewed studies of the behaviour of captive elephants they found just 10 that quantified behaviour. In these studies, only 9 (43%) of the 21 zoo

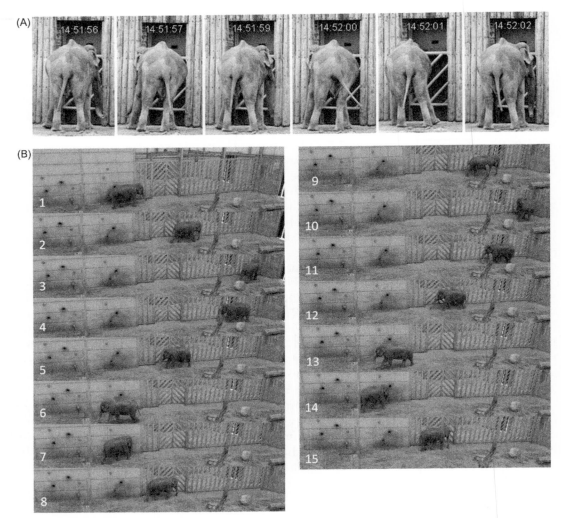

FIGURE 7.4 Stereotypic behaviour in Asian elephants. (A) A female Asian elephant exhibiting stereotypic behaviour (weaving) at the entrance to the elephant house in a zoo. The sequence shows the positions of the body over a period of 6 seconds. Note that the head and trunk are swaying from side, the weight of the body is being transferred from the left legs to the right legs, and the spine is twisting. (B) A female Asian elephant repeatedly pacing (route-tracing) along a feeder wall and fence inside an elephant house. The sequence of images 1–15 represents approximately 1.5 minutes. Source: *(A) Reproduced with permission from Wiley-Blackwell. First published in Rees, P.A. 2015. Studying Captive Animals: A Workbook of Methods in Behaviour, Welfare and Ecology. Wiley-Blackwell, Chichester; (B) Author's original photographs.*

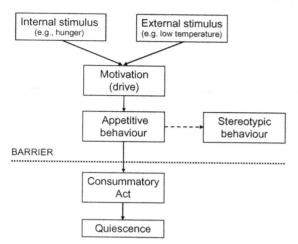

FIGURE 7.5 A theoretical model of one possible cause of stereotypic behaviour. Source: *Modified from Rees, P.A., 2004. Low environmental temperature causes an increase in stereotypic behaviour in captive Asian elephants* (Elephas maximus). *Journal of Thermal Biology* 29, 37–43.

elephants studied were observed stereotyping, but all of the 29 circus elephants studied by Schmid (1995) exhibited stereotypies. A questionnaire survey of British zoos found that stereotypic behaviour occurred in 20% of bulls, 47% of Asian females and 23% of African females (Stevenson and Walter, 2006). Overall, this represented a third of the 85 animals in the collections surveyed. Such data are almost meaningless because they do not distinguish between individuals who spend 1 minute in a 24-hour period stereotyping and those who perform such behaviour for most of each day.

Studies of stereotypic behaviours in elephants have frequently examined short-term phenomena, for example the effects of changes to a social group (Schmid et al., 2001), the response to a particular environmental enrichment technique (Stoinski et al., 2000; Wiedenmayer, 1998) or the effects of pregnancy (Szdzuy et al., 2006). Other studies have examined the effects of different husbandry methods (Gruber et al., 2000; Schmid, 1995; Schmid and Zeeb, 1994), behavioural sleep (Tobler, 1992), nocturnal behaviour (Brockett et al., 1999; Weisz et al., 2000), seasonal variation (Elzanowski and Sergiel, 2006) and the relationship between stereotypic behaviour and serum cortisol levels (Wilson et al., 2004).

Stereotypies have a wide range of origins and proximate causes (Mason, 1991a). Some occur when an animal is consistently unable to reach a particular goal by performing appetitive behaviour, unable to reach a desired place or unable to escape from disturbance (Carlstead, 1996) (Fig. 7.5). Although stereotypic behaviour is often considered to be an indicator of poor welfare, there is evidence that some stereotypic behaviours may function to regulate arousal and possibly reduce distress, thereby alleviating the effect of a suboptimal environment (Mason and Latham, 2004; Rushen, 1993). Nevertheless, the public perception of this behaviour is that it is unnatural and anything that reduces its frequency may benefit the elephants themselves and improve the public perception of elephant exhibits.

There have been relatively few extensive studies of stereotypic behaviours in elephants. Attempts to determine the factors affecting the occurrence and frequency of stereotypic behaviours have suffered from small sample sizes (in terms of the number of animals studied, number of samples collected and the duration of the study) and the compounding factors associated with differences in social group sizes, the individual histories of the animals (captive born, wild born, former circus animals, etc.), variation in husbandry techniques between zoos and a host of other possible influences.

Recently an epidemiological approach has been taken to the study of stereotypic behaviour. Greco et al. (2016a) studied stereotypic behaviour in 89 elephants in 39 North American zoos and found that it was the second-most commonly performed behaviour after feeding: 15.5% of all observations made during the day and 24.8% at night. Asian elephants were at greater risk of exhibiting stereotypic behaviour than African elephants. For both species, the risk of performing higher rates of stereotypic behaviour during the day increased for individuals who spent time housed separately and those who experienced interzoo transfers. This risk was reduced in individuals who spent more time with juveniles and engaging with zoo staff. At night, reduced risk corresponded with spending more time in environments with both indoor and outdoor areas, and in large social groups. The risk at night was increased for those elephants who had experienced interzoo transfers. Greco et al. concluded that social factors were the most influential in predicting rates of stereotypic behaviour.

As part of the same research project, Greco et al. (2017) made video recordings of 77 elephants that exhibited stereotypic behaviour at 39 North American zoos to determine the factors that affected the extent to which locomotion was incorporated into this behaviour. They characterised each individual according to whether it incorporated locomotion into its stereotypic behaviour or only exhibited whole-body movements. The study found that the odds of having locomotion present in stereotypic behaviour increased when elephants spent time housed separately, when they spent time indoors and when they were members of several social groups.

During my study of eight (2.6) Asian elephants at Chester Zoo, United Kingdom, only the five adult cows exhibited stereotypic behaviour during the day, with frequencies ranging from 3.9% to 29.4% of all observations. These elephants exhibited individual, diurnal and seasonal variation in stereotypic behaviour (Rees, 2009b). This variation has clear implications for studies that use short sampling periods and may make comparisons of data collected at different times of the day and different times of the year invalid. In adult cow elephants, stereotypic behaviour increased in frequency towards the end of the day. The elephants spent more time stereotyping during the winter months than at other times. The frequency of stereotypic behaviour was negatively correlated with the frequency of feeding behaviour in adult cows, suggesting that the provision of additional food throughout the day might reduce stereotyping. Creating more opportunities for elephants to exhibit foraging behaviour and introducing greater unpredictability into management regimes, especially feeding times, may reduce the frequency of stereotypic behaviour and increase general activity levels.

Wilson et al. (2006) recorded that three unchained female African elephants at Zoo Atlanta, Georgia, United States, exhibited stereotypic behaviour during 3.5% of scans made at night, but a similar study of the same elephants made 8 years earlier detected no stereotypic behaviour at night (Brockett et al., 1999). During a study of five Asian elephants at Busch Gardens Tampa Bay, Florida, United States, Lukacs at al. (2016) recorded no incidences of stereotypic behaviour during 270 elephant hours of observations and did not include stereotypic behaviour in their ethogram.

Caution is required when attempting to interpret data collected on stereotypic behaviour in elephants over a short period of time. Harris et al. (2008) examined the relationship between stereotypic behaviour and space (enclosure size) — and other variables — but the validity of this work depends heavily on how the frequencies of stereotypic behaviours were measured. In common with many other studies of the behaviour of elephants living in zoos, they collected data during relatively short sampling periods (30 minutes and 1 hour) spread over a small number of days (a maximum of nine) and amounting to a mean of approximately 21 hours of study per elephant.

Harris et al. admitted that their database was not amenable to the conventional statistical approach for some of their analysis and that there was an increased probability of a Type I error (i.e. false-positive error) due to repeated analysis on a single dataset. The authors noted that the two elephant species 'can differ markedly in their physiology and behaviour', and yet they pooled the data for both species. The study only examined 40 Asian elephants and 36 African elephants, their ages ranged from 0.5 to 50 years (but in some cases ages were not known), some were captive born and some were wild-caught, and many had lived in several captive environments including zoos, circuses and logging camps. During the study, some cows were pregnant and changes occurred to the size and composition of six of the herds. With so many variables and so few elephants it is doubtful whether any of the conclusions drawn concerning behaviour have any validity.

The total number of hours of observation in the study conducted by Harris et al. was 1634.5 on 77 elephants spread over three visits of 1–3 days each. This amounted to a mean of about 21.2 hours for each elephant. Some elephants may exhibit stereotypic behaviour for many hours continuously on 1 day and not at all the next day, for no obvious reason (see Box 7.1). This being the case, it is inconceivable that studying an elephant

---

BOX 7.1 *Kumara*: a case study in stereotypic behaviour

In some cases, the design of elephant accommodation can induce stereotypic behaviours. *Kumara* was a female Asian elephant born in the wild in 1966. She arrived in ZSL Whipsnade Zoo in 1967 and was moved to Chester Zoo in 1989.

At Chester, *Kumara*'s stereotypic behaviour consisted of route-tracing which involved walking repetitively in a circle in an anticlockwise direction, often while squinting (Fig. 6.8). If another elephant blocked her path she would stop until it moved and then carry on. This behaviour was performed in one corner of the outdoor enclosure which formed a quadrant. She followed the edge of this quadrant until the edge of the enclosure became straight and then completed the circle, walking back to the beginning of the quadrant. On one occasion I recorded *Kumara* walking in circles for 63% of the time between 1000 and 1400 (Fig. 7.6). On other days she did not perform the behaviour at all between these hours (Fig. 7.7). There was no obvious reason for the variability in the time she spent stereotyping.

(Continued)

BOX 7.1 (Continued)

FIGURE 7.6 A female Asian elephant (*Kumara*) at Chester Zoo, United Kingdom, route-tracing by walking in an anticlockwise circle. She had previously been kept at ZSL Whispnade Zoo where the elephant house consisted of individual circular 'silos'. This behaviour appears to have developed after walking continuously around her circular space.

The aetiology of this behaviour appears to have had nothing to do with the accommodation and husbandry practices that prevailed at Chester at that time. *Kumara* had previously lived at ZSL Whipsnade Zoo. When she lived there she was housed in an elephant house made of reinforced concrete and designed by Lubetkin and the Tecton Group in 1935; it is of some architectural importance and it is a listed building, but was entirely unsuitable for elephants. They were housed in four separate 'cylindrical cubicles', so the floor space for each animal was circular. While she lived at Whipsnade, *Kumara* apparently walked repeatedly around the edge of her circular pen. When she was moved to Chester, she brought this stereotypic behaviour with her and continued to perform it for at least 10 years.

When the enclosure at Chester was redeveloped, the shape of the corner of the enclosure where *Kumara* had performed her stereotypic behaviour was changed so that the curved section no longer existed. This prevented her from walking her normal circuit of the area, possibly because the stimulus of the curved edge had been removed. However, on one occasion I observed *Kumara* within a group of other elephants turning her body — more or less on-the-spot — in an anticlockwise direction near the area where she had formerly walked a circular anticlockwise path (Fig. 7.8).

(Continued)

BOX 7.1 (Continued)

TIME	1	2	3	4	5	6	7	8	9	10	11	12	13	14	15	16	17	18	19	20	21	22	23	24	25	26	27	28	29	30	31	32	33	34	35
10:00	L	F	L	F	F	F		L	F	F	F	F	F	F	F	F	F		F	F	DT	F	F	F	L	F	F	F	F	F	F	F	F	F	F
10:05	F	F	DT	F	F	F		S	F	F	F	F	F	F	F	F			F	F	F	F	F	F	F	F	F	F	F	S	F	F	F	F	F
10:10	F	F	S	F	F	F		F	F	F	F	F	F	F	F	F	L	S	F	S	F	F	F	F	F	F	F	F	F	F	F	F	F	F	F
10:15	F	F	S	F	F	F		DK	F	F	F	F	DT	S	F	F	F		F	S	F	F	F	F	F	F	F	F	F	S	F	F	F	F	F
10:20	F	F	S	F	F	F		L	F	F	F	L	S	S	F	F	DK	S	F	S	L	F	F	F	F	F	F	F	F	L	DK	F	F		
10:25	F	F	S	F	F	F		S	F	F	F	DT	F	F	S	F	F	S	S	F	F	F	F	DK	F	F	F	S	F	S	S	L	F	F	
10:30	F	F	L	DG	F	F		L	F	F	F	F	F	S	DT	F	F	F	S	F	L	F	F	F	F	F	F	F	F	S	F	F	F		
10:35	DT	ST	S	DT	F	F	F	S	F	F	F	F	L	F	L	F	F	F	DK	F	F	F	DT	F	F	F			F	F	S	F		F	
10:40	F	S	L	DT	L	F	F	S	L	F	F	F	S	S	F	S	F	L	F	L	F	F	F	F	F				L	F	S	F			F
10:45	F	S	S	S	F	F	F	S	L	F	S	F	DT	DT	F	F	F	F	F	F	DT	L	F	F	F	F	L	DT	L	S			F	F	
10:50	F	S	ST	S	F	F	F	S	F	F	F	F		S	F	PY	S	F	F	L	F	F	F	F	F	F	F	S	F	F	S	S	F	S	
10:55	L	L	S	F	F	F	F	F	F	F	F	DT	F	L	S	L	F	F	F	S	F	L	S	S	F	S	F	F	S	L	DT	F	S	S	S
11:00	S	S	ST	S	F	F	F	S	F	F	S	F	ST	S	S	S	F	L	S	F	L	S	L	F	S	DT	F	L	S	S	F	S	S	S	S
11:05	F	ST	S	S	F	S	F	S	L	F	F	F	DT	S	S	F	S	F	DT	L	S	F	DT	F	S	F	L		L	S	L	S	S	F	F
11:10	F	S	S	S	F	S	F	S	L	F	L	F	F	S	F	F	F	DT	S	S	S	S	F	F	F		S	S	L	F	S	F	S	F	S
11:15	F	S	L	ST	F	L	F	S	S	F	S	F	L	S	PY	S	F	F	L	ST	L	S	S	S	DK	DT	S	S	F	S	F	S	F	S	
11:20	S	L	ST	S	L	DT	L	S	L	F	S	F	S	S	S	S	F	S	ST	F	L	DT	S	S	S	S	F	S	F	S	S	F	S	S	
11:25	F	S	ST	L	S	S	F	S	S	S	DT	F	F	S	S		L	L	F	ST	S	S	S	F	S		S	F	S	S	S	F	S	F	F
11:30	DT	F	ST	S	L	S	F	S	S	S	S	F	DK	S	DT	S	F	S	F	L	F	DT	S	S	S	F	S	S	DT	F			S	S	S
11:35	L	F	S	S	L	L	F	S	L	S	S	F	S	ST	S	F	F	F	L	ST	L	S	F	S	F	S	F	S	S	S	F		S	S	S
11:40	F	F	ST	AG	S	SEX	F	S	S	S	ST	DT	ST	S	S	F	F	S	ST	S	S	S	S	S	DT	S	S	S		F	F	S	S	S	F
11:45	F	S	ST	S	L	F	S	S	S	F	DT	S	ST	S	S	S	F	L	DT	ST	S	DT	S	S	S	F	S	S	F		F	F	F	S	F
11:50	F	L	ST	L	S	F	F	S	L	S	ST	S	ST	S	S	S	F	DK	S	S	S	S	DT	S	S	DT	S	S	L	F	DT	S	F	S	
11:55	ST	L	ST	S	S	F	F	F	L	S		S	ST	S	S	F	F	ST	L	L	S	S	DT	S	S		S	S	L	F	F	S	F	S	S
12:00	ST	L	ST	S	S	S	F	F	L	S		S	ST	S	S	F	F	DT	L	S	S	S	L	F	F	S	DT	S	S	DT	F	S			S
12:05	ST	L	ST	S	L	L	F	L	S	L	ST	S	ST	S	S	F	F		ST	ST	S	S	S	F	S	S	S	F	DT	F	S	L	S	F	
12:10	L	F	ST	L	L	S	F	F	S	S	L	S	S	S	F	S	L	ST	ST	S	S	DT	S	F		S	S	DT		S	S	S	S	S	
12:15	ST	F	ST	S	S	S	S	S	S	F	ST	S	ST	F	S	F	S	S	ST	S	S	DT	S	S	DT	L	S	S	S	F	DK	S	S	L	F
12:20	ST	F	ST	L	S	L	F	F	S	F	S	S	ST	S	F	F	S	ST	ST	S	S	S	S	F			S	F	S	S	F	S	S	F	
12:25	ST	L	F	S	S	F	F	F	S	S	ST	S	S	F	S	F	S	F	L	S	ST	S	F	L	F		S	S	DT	DK	S	S	F	F	
12:30		L	F	S	S	S	F	S	L	S	S	S	F	F	S	S	S	S	ST	ST	S	L	S	S		S	S		F	F	S	S	F	F	
12:35	F	S	ST	S	L	F	F	S	L	S	S	S	DT	F	F	S	S	S	ST	ST	S	F	S	S	S		S	S	F	F	F	S	S	S	
12:40	ST	F	ST	L	F	F	F	S	L	F	ST	S		F	F	S	F	S	F	ST	S	S	F	S		S	S		F	F	S	S	S	S	
12:45	S	ST	ST	L	F	F	F	L	L	S	S	S		F	F	S	F	S	F	S	F	S	F	L	L	S	S		F	F	S	S	S	S	
12:50	ST	L	ST	S	S	F	F	S	ST	S		S	F	ST	L	F	F	S	S	S	S	DT	S	S	DT		F	F	S	S	S	S			
12:55	S	ST	ST	L	L	F	F	S	F	S	ST	S	ST	S	F	S	F	F	S	S	S	S	S		DT	S	F	S	F	S	S	S	S	L	
13:00	ST	S	ST	S	S	S	F	S	L	S	ST	S	ST	S	S	S	F	F	S	S	S	DT	S	S	ST	S	F	S	F	S	S	S	L	F	
13:05	ST	S	ST	S	L	S	ST	ST	S	S	ST	S	S	S	S	F	L	F	S	S	S	S	S	F		S	F	F	S	S			L		
13:10	ST	S	ST	S	L	S	L	ST	S	S	ST	F	ST	S	DT	S	S	L	ST	S	S	DT	S	S	DT	L	S	L	S	S	S	L		F	
13:15	ST	S	ST	S	S	S	L	ST	S	S	S	F	ST	S	S	S	S	S	ST	S	S	F	S	S	DT	S	ST	S	S	S	L		S	L	F
13:20	S		ST	ST	S	S	L	L	S	S	L	F	ST	S	L		S	ST	ST	ST	L	S	S		S	S	L	S	F	F	S	F	S	F	
13:25	ST		ST	S	S	L	ST	ST	L	S	ST	F	ST	S	S		S	ST	ST	S	S	S	L	F	F	F		L		F	F	S	F	S	
13:30	ST	L	ST	S	L	S	L	L	L	S	ST	S	S	S	L		S	ST	ST	S	S	ST	S	S	S	S		F	F	S	F	S	S	L	
13:35	ST	S	ST	S	S	ST	S	S	ST	S	S	S	DT	S	S	S	S	ST	ST	S	S	L	S		DT	S	S	DT	S	DT	S	L	S	S	
13:40	ST	L	ST	S	S	L	L	ST	S	S	S	S	S	S	S	S	ST	DT	DK	S	S	S	AG		S	S	S	DT	S	L	S	S	DT	S	
13:45	ST	S	ST	S	S	S	ST	ST	L	S	F	S	ST	S	S		L	ST	ST	S	S	S	S	L		S	L	S	L	S	S	S	S	S	
13:50	ST	S	ST	S	S	ST	S	S	S	S	ST	S	S	S	S	L	ST	S		L	L	S	S		S	S	L	F			S	S	S		
13:55	S	L	S	S	L	S	L	DT	L	ST	S	ST	S	S	S	L	ST	S	ST		S	ST	S			S	L	L	F	S	L	F	S		
% stereo	38	9	63	10	0	0	12	13	0	0	30	0	52	0	2	2	0	13	31	38	0	0	4	0	2	0	0	0	2	0	0	0	0	0	0

FIGURE 7.7 The diurnal pattern of stereotypic behaviour of an adult cow Asian elephant, *Kumara*, during a 35-day study over a period of 8 months. Each cell represents one scan sample (taken at 5-minute intervals). Light grey cell labelled ST = stereotypic behaviour; dark grey cell = no recording made due to disturbance, absence from enclosure, etc.; white cell = other behaviour recorded [e.g. feeding (F), walking (L), standing (S), dusting (DT), drinking (DK), aggression (AG), sexual behaviour (SEX); playing (PY)]. Source: *Rees (unpublished data)*.

(*Continued*)

BOX 7.1 (Continued)

FIGURE 7.8 *Kumara* turning on-the-spot after the elephant enclosure was redesigned. Sequence runs from top left to bottom left (1–4) then top right to bottom right (5–8).

for just 21 hours over a small number of days will provide anything other than an extremely superficial picture of how the animal spends its time. The frequency of stereotypic behaviour has been shown to be affected by the time of day (increasing when elephants expect to return to their indoor accommodation at the end of the day) and temperature (increasing when it is cold) (Rees, 2002a, 2009b). Harris et al. did not appear to take either of these factors into account in their data collection or analysis.

Wild Asian elephants have a fragmented distribution across the Indian subcontinent and Southeast Asia between approximately latitude 30°N and the equator. They are adapted to warm, moist conditions but may also experience cold temperatures (below 0°C) in parts of their range. In European zoos, Asian elephants may potentially be exposed to much lower temperatures [e.g. in Moscow minimum average daily temperatures may fall to −16°C in January (Pearce and Smith, 2000)]. However, zoo elephants are normally confined to heated indoor quarters in severe weather. A number of welfare scientists and organisations have drawn attention to the unsuitable climates experienced by elephants that have been kept in Alaska (until 2007) and Moscow (which, at the time of writing, still keeps Asian elephants) without any clear scientific evidence of specific harm caused by low temperatures. However, there is some evidence that low temperatures may increase stress in elephants.

In my study of Asian elephants living in Chester Zoo, I found a strong negative correlation between maximum daily temperature and the frequency of stereotypic behaviour in those animals predisposed to exhibit stereotypies (Rees, 2004a) (Fig. 7.9). The frequency of stereotypic behaviours increased throughout each study day (1000–1400), reaching maximum values at the end of the day as feeding time approached. The frequencies of stereotypic behaviour were higher at all times of the day on the 10 coldest days (mean maximum daily temperature of 9.0°C) than on the 10 warmest days (mean maximum daily temperature of 23.2°C) of the study (Fig. 7.10). There was a strong negative correlation between body mass and the mean frequency of stereotypic behaviour, possibly because smaller animals lost heat faster than larger animals. The aetiology of stereotypic behaviour was unclear, but it was likely to be the result of poor husbandry experienced in early life, such as chaining and inappropriate housing. Hunger and the physical thwarting of attempts to reach food and shelter may have been the proximate cause of individual episodes of stereotypic behaviour, with temperature acting as a compounding factor. There was no evidence that stereotypic behaviour developed in response to exposure to cold.

Higher glucocorticoid levels have been found during colder seasons among small numbers of zoo-housed Asian (Marcilla et al., 2012; Menargues et al., 2012) and African (Casares et al., 2016) elephants. Carlstead et al. (2018) found that cortisol levels increased with latitude in elephants kept in zoos in North America. They suggest that colder temperatures may have been responsible for this, but it could also have been due to shorter day length resulting in more time spent indoors.

I found that the frequency of stereotypic behaviours in five adult cow elephants at Chester Zoo was negatively correlated with the frequency of feeding behaviour (Rees, 2009b) (Fig. 7.11). In their study of a solitary female Asian elephant at Higashiyama Zoo in Japan, Koyama et al. (2012) also found a negative correlation between stereotypic pacing and feeding. At Chester Zoo, stereotypic behaviour increased in frequency towards the end

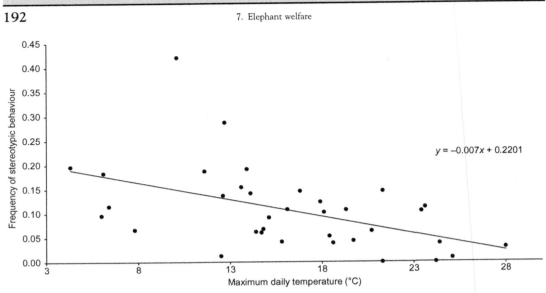

FIGURE 7.9 The effect of environmental temperature on the frequency of stereotypic behaviour in five female Asian elephants living in a zoo. Source: Rees, P.A., 2002. Asian elephants (Elephas maximus) *dust bathe in response to an increase in environmental temperature. Journal of Thermal Biology 27, 353–358.*

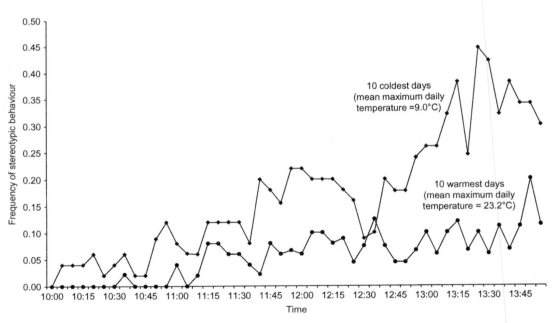

FIGURE 7.10 The effect of time of day on the frequency of stereotypic behaviour on the 10 hottest and 10 coldest days of the study. Source: Rees, P.A., 2002. Asian elephants (Elephas maximus) *dust bathe in response to an increase in environmental temperature. Journal of Thermal Biology 27, 353–358.*

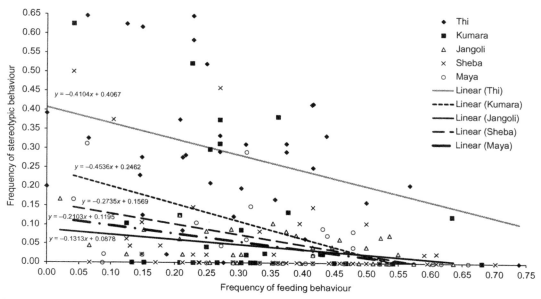

FIGURE 7.11 The relationship between the frequency of feeding behaviour and the frequency of stereotypic behaviour in five Asian elephants. Source: Rees, P.A., 2009. *Activity budgets and the relationship between feeding and stereotypic behaviors in Asian elephants* (Elephas maximus) *in a zoo. Zoo Biology* 28, 79–97.

of the day — while waiting to return to the elephant house for food — and elephants spent more time stereotyping during the winter months than during the summer months.

Hasenjager and Bergl (2015) studied the environmental conditions associated with repetitive behaviours in a group of six African elephants at North Carolina Zoo, United States. They found that most time was spent in feeding, locomotion, resting and repetitive behaviour. Patterns of repetitive behaviour were best explained by time of day in conjunction with location on- or off-exhibit. These behaviours occurred at a lower rate in the morning when the animals were on-exhibit compared with in the afternoon while on-exhibit or at any time in the day while off-exhibit. Hasenjager and Bergl suggested that the increased repetitive behaviour rates recorded in the afternoon on-exhibit may have been anticipatory responses prior to the transfer to indoor accommodation and evening feeding. This variation in the rate of repetitive behaviour during the day on-exhibit concurs with my observations made of the Asian elephant herd at Chester Zoo, which also appeared to be an anticipatory response (Fig. 7.10). The elevated rates of repetitive behaviour off-exhibit at all times of day at North Carolina Zoo were thought to have been caused by the differences in exhibit complexity between off-exhibit and on-exhibit areas and also the lack of additional foraging opportunities off-exhibit.

The effect of restraining method on stereotypic and other behaviours in Asian and African elephants ($n = 10$ and $n = 3$, respectively) in a travelling circus was studied by Gruber et al. (2000). Video recordings were made for at least 24 hours at each of six locations where they performed. The elephants were held on turf in electric-wire pens at four of the locations and at the other two they were restrained by leg chains while standing on a tar macadam surface. The frequencies of seven behaviours were measured, namely aggression, comfort, ingestion, locomotion, resting, social and stereotypy. The authors found that younger elephants were more likely to exhibit stereotypic behaviour than older elephants. Chained restraint resulted in higher frequencies of stereotypic and social behaviour and lower frequencies of comfort behaviour, ingestion and locomotion than observed in pens. Frequencies of aggression and resting were unaffected by restraint type.

Giving female Asian elephants housed at the Bronx Zoo, New York, the option of spending time outside during the night appeared to reduce significantly the amount of time they spent engaged in stereotypic swaying, even though one of the animals chose to spend most of her time inside (Powell and Vitale, 2016). In a study of the nocturnal behaviour of 34 orphaned elephants kept in a 4 ha paddock at night, Stokes et al. (2017) recorded no stereotypic behaviour.

Schiffmann et al. (2019) studied stereotypic behaviour in a 44-year-old female African elephant and at Zoo Basel, Switzerland, for 15 days during reconstruction of, and transfer to, a new enclosure. They found that social factors and the complexity of an enclosure providing appropriate resting locations to be more relevant for the decrease of this zoo elephant's stereotypic behaviour than exhibit size or dietary enrichment.

It is clear that stereotypic behaviour in elephants is affected by a number of aspects of their physical and social environment. Very short studies with few subjects cannot adequately investigate these factors.

7.3 Environmental enrichment

7.3.1 Defining environmental enrichment

Environmental enrichment has been defined by Shepherdson (1998) as:

> ...an animal husbandry principle that seeks to enhance the quality of captive animal care by identifying and providing the environmental stimuli necessary for optimal psychological and physiological well-being.

Other definitions have emphasised the importance of promoting natural (species appropriate) behaviours, for example the Association of Zoos and Aquariums (AZA) (2009) has defined enrichment as:

> ...a process for improving or enhancing zoo animal environments and care within the context of their inhabitants' behavioral biology and natural history. It is a dynamic process in which changes to structures and husbandry practices are made with the goal of increasing behavioral choices available to animals and drawing out their species appropriate behaviors and abilities, thus enhancing their welfare.

In 2006 a survey of elephant facilities in North American zoos found that 95% of AZA institutions operated a structured enrichment programme (Lewis et al., 2010). Lukacs et al. (2016) found that the five Asian elephants at Busch Gardens Tampa Bay, Florida, used enrichment items for 1%—20% of the time during the day but much less at night (0.1%—10%). They defined this as: 'Interacting with any food item other than hay or concentrate and other nonfood enrichment objects; may include walking'.

If the purpose of enrichment is to promote natural behaviours, it follows that the use of a man-made device that requires an animal to perform behaviours that do not occur in nature, for example operating a mechanical feeding device, painting or playing a musical instrument, cannot be enriching (Fig. 7.12). Zoo enclosures abound with potentially enriching devices that are broken, and in some cases, apparently unused. Relatively few of these have been scientifically assessed for their effectiveness.

Zoo elephants have been the subject of many attempts at environmental enrichment, but unfortunately many of these have been reported in the grey literature, contained little or no scientific data and have not been peer-reviewed. These include reports of the use of large blocks of ice containing fruit ('fruitsicles') produced using a garbage can as a mould at San Diego Zoo (Hartnett, 1995) and other projects using feeder balls, logs, tyres, pumpkins, browse, sand and earth, and carrots hidden in sand (Gilbert, 1994; Green, 1993; Haight, 1993) (Fig. 7.13).

Zoo managers and keepers face a dilemma when designing elephant enclosures containing enrichment devices. However desirable it is to create naturalistic enclosures, some enrichment devices inevitably look incongruous. It is difficult to disguise devices used to suspend food high above the ground and even more difficult to make some enriching activities look like natural behaviour (Fig. 7.14).

The use of carefully controlled scientific studies to assess the effectiveness of enrichment is important. Posta (2011) retrospectively examined zookeeper documentation for two African elephants kept at Toledo Zoo, Ohio, United States, and concluded that,

FIGURE 7.12 Suspended heavy objects can be useful toys for elephants but look out of place in naturalist exhibits: (A) a tyre in the old elephant enclosure at Blackpool Zoo, United Kingdom; (B) a block of wood at Berlin Zoo, Germany.

FIGURE 7.13 *Kumara* breaking a fruitsicle at Chester Zoo, United Kingdom: (A) a keeper floats the fruitsicle in the pool; (B) *Kumara* scoops it out of the water with her trunk; (C) she crushes it with her foot; (D) elephants collect the pieces of ice and eat the fruit.

although it was adequate for monitoring the daily care and well-being of elephants, it did not allow for a robust analysis of the effectiveness of enrichment.

7.3.2 Food and foraging as enrichment

Elephants living in zoos tend to spend less time than their wild counterparts foraging, so any change to their husbandry or enclosure design that promotes foraging should be beneficial. The AZA's *Standards for Elephant Management and Care* (Section 2.1.1.5) require

FIGURE 7.14 'Football' as enrichment: (A) The herd of Asian elephants at Chester Zoo, United Kingdom, 'playing football'; (B) *Chang* put an end to the fun by knocking the ball into the dry moat.

that elephants should be provided with opportunities to process food in ways similar to their wild counterparts and mechanisms that enable them to work for food (AZA, 2019a):

2.1.1.5

…Opportunities for searching, browsing, grazing, reaching, opening, etc. can be provided by scatter-feeding, hiding foods in crevices and substrates around the exhibit, or by using elevated feeders such as hanging hay nets that encourage an elephant to reach for and manipulate its trunk to gain access to the food.…

It cannot be assumed that simply hiding food will necessarily have an enriching effect. In an attempt to increase the amount of time a group of Asian elephants living at Zürich Zoo, Switzerland, spent foraging, peanuts were hidden above the stone border in their outdoor enclosure (Wiedenmayer, 1998). After training, the elephants learned the association between the border and the presence of food and they searched more above these borders than previously. However, hiding food did not enhance searching behaviour overall and did not, therefore, constitute an environmental enrichment.

The use of browse as food for elephants acts as a behavioural enrichment and also enhances the experience of zoo visitors. Stoinski et al. (2000) used an A—B—A design to test the effect of substituting an equal dry weight of browse for hay on the behaviour of three African elephants. When browse was present the elephants showed a significant increase in feeding and significant decreases in inactivity and drinking compared with when hay was present. When browse was present the elephants exhibited species-typical feeding behaviours and their visibility to zoo visitors was increased.

Elephants will spend a great deal of time interacting with tree trunks and tree branches: stripping leaves from branches by pulling the curled end of the trunk along the branch, breaking off pieces of bark with the toes or teeth, and breaking large branches into pieces small enough to eat (Fig. 7.15). Browse also provides young elephants with objects to manipulate, carry and use as 'toys'.

A number of zoos now hide food for their elephants in cavities in walls (Fig. 7.16) and feed them at a higher level, often using cargo nets or suspended containers, so the elephants have to reach above their heads to obtain food (Fig. 7.17). In the wild, feeding at high levels has been reported, but accounts for a relatively small proportion of time spent feeding. African elephants in Uganda studied by Wyatt and Eltringham (1974) fed at ground level during 75% of the time spent feeding. In Zimbabwe, elephants studied by Guy (1976) obtained most of their food from within 6 m off the ground and did not make full use of their reach. However, some 8.5% of food was obtained from above the head height of an adult elephant.

Some zoos provide their elephants with feeding devices that dispense small food treats (Fig. 7.18). The extent to which these act as enrichment depends, at least in part, upon whether or not they are filled with food as busy keepers do not always have sufficient time in the day to chop up food and refill them. In some cases, individual elephants may

FIGURE 7.15 Elephants at Chester Zoo, United Kingdom, spent considerable periods of time breaking branches and stripping bark from large tree trunks when these were available.

FIGURE 7.16 Feeder walls: (A) in the outdoor enclosure at Chester Zoo, United Kingdom; (B) in the elephant house at Blackpool Zoo, United Kingdom.

monopolise feeders. The only adult bull at Chester Zoo, United Kingdom, was found to use a large metal feeder 'ball' more than any of the other herd members (Rees, 2009b) (Fig. 7.19). The AZA's *Standards for Elephant Management and Care* recommend that feeders should be placed in several locations to discourage undue competition for food (AZA, 2019a).

7.3.3 Substratum and trees as enrichment

Loose soil or sand provides an important source of enrichment for captive elephants because, given the opportunity, they will spend a considerable amount of time using it to dust their bodies or digging in it (Figs 2.11 and 6.14).

FIGURE 7.17 (A) A female African elephant feeding high in a tree in Lake Manyara National Park, Tanzania. (B) A bull Asian elephant feeding from a cargo net at Chester Zoo, United Kingdom.

FIGURE 7.18 A female Asian elephant hitting a feeder tub in the elephant house at Blackpool Zoo, United Kingdom.

FIGURE 7.19 *Chang*, the only adult elephant bull at Chester Zoo, often monopolised the feeder 'ball'.

The AZA's *Standards for Elephant Management and Care* (Section 1.4.5.1) require that outdoor habitats contain a variety of substrates and that surfaces consist primarily of natural substances such as soil, sand and grass, with good drainage. Soft substrates promote foraging, wallowing, bathing, digging and resting. If the ground is landscaped, this will provide mounds where elephants can lie down. Some hard substrate is necessary to promote the normal wear of footpads.

Facilities that hold elephants are increasingly using sand on the floor of their elephant houses, although some use stall mats, straw or shavings for insulation. Sand on the floor of an elephant house provides opportunities for elephants to dust and dig for hidden food, and, as in the outdoor enclosure, if the floor is landscaped elephants have the opportunity to climb and places to lie down. This is particularly important in certain climates and at times of the year when elephants are kept indoors for extended periods (Fig. 7.20). The AZA (2019a) recommends that institutions renovating existing facilities or creating new elephant houses should consider using natural, changeable substrates indoors (standard 1.4.5.2).

Although elephants in zoos will quickly destroy any trees to which they have access by breaking off branches and removing bark, real and artificial trees and tree trunks provide useful opportunities for scratching (Fig. 7.21).

FIGURE 7.20 Sand substratum in the elephant house at Dublin Zoo, Ireland, is landscaped by keepers.

7.3.4 Water as enrichment

Hancocks (2008) discussed the importance of access to water to elephants living in zoos and believes that they should be treated as 'almost essentially amphibious creatures'. Well-designed elephant enclosures contain deep pools and high-powered hoses so that elephants can be cooled down on hot days (Fig. 7.22B and D). In Southeast Asia working elephants and those kept in orphanages are routinely washed or allowed to bathe in rivers (Fig. 7.22A and C). Indeed, in the state of Kerala, in India, 'it is the duty of each mahout to ensure that [his] elephant gets a thorough bath every day' [Kerala Captive Elephants (Management and Maintenance) Rules 2012, Rule 4(3)].

The recommended sizes of pools for zoo elephants are discussed in Section 8.3.3.

7.3.5 Sleep, rest and enrichment

The ability to obtain adequate rest and sleep has long been recognised as important to the welfare of elephants (Evans, 1910). It has recently been suggested that excessive drowsiness occurs in elephants as a number of cases have been reported whereby falling bouts occurred after a period of lying rest deprivation (Schiffmann et al., 2018). This phenomenon has been recorded in horses (e.g. Bertone, 2006). It is possible that elephants are reluctant to lie down — thereby allowing REM sleep — when they feel unsafe, vulnerable or anticipate being unable to get up again. A shift from lying to standing rest has been reported in elephants who were ill, stressed or geriatric (Holdgate et al., 2016b; Laws et al., 2007; Walsh, 2017).

Schiffmann et al. (2018) have discussed the usefulness of providing elevated and horizontal structures in enclosures allowing elephants to relieve their limbs and supporting standing rest, along with a substratum shaped so that it facilitates lying down by making it easier for them to return to a standing position (Fig. 7.23). This is particularly important for pregnant elephants.

FIGURE 7.21 A 4-year-old female Asian elephant at Blackpool Zoo, United Kingdom, using a tree for scratching. Sequence runs top left to bottom right.

FIGURE 7.22 Elephants enjoying water: (A) elephants from the Pinnawala Orphanage, Sri Lanka, visiting the Kuda Oya River; (B) *Thi* and her daughter *Sithami* bathing at Chester Zoo, United Kingdom; (C) mahouts washing a Forest Department elephant in the Cauvery River, Karnataka, India; (D) washing the elephants at Dublin Zoo, Ireland.

FIGURE 7.23 *Jangoli* resting while pregnant at Chester Zoo, United Kingdom, in June 2000: (A) against a steel bar; (B) against the base of the waterfall.

7.3.6 Sound, music and art as enrichment

Sound is a very important component of the sensory environment of wild elephants. Some of the Asian elephants housed at the Thai Elephant Conservation Center near Chiang Mai, Thailand, play musical instruments in the *Thai Elephant Orchestra*. They play large percussion and wind instruments specially designed for their physical abilities such as gongs, bells, drums, xylophones, harmonicas and single-string instruments. The elephants appear to be capable of playing complex patterns and are given a minimal number of commands indicating when they should start and stop, and occasionally the number of times they should strike an instrument. The purpose of this project was to raise money for the elephant centre by attracting tourists and selling audio CDs of the music. It has been suggested that, rather than having an enriching effect on the animals' lives, the project may have had a negative impact because the focus was on training the elephants to produce human-like music for a human audience rather than allowing them to use the musical instruments as they pleased (Gupfinger and Kaltenbrunner, 2018).

At the National Zoological Gardens (Dehiwala Zoo) in Colombo, Sri Lanka, Asian elephants have been taught to 'dance' to music to entertain visitors seated in an outdoor arena. This is inappropriate in a modern zoo, but may have some enrichment value for elephants who spend much of their day chained to a concrete floor (Fig. 7.24).

A more elephant welfare-focussed attempt to engage captive elephants with music has recently been made by French et al. (2018) by designing and building interactive musical 'smart toys'. Working with keepers and elephant welfare experts the authors built

FIGURE 7.24 The elephant shelter at the National Zoological Gardens, Colombo, Sri Lanka (2003). Elephants were chained here during the day.

instruments equipped with hidden sensors that enabled elephants to control sounds using trunk interactions with tactile surfaces mounted in wooden frames.

Wells and Irwin (2008) examined the use of auditory stimulation as enrichment for four Asian elephants at Dublin Zoo, Ireland, using an A−B−A experimental design. They found that elephants spent significantly less time exhibiting stereotypic behaviour when a CD of classical music was played than when no music was played, and that other behaviours were not affected.

Elephants in some zoos have been encouraged to produce paintings as a presumed enrichment and sometimes to raise money for elephant conservation. English et al. (2014) investigated the relationship between painting and stress-related behaviours in four Asian elephants at Melbourne Zoo, Australia. If the activity of painting is enriching for elephants, the authors argued, they would expect stress-related behaviours to be reduced. However, they found no evidence of the elephants anticipating the painting activity or any effect of painting on stereotypic or other stress-related behaviours, either before or after a painting session. Nevertheless, elephants not selected for painting on any particular day exhibited higher levels of noninteractive behaviour: standing still, separate from other elephants with the trunk tip touching the ground, eyes open, with ears and tail in a neutral position. This is an indicator of increased stress in some species (Carlstead, 1996). English et al. concluded that, while the paintings produced by elephants have an aesthetic appeal to humans and their sale generated funds for elephant conservation, the activity of painting itself did not improve elephant welfare.

7.3.7 Interactive toys

A particular challenge for researchers developing interactive toys for elephants is constructing them so that they are sufficiently robust to withstand the destructive abilities of these animals. In the process of designing an interactive shower system for an Asian elephant called *Valli*, French et al. (2016) noted that:

> ...Valli has trashed several of our systems when she has been left unattended. While the systematic destruction of a novel object in the enclosure also provides her with cognitive and sensory enrichment, the effect is transient and leaves little scope for progress. Interestingly, she has not attempted to destroy the controls themselves, but has targeted the wires and pipes that facilitate the deployment of a shower system.

7.3.8 Improving elephant welfare through breeding

In the United Kingdom, the Federation of Zoos' [now British and Irish Association of Zoos and Aquariums (BIAZA)] *Management Guidelines for the Welfare of Zoo Animals: Elephants* claimed that zoos 'improve elephant welfare through captive breeding' (Stevenson and Walter, 2002). The Federation asserted that:

> Giving birth and raising young is one of the most enriching experiences possible for an individual animal...

This is almost certainly true, but is it legitimate to breed elephants as an enrichment for other elephants? Some zoos allow certain species [e.g. African lions (*Panthera leo*)] to breed as an enrichment and to produce young animals to attract visitors, and then have to euthanise young animal surplus to their requirements when they cannot be transferred to other collections. This claim — improving welfare through breeding — appears to have been removed from subsequent editions of BIAZA's guidelines.

7.3.9 Social contact: the ultimate in enrichment

Even the best enrichment devices inevitably offer a very limited number of responses to interactions with an elephant. The ultimate elephant enrichment is social interaction, preferably with conspecifics. In spite of what we know about elephant social behaviour, some elephants living in zoos are still being kept alone or in small groups and their only social interaction is with humans The complexity of elephant society and the size of elephant groups in zoos have been discussed in Chapter 3, Elephant social structure, behaviour and complexity.

Vanitha et al. (2011) suggested a link between social isolation and abnormal elephant behaviour in captive Asian elephants in Tamil Nadu, India. They found that the proportion of elephants exhibiting stereotypic behaviours was highest in those kept in temples (49%) followed by those owned privately (26%), with only 6% showing this behaviour in forest department facilities. Single elephants were kept by 95% of temples and 82% of private facilities. Vanitha et al. suggested that elephant well-being could be improved if patchily distributed temple and privately owned elephants free from contagious diseases were kept in small groups of socially compatible individuals in 'common elephant houses'.

Young calves represent perhaps the most significant behavioural enrichment for elephants in captivity, providing opportunities for adults to exhibit an extended range of behaviours including allomothering, chastisement and play (Lee, 1987; Rapaport and Haight, 1987). Keeping bull elephants with cows and their calves offers opportunities for adults to engage in sexual behaviour and for calves to take part in the occasional 'mating pandemonium'. Adults in zoo herds containing calves will also form protective circles around the young when they perceive a threat from unusual noises or events, just as they would in the wild (Douglas-Hamilton and Douglas-Hamilton, 1975; Moss, 1988) (see Section 3.5).

Feeders and other enrichment devices may reduce stereotypic behaviours, but these may be replaced by equally 'abnormal'' behaviours such as spending long periods at a feeding device. Social contacts such as courtship, playing and fighting also reduce the frequency of undesirable stereotypic behaviours, but are only available to elephants kept in groups. Many elephants kept in zoos are either completely denied contact with their own species or have very limited social contacts.

7.4 Training

The training of elephants has been practised since ancient times. It was mentioned by Aristotle in his *History of Animals* in 350 BCE (Creswell, 1883) (see Box 1.1).

Training is widely used to assist in veterinary procedures and to control working elephants. It may also have a function as enrichment for some captive elephants and to give

TABLE 7.6 Checklist of AZA standard elephant programme behavioural components.

Bathe/scrub skin	Tusk exam	Transrectal ultrasound	Trunk wash for TB testing
Treat skin	Tusk trim	Accepts injections	Foot X-ray
Trim all feet	Blood collection (note frequency of collections)	Accepts oral medications	Separation
Eye exam	Urine collection	Enters chute (remains inside with doors closed)	Leg restraint
Ear exam	Vaginal exam	Allows chute walls to move	Reproductive assessment completed
Mouth exam	Rectal palpation	Allows husbandry procedures to be performed by staff	
Tooth exam	Enema	Allows veterinary procedures to be performed by vet	

Source: *Based on AZA, 2019. Standards for Elephant Management and Care. In: The Accreditation Standards and Related Policies, 2019 edition. Association of Zoos and Aquariums, pp. 38–66.*

demonstrations to the public, some of which have conservation messages (Figs 10.7 and 10.9). However, if elephants in zoos are trained to perform behaviours that are not 'natural' this may detract from the conservation and welfare messages that zoos aspire to send to the public. Training a bull elephant to allow a caretaker to inspect his feet and ears, or cooperate while being washed is clearly for the animal's benefit but zoo visitors — or at least some of them — may see little value in training an elephant to play a musical instrument.

Elephant facilities in North America that are AZA-accredited use the AZA Elephant Profile Form (AZA, 2013) — which includes a 'Checklist of AZA Standard Elephant Program Behavioral Components' — to record those behaviours and procedures in which the individual has, and has not, been successfully trained (Table 7.6).

Modern training techniques in zoos utilise positive reinforcement, providing small food rewards and encouragement to reinforce appropriate responses. Handlers of captive elephants in Nepal have traditionally relied on punishment and aversion-based methods in training. Fagen et al. (2014) have shown that it is possible to train juvenile, free contact, traditionally trained Asian elephants in Nepal to participate in a trunk wash — for the purpose of tuberculosis (TB) testing — using only secondary positive reinforcement techniques.

The training of elephants so that they may be handled under protected contact is discussed in Section 8.5.4.

7.5 Locomotion and gait

Locomotion in elephants has been of great interest to scientists due to its unusual characteristics. The general consensus has always been that elephants are not capable of running and the stiff manner of their gait has always attracted interest. The English

FIGURE 7.25 A sequence of consecutive photographic images of an Asian elephant walking, created by the English photographer Eadweard Muybridge c.1887. Source: *US Library of Congress: Reproduction No. LC-USZ62-107505 (b&w film copy neg.)*

photographer Eadweard Muybridge pioneered the use of photography to study movement in animals. He is famous for using a series of large cameras to determine whether or not all four of a horse's hooves leave the ground during a gallop. Around 1887 he used a similar technique to record the gait of an Asian elephant by producing a series of consecutive images of the animal walking (Fig. 7.25) and a second series of two Asian elephants walking side by side away from the camera.

Most captive elephants — at least those living in zoos — do not walk as much as wild elephants because they do not need to search for food, water, mates or other resources. Zoo elephants have traditionally been kept on concrete floors and these factors combined have led to skeletal and foot problems.

Some zoo elephants and those kept in elephant facilities in Asia can be ridden or led by their keepers and mahouts. This makes them perfect for the study of the dynamics and energetics of locomotion. Genin et al. (2010) performed experiments on the biomechanics of locomotion using 34 Asian elephants at the Thai Elephant Conservation Center near Lampang, northern Thailand. This subject is of particular interest in elephants because they are relatively fragile compared with smaller animals since forces are proportional to body volume while supportive tissue strength depends on their cross-sectional area. The authors showed that the mass-specific mechanical work required to maintain the movements of the centre of mass (COM) per unit distance is approximately 0.2 J/kg per metre [about one-third of the average of other animals ranging in size from a kangaroo rat (35 g) to a human (70 kg)]. At low speeds a pendulum-like exchange between the kinetic and potential energies of the COM reduces this work. At high speeds, a bouncing mechanism is used, with little exchange between kinetic and potential energies of the COM, and with no aerial phase.

Ren and Hutchinson (2008) studied the dynamics of gait in elephants using two adult Asian elephants from Woburn Safari Park (Bedfordshire, England) and three adult African elephants from the West Midlands Safari Park (West Midlands, England). This study could not have been conducted in the wild because it required that each elephant be equipped with two miniature integrated inertial sensors (attached to the back), two foot-mounted accelerometers and a stand-alone global positioning system device (also attached to the back). While this study added to our knowledge of locomotion in healthy adult elephants, no gait-related welfare issues were discussed by the authors.

Ren et al. (2010) studied locomotor performance in elephants using six Asian elephants at the Thai Elephant Conservation Center in Thailand and dissections of cadavers supplied by zoos in the United Kingdom. The elephants were led or ridden by their mahouts and movements were analysed using an infrared camera motion analysis system to record limb marker motions and custom-made force platforms to measure ground reaction forces. Once again, the study made interesting discoveries about the nature of locomotion of elephants. The authors described the limb function as analogous to the four-wheel-drive function in vehicles. However, no mention was made of any relevance of the study to elephant welfare.

Elephants have attracted the attention of nonbiologists because of their ability to maintain stability in spite of their weight. Wijesooriya et al. (2012) developed a methodology for studying movement in large mammals without needing sophisticated apparatus and instead used video and still images to analyse the biomechanics of gait in three captive Asian elephants. Their interest was not in the elephant per se, but in the value of studying these animals to assist the development of robotic devices because of their ability to maintain the stability of a body weighing up to 4500 kg in wide range, of unstructured terrain conditions.

Although much of this work is concerned with increasing basic knowledge, nevertheless, the accumulation of data on walking in healthy elephants may be of use in improving elephant welfare in captivity. Hutchinson et al. (2006) analysed locomotor kinematic data from over 2400 strides for 14 African and 46 Asian elephants from facilities in California, Indiana, Germany, the United Kingdom, and Thailand. This large dataset was used to establish the normal kinematics of elephant locomotion and the authors suggest it can be utilised to identify gait abnormalities that may be indicative of musculoskeletal pathologies.

7.6 Obesity

Captive elephants, particularly those kept in zoos, are prone to becoming obese due to lack of exercise and excessive calorie intake. Obesity appears to be an important factor in the poor reproductive performance of elephants living in zoos (Morfeld and Brown, 2014).

Das et al. (2015) studied the effect of restricting concentrates in the diet of a group of captive Asian elephants on feed consumption, diet digestibility and nitrogen utilisation. When the amount of concentrate was reduced by 50% the elephants ate more roughage, digestibility of dry matter and gross energy decreased, but the diet still provided sufficient digestible energy and crude protein for the elephants' needs during the trial. The authors of the study concluded that concentrates should be restricted in the diets of elephants to reduce excessive calorie intake and the risk of obesity.

Morfeld and Brown (2014) compared body condition, glucose, insulin and leptin concentrations, and the glucose-to-insulin ratio in 23 cycling and 23 noncycling African elephants. They used a body condition score (BCS) index with a 5-point scale: 1 = thinnest, 5 = fattest. The mean score of noncycling elephants was higher than that of cycling elephants. Among the fattest elephants (BCS = 5), noncycling animals had higher leptin and insulin concentrations and a lower glucose-to-insulin ratio than cycling animals. High BSC was a strong predictor of noncycling status and this study provided the first evidence that ovarian acyclicity in African elephants is associated with obesity and altered circulating metabolic hormone concentrations.

A study of 22 female African elephants housed in eight North American zoos recorded body fat percentages ranging from 5.24% to 15.97% (Chusyd et al., 2018). However, it found that fat was not significantly associated with cyclicity, and concluded that age was a stronger predictor of cyclicity status.

Elephants living in zoos tend to be more sedentary than those living wild because they do not need to move far to find food, water and mates. However, it is not inevitable that zoo-living elephants will engage in little exercise. The walking rates of eight African elephants living in San Diego Zoo Safari Park in California, United States, were compared with those of 11 wild elephants living in different habitats and locations in Botswana by Miller et al. (2016). The methods used to determine walking rates had been previously validated by Miller et al. (2012). The 2016 study concluded that there was no difference in the average walking rates of the animals in the two groups and, therefore, the elephants living in the zoo were meeting their exercise needs. However, my study of eight Asian elephants at Chester Zoo – in a much smaller enclosure - found that individuals were inactive (i.e. exhibited behaviours other than locomotion) for between 70.1% and 93.9% of the time (Rees, 2009b).

7.7 Disease

7.7.1 Introduction

A great deal has been published about the diseases of elephants, most notably in the work edited by Fowler and Mikota (2006) entitled *Biology, Medicine, and Surgery of Elephants*. It is not my intention to summarise the information contained therein – or in work published thereafter – but rather to indicate the role that captive elephants have played in the accumulation of knowledge about elephant diseases and medicine. I shall also briefly consider particular diseases, especially emerging diseases, that are of particular concern to those who care for elephants living in zoos and other captive environments.

The long-standing practice of using elephants in the logging industry, and more recently in tourism, in Southeast Asia has led to the development of considerable expertise in the care and treatment of elephants by veterinarians working in this region.

FIGURE 7.26 An elephant being fed a ragi ball at an elephant camp operated by the Forest Department, Karnataka, India.

Krishnamurthy and Wemmer (1995) have provided an historical account of the veterinary care of elephants in Asian timber camps. They noted that both indigenous and European practices were incorporated into the veterinary management of timber elephants by the British during the colonial period, and that considerable time and energy were invested to improve care and humane standards. They examined aspects of the preventive care and clinical treatment of elephants based on records from the Madras Presidency (an administrative division of British India) and the modern-day Tamil Nadu state, including the administration of broad-spectrum anthelmintics to treat parasitic infections by disguising them in ragi balls. Ragi is a type of cereal from which flour can be made (Fig. 7.26).

The difficulty of determining appropriate drug doses for elephants — due to the absence of similar wild or domestic species upon which to base them — has been discussed by Cheeran et al. (1995) of the College of Veterinary and Animal Sciences at Kerala Agricultural University, India. They suggested suitable doses of drugs used to treat a wide range of maladies, including parasitic infections based on their experience in clinical practice in India.

An analysis of diseases recorded in captive and wild Asian elephants (especially in Kerala, India) over a period of some 30 years has been provided by Chandrasekharan et al. (1995), also of the College of Veterinary and Animal Sciences in Kerala. In total more

FIGURE 7.27 A visually impaired bull elephant from the Pinnawala Orphanage, Sri Lanka, returning from bathing in the local river.

than 1000 cases were recorded, most of which related to captive animals (93%) and 58% of which were parasitic infections. As part of this record they included details of anthelmintic drugs and dose rates they had found to be effective against a range of elephant parasites.

A visual health assessment of 81 captive elephants in 10 facilities in India was conducted by Ramanathan and Mallapur (2008) in 2004—05. They found that the overall condition of elephants housed at tourist camps was poor compared with those kept by zoos and a forest camp. Some 43.2% of the elephants studied exhibited hyperkeratosis, and skin condition was generally poorer among elephants kept by tourist camps compared with those in zoos and a forest camp. Captive-born animals had better skin condition than did wild-caught elephants. Foot pad fissures were found in 74.1% of elephants and were more common in females than in males; 20% of these were severe. Vertical and horizontal toenail cracks were more common in tourist camp elephants (see Fig. 7.28, below) (see Section 7.7.3). Approximately 76.9% of the 'wounded' animals and 80% of those with abscesses were housed in tourist camps and temples, as were all the elephants with eye problems (8.5% of the total studied) (Fig. 7.27).

7.7.2 Histology

Knowledge of the histology of elephants may assist in the diagnosis and understanding of pathogenesis in sick elephants or assist in determining cause of death. Captive elephants continue to be a source of specimens for such studies. Thitaram et al. (2018) examined the histology of 24 organs from two Asian elephant calves provided by Chiang Mai elephant camps in Thailand. Although most structures were found to be similar to those found in adult elephants or other mammalian species, some structures were different. Plexiform bone was found in flat bone only (bones whose primary function is protection or the provision of broad surfaces for muscular attachment); in the trachea a thin trachealis muscle was observed (whose function is to constrict the trachea allowing air to be expelled with increased force); and the submucosa of the trachea appeared to contain no serous or mucinous glands.

7.7.3 Foot health

Foot problems are common in elephants living in zoos and it is essential that they are trained to allow caretakers to inspect their feet so that damage and disease can be detected at an early stage (Fig. 7.28). Elephants have evolved a specialised foot morphology — including a large fat pad — that helps reduce pressures during walking. Peak pressures that could cause tissue damage are mitigated passively by the anatomy of the foot but this mechanism does not work well in some captive elephants. Variation in pressure patterns may be related to differences in husbandry practices including trimming and the nature of the substrate.

In 2006 Lewis et al. (2010) conducted a survey of elephant husbandry and foot health in North American zoos. They found that 69% of indoor areas had concrete floors and 85% of outdoor areas consisted of natural substrates. Elephants in AZA facilities had an

FIGURE 7.28 An Asian elephant trained in free contact to allow the examination of the base of the foot.

average of 45.5 minutes exercise per day; those facilities with a structured exercise plan provided significantly more exercise than those without such a plan. Preventative foot care was almost universal at AZA facilities and all routinely trimmed nails and pads (Figs 7.29 and 7.30). Nevertheless, 33% of institutions reported at least one foot pathology in the year preceding the survey. A strong inverse relationship was identified between exercise and foot pathology. Younger herds were less likely than older herds to have a member diagnosed with arthritis. Lameness was unrelated to pathology or age, but could be

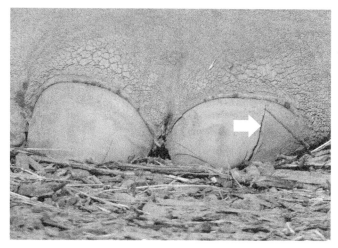

FIGURE 7.29 A cracked toenail in an Asian elephant living in a zoo (indicated by white arrow).

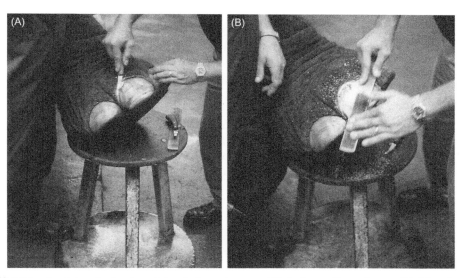

FIGURE 7.30 An Asian elephant receiving foot care in a zoo: (A) dealing with an abscess; (B) filing down nails.

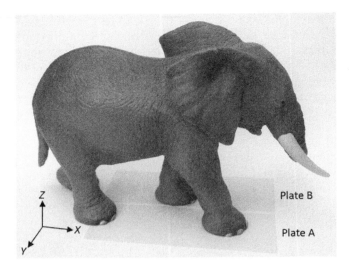

FIGURE 7.31 The arrangement of two pressure platforms (each of which contained 7040 sensors) in the experiments performed by Panagiotopoulou et al. (2016) on foot pressure distributions in elephants.

explained by the presence of arthritis. Although Lewis et al. found lower rates of foot pathology and arthritis in African elephants than in Asian elephants, this was explained by differences in age rather than any real species-specific difference.

Foot pressure distributions in African elephants were examined by Panagiotopoulou et al. (2016) using five elephants kept by Adventures with Elephants in Bela Bela, South Africa. Experiments were conducted using two pressure platforms, each of which contained 7040 sensors (Fig. 7.31). The results of this study were compared with a similar study of Asian elephants (Panagiotopoulou et al., 2012) at ZSL Whipsnade Zoo and Woburn Safari Park, both in the United Kingdom. Natural substrates predominated in both studies, but foot care protocols differed. Both species exhibited the highest concentration of peak pressures on the lateral digits of their feet — which are more prone to disease — and lower pressures around the heel. When walking at similar speeds (a mean of 1.2 m/s) both species load their feet laterally at impact with the ground and then shift their weight medially until toe-off. This method of locomotion works well for natural substrates, but may not be as effective on hard substrates such as concrete and asphalt. Elephants confined for long periods on hard substrates are more prone to foot disease than their wild conspecifics. Panagiotopoulou et al. (2016) found differences in foot pressure patterns that might, in part, be attributable to husbandry. The authors speculated that captivity-induced foot infections around the fat pad may reduce its ability to dissipate locomotory pressures. A cracked or infected foot pad may potentially increase the loading on the distal part of the lateral digits and accelerate foot disease.

7.7.4 Tuberculosis

Bovine TB emerged as a disease of concern in captive elephants in the mid-1990s. Screening for the disease was introduced in the United States by the *Guidelines for the Control of TB in Elephants* which were first produced in 1997 (USDA, 1997) and have been revised several times.

Mikota et al. (2000) isolated *Mycobacterium tuberculosis* from 18 of 539 elephants (3.3%) in North America between August 1996 and May 2000. Reports of TB in captive elephants are not new. A case was reported in an Asian elephant at London Zoo as early as 1875 (Garrod, 1875) and Baldrey (1930) reported a case of TB in an elephant who had been landed at Port Said in Egypt and subsequently died. Another case was reported by Bopayya (1928).

The use of trunk wash cultures in screening for, and the diagnosis of, TB in elephants is limited by variable sample quality, long turnaround time and low test sensitivity. Responding to the need for improved diagnostic tools, Greenwald et al. (2009) described three novel serologic techniques for the early and rapid detection of TB antibodies in elephants, the ElephantTB Stat-Pak kit, multiantigen print immunoassay, and dual-path platform VetTB test. They tested serum samples from 236 captive Asian and African elephants from 53 locations in Europe and North America. Greenwald et al. emphasised the importance of rapid and accurate antibody tests to identify infected elephants and facilitate earlier and more efficient treatment, thereby reducing the chances of transmission.

7.7.5 Elephant endotheliotropic herpesviruses

The risks associated with the movement of elephants between collections have increased over the past two decades since the appearance of new fatal elephant endotheliotropic herpesviruses (EEHV) in the zoo population (Montali et al., 1998; Richman et al., 1999). One is fatal for African elephants and the other for Asian elephants. The disease affects predominantly young elephants, and has been recorded in North America, Europe and elsewhere. Richman et al. (2000) have suggested that otherwise healthy African elephants may have been the source of the herpesvirus that causes death in both species. This has implications for the management of zoo elephants and the EEP species coordinators. Dorresteyn and Terkel (2000a) recommended that Asian and African elephants should not be kept in mixed groups some two decades ago and this practice is now widely recognised as unwise.

Managers of elephant breeding programmes and reintroduction projects have shown great concern about the risk of disease transmission when individuals are transferred among facilities and between facilities and the wild. Data from studbooks have been used by Ryan and Thompson (2001) to estimate the transmission risk between the elephants in the AZA's Species Survival Plan in North America. They found that, although elephants experience few transfers between zoos compared with many other species, the number of direct contacts with other elephants born during the study period (1983–96) was much higher than anticipated, and the number of indirect contacts was also large. This made infection pathways for herpesvirus difficult to determine. However, the authors were able to propose potential transmission routes for the cases of the virus identified histologically and the studbooks also allowed the identification of two other groups, namely elephants who had been exposed to the virus but had not succumbed, and those elephants who should undergo histological testing for the disease.

The management challenges associated with EEHV and TB in relation to disease control and the transportation of elephants between collections within the Asian elephant European Association of Zoos and Aquaria Ex-situ Programme (EEP) has been recently discussed by Schmidt and Kappelhof (2019).

CHAPTER 8

Housing and handling elephants

...but when somebody told you he was an elephant handler, it was always best to wait and see. Maybe he'd fed an elephant peanuts at some zoo.

George 'Slim' Lewis (Lewis and Fish, 1978)

8.1 Introduction

This chapter is concerned with the construction of housing for elephants in zoos and the cost of meeting modern welfare standards. It also discusses the relationships between elephants and their caretakers — especially keepers in zoos and mahouts in logging camps — and the safe transportation of elephants.

8.2 Wild elephant decline and the establishment of ex situ breeding programmes

Humans have been responsible for a spectacular decline in elephant numbers over a period of just a few decades. The professional hunter T. Murray Smith remarked in 1963:

There must be fifty thousand elephants in Kenya, Tanganyika and Uganda alone. I do not think they will decrease.

Smith (1963)

Assuming his original estimate was correct, he was right, although, this can have been no more than an educated guess as no accurate figures were available in the 1960s. In 1998 it was estimated that these three territories held between 82,000 and 125,000 elephants [IUCN/SSC African Elephant Specialist Group database data for Kenya, Tanzania and Uganda, quoted in Sukumar (2003)]. However, the whole continent probably only supported about 609,000 elephants in 1989, compared with perhaps 1.3 million in the late 1970s (Spinage, 1994). In the 1970s, Dr Keith Eltringham warned the world of the risk that the African elephant could soon become extinct (see Section 10.2). In 2016 a continent-wide survey produced an estimated population of 352,271 savannah elephants on study sites in 18 countries (Chase et al., 2016). This represented approximately 93% of all savannah elephants in these countries. The population of Asian elephants in 2017 was estimated to be 47,602—50,324 wild individuals and 14,022—14,222 captive elephants within 12 range states (AsERSM, 2017).

In spite of decades of evidence of the decline in wild elephant populations, it was not until the mid-1980s that zoos began developing coordinated captive breeding programmes for elephants in zoos. The Species Survival Plan (SSP) for the Asian elephants was established by the American Association of Zoos and Aquariums (AZA) in 1985 and combined with the African elephant SSP in 1990. The European Endangered species Programme (EEP) for the Asian species was founded in 1991 by the European Association of Zoos and Aquaria. The European programme is now known as the European Association of Zoos and Aquaria Ex-situ Programme (EEP).

When I examined the global distribution of elephants living in zoos in 2006 using the database of the International Species Information System (ISIS) — now the Zoological Information Management System (ZIMS) of Species 360 — I identified 495 Asian elephants and 336 African elephants in 194 zoos, most of which were located in Europe (49.1%) and North America (32.6%) (Rees, 2009a). More recent data from ZIMS show a similar geographical pattern. By 2018 there were 397 African elephants (including 3 whose sex was not recorded) and 750 Asian elephants listed by ZIMS, making a total of 1147 animals (Tables 8.1 and 8.2). During this period some zoos stopped keeping

TABLE 8.1 The geographical distribution of zoos holding elephants recorded by ISIS in 2006 and ZIMS in 2018.

Region	*Loxodonta africana*		*Elephas maximus*	
	2006	2018	2006	2018
	Zoos (%)	Zoos (%)	Zoos (%)	Zoos (%)
Europe	42 (40.4)	45 (43.7)	54 (47.4)	76 (46.9)
North America	53 (51.0)	40 (38.8)	44 (38.6)	30 (18.5)
Africa	2 (1.9)	4 (3.9)	0 (0)	1 (0.6)
Central America	2 (1.9)	0 (0)	2 (1.8)	0 (0)
South America	2 (1.9)	6 (5.8)	1 (0.9)	3 (1.9)
Southeast Asia	2 (1.9)	8 (7.8)	8 (7.0)	47 (29.0)
Australia	1 (1.0)	0 (0)	5 (4.4)	5 (3.1)
Total	104 (100)	103 (100)	114 (100)	162 (100)

Source: *Data from ISIS. Anon., 2006a. ISIS Abstracts:* Loxodonta africana. *International Species Information System, https://app.isis.org/abstracts/abs.asp (accessed 27.10.06.); Anon., 2006b. ISIS Abstracts:* Elephas maximus. *International Species Information System, https://app.isis.org/abstracts/abs.asp (accessed 27.10.06.); Rees, P.A. 2009. The sizes of elephant groups in zoos: implications for elephant welfare. Journal of Applied Animal Welfare Science, 12, 44-60; ZIMS, 2018a. Species holding report:* Elephas maximus/*Asian elephant. Zoological Information Management System. Species 360. https//www.species360.org (accessed 26.03.18.); ZIMS 2018b. Species holding report:* Loxodonta africana/*African elephant. Zoological Information Management System. Species 360. https//www.species360.org (accessed 26.03.18.).*

TABLE 8.2 The global distribution of elephants in zoos recorded by ISIS in 2006 and ZIMS in 2018.

Region	Loxodonta africana				Elephas maximus			
	2006		2018		2006		2018	
	♂.♀	Sex ratio	♂.♀	Sex ratio	♂.♀	Sex ratio	♂.♀	Sex ratio
Europe	40.126	1:3.15	47.135	1:2.87	51.188	1:3.69	89.216	1:2.43
North America	22.125	1:5.68	40.138	1:3.45	23.101	1:4.39	17.83	1:4.88
Africa	2.4	1:2.00	2.5	1:2.50	0.0	–	0.1	0:1.00
Central America	1.3	1:3.00	0.0	–	0.4	0:4.00	0.0	–
South America	1.1	1:1.00	4.8	1:2.00	1.0	1:0.00	0.5	0:5.00
Southeast Asia	2.5	1:2.50	6.9	1:1.50	40.67	1:1.68	112.203	1:1.81
Australasia	0.3	0:3.00	0.0	–	4.14	1:3.50	8.16	1:2.00
Total	68.267	1:3.93	99.295	1:2.98	119.374	1:3.14	226.524	1:2.32
	335*		394⁺		493**		750	
	Totals for both species: **2006** = 828 **2018** = 1144							

One *Loxodonta** and two *Elephas*** specimens whose sex was not recorded have been omitted from the 2006 data. Three *Loxodonta*⁺ whose sex was not recorded have been omitted from the 2018 data.
Source: *Data from ISIS. Anon., 2006a. ISIS Abstracts:* Loxodonta africana. *International Species Information System, https://app.isis.org/abstracts/abs.asp (accessed 27.10.06.); Anon., 2006b. ISIS Abstracts:* Elephas maximus. *International Species Information System, https://app.isis.org/abstracts/abs.asp (accessed 27.10.06.); Rees, P.A. 2009, The sizes of elephant groups in zoos: implications for elephant welfare. Journal of Applied Animal Welfare Science, 12, 44–60; ZIMS, 2018a. Species holding report:* Elephas maximus/*Asian elephant. Zoological Information Management System. Species 360. https//www.species360.org (accessed 26.03.18.); ZIMS 2018b. Species holding report:* Loxodonta africana/*African elephant. Zoological Information Management System. Species 360. https//www.species360.org (accessed 26.03.18.).*

elephants while others have increased the size of their elephant groups due to births, transfers between institutions and importation from range states. The data for the 2 years are not directly comparable because zoos are not compelled to engage with ZIMS and the list of participating zoos has changed over time. Nevertheless, there are more zoos reporting that they keep elephants in Europe (especially Asian elephants) but apparently fewer keeping elephants of either species in North America. The total number of elephants in zoos reported to ISIS/ZIMS has increased from 828 in 2006 to 1144 in 2018 (omitting a small number of animals whose sex was not published), largely due to increases in the numbers of Asian elephants reported from facilities in Europe and Southeast Asia and much smaller increases in African elephants held in Europe and North America. The number of Asian elephants reported from North American facilities has decreased over this period.

All these elephants must be housed and the nature of their accommodation varies between facilities and, of course, has varied over time. The remainder of this chapter considers how humans contain, house and handle elephants and the animals' relationships with their caretakers.

8.3 Elephant enclosures

8.3.1 Housing and containment

Elephants in zoos have traditionally been kept in an elephant house — or elephant barn — at night, with access to an outside enclosure in the day. Victorian zoos typically kept elephants behind vertical metal railings within reach of the public (Fig. 8.1). Railings were replaced by dry moats (e.g. at ZSL London, Chester, Berlin and Stuttgart zoos), sometimes bounded by small sharp rocks, a low concrete wall or lip on the edge of an 'island' (Fig. 8.2) and later by ha-has and wet moats. At Amsterdam Zoo elephants were previously kept behind a barrier consisting of rows of small concrete pyramids that created an angular, uneven surface. Elephants are also kept behind steel fences (especially indoors) and behind fences consisting of steel posts connected by steel cables. In safari parks elephants were once allowed to roam freely among visitors' vehicles, but are now often kept behind electric fences made of a series of parallel wires suspended just above the surface of the ground.

An important principle of enclosure design for elephants has historically been based on the assumption that they cannot traverse a trench approximately 2 m (6 feet) deep and 2 m (6 feet) wide because they are anatomically incapable of jumping.

A letter published in the *Journal of the Bombay Natural History Society* made reference to the ability of wild Asian elephants to cross trenches (Morris, 1958):

> 6. STRIDE OF ELEPHANT CROSSING TRENCH
>
> On page 933, Vol. 50 of the Journal I recorded elephants crossing a 5½-ft. trench. This month an elephant on two successive nights, in a single stride, crossed and re-crossed an 8-ft. V-shaped trench bordering our raghi farm. From the tracks he would appear to have lurched himself over.
>
> HONNAMETI ESTATE
> ATTIKAN P.O.,   C. MORRIS
> VIA MYSORE (S. INDIA),
> October 29, 1953.

In some parts of the world standards have been published for the design of elephant barriers (Box 8.1). Barriers clearly need to be safe and secure for the benefit of the elephants, zoo staff and visitors. Some barriers can be very dangerous for elephants. In 1974, *Sheba*, an adult Asian cow, was pushed into the concrete dry moat at Chester Zoo, United Kingdom, by one of the other elephants. She was successfully recovered but 4 days later she miscarried a full-term calf (Rees, 2001a).

When I was working as a keeper at Knowsley Safari Park, United Kingdom, in 1976 it was the practice to keep the seven African elephants on a hard standing area outside the elephant house for an hour or so before the park opened and while keepers were cleaning the floor of the house. On one occasion a passing keeper came into the house to

FIGURE 8.1 Adelaide Botanic and Zoological Gardens, South Australia, 1920. Source: *The University of Queensland, Australia, UQ eSpace, Henry William Mobsby Collection, UQFL181, Box 2, Folder 6, photo 157.*

inform us that the elephants were standing in the service road outside. Somehow they had managed to open the gate to the hard standing area and quietly left the premises. At this time there was no impediment to their walking through the car park and onto the dual carriageway into the town of Prescot because the elephant house was located outside the perimeter fence of the rest of the park. Thankfully, the elephants remained together as a group and, having escaped from their enclosure, did not appear to know what to do next.

An elephant escape that occurred in 1975 some 40 km to the south, at Chester Zoo, did not have such a happy ending. *Nobby*, an Asian bull, escaped from the zoo to a nearby residential area. After a failed attempt to tranquillise him he was shot dead by zoo officials (Rees, 2001a).

8.3.2 Early elephant houses

Early elephant houses were utilitarian and often little more than brick-and-concrete shelters and sometimes intended to hold one elephant. Others were more elaborate and decorative.

The first elephant house at Berlin Zoo, Germany, was designed by the architect Gustav Herter and built in 1859. It was castle-like in form with Italian and other exotic elements, and intended to hold a single elephant along with zebras and giraffes. The house at Antwerp Zoo was built like an Egyptian temple (Strehlow, 1996). At the Bronx Zoo in New York, the general director of the New York Zoological Society described their elephant house as 'a palace in praise of elephants and rhinos' (Veltre, 1996) (Fig. 8.3).

FIGURE 8.2 Examples of containment barriers used for elephants: (A) African elephants behind an electric fence at Knowsley Safari Park, United Kingdom; (B) Asian elephant standing near the edge of a dry moat (Berlin Zoo, Germany) 2019. Note the hot wire on the visitor side of the moat; (C) Asian elephant climbing a steel cable fence (Twycross, United Kingdom) 2008; (D) electric fence used to separate visitors' cars from elephants at West Midlands Safari Park, United Kingdom, c.2008; (E) line of artificial trees (Chester Zoo, United Kingdom) 2019; (F) ha-ha (Chester Zoo, United Kingdom) 2019. The vegetation at the top of the wall is protected by a hot wire.

BOX 8.1 Examples of guidelines for the design of elephant barriers in zoos

In India, the Central Zoo Authority produces guidelines for zoo barrier designs. It groups together Asian elephants with other taxa — 'Gaur, Wild Boar, Rhinoceros, Asian Elephant' — and recommends the following (Anon., n.d.):

V-shaped dry moats, or low walls (clay banks), cattle grids (for gaur) or 5 meters away a sunken B.G. Rail Barrier with 1 to 1.5 m high or hot wire fence made in depression, created by excavating earth for camouflaging it from viewer.

It should be noted that Asian elephants are no longer kept in Indian zoos.

In New Zealand, standards have previously been published by the government [MAF Biosecurity New Zealand Standard 154.03.04: Containment facilities for zoo animals (MAF, 2007)]:

14.2 Enclosure standards for Pachyderms (elephant, rhinos, hippos)

14.2.1 Physical containment.

The perimeter of the enclosure shall be either:

1. Vertical and unclimbable, elephant 0.8x mean species body height, rhino shoulder height, hippo body length in height.

OR,

2. Horizontal, elephant and rhino 1.2x mean species body length, and hippo 1.2x mean species body length with a failsafe.

AND

0.8 x body height in depth for all animals above.

Other specifications related to the strength of materials to be used, spacing of railing, engineering performance of doors and so on. This standard was replaced in July 2018 and the new standard makes no specific mention of elephants (EPA, 2018).

In England the *Secretary of State's Standards of Modern Zoo Practice* (SSSMZP, 2017) lay down requirements for elephant barriers:

Boundaries

8.8.27 Barriers must prevent escapes and direct contact with the public and must also ensure the safety and well-being of both the elephants and staff. Methods of quick escape must be provided for keepers.

8.8.28 Barriers and gates should not have horizontal bars, which would allow elephants to climb. The minimum height is 1.9 m for cows and 2.5 m for bulls. A large bull may require a 3 m barrier. Designated safety areas for keeping staff must be clearly identified in line with the management system employed. Stand-off areas must be designed to ensure no contact between elephants and public.

8.8.29 Gates should be robust and any hydraulic system should have manual back-up and/or alternative power. Hydraulic gates must be capable of being operated remotely by staff, i.e. outside the area within elephant reach, and must be able to be opened and closed quickly with a stop facility to ensure trunks/tails are not crushed. The safe operation of any manual gates must be able to be demonstrated to inspectors, including safe systems of work for scenarios where direct access in free contact (FC) situations (see 8.8.51) is not possible.

8.8.30 Electric fences used as a secondary barrier must be of sufficient voltage to deter elephants and must have a failsafe alarm system. Fences should be checked daily and recorded. Electric fences should not normally be used as a

(Continued)

BOX 8.1 (Continued)

main barrier, but where they are, such as to allow access to large grassed areas, suitably trained staff must be present, directly supervising the animals. Moats are not suitable as barriers and should not be used.

In North America the AZA's *Standards for Elephant Management and Care* specify suitable containment (AZA, 2019a):

1.4.9.1   Containment

...A recommended minimum height of walls, cables and horizontal railings for adult elephants is 8 ft (2.4 m). The use of electric fences in not sufficient as a primary containment barrier...The barriers must be safe for the elephants, must be able to withstand an elephant's strength, must contain the elephant in a specific space, and must prohibit direct contact between elephants and visitors. Recommended material for barriers include solid concrete, rock walls or horizontal steel rails, pipe or cable.

1.4.9.3.   Dry moats

...The use of dry moats with steep sides and hard bottoms as primary containment should be limited.

...Dry moats can pose a substantial threat to elephants, especially those out of which an elephant cannot easily climb. Where present, moats should be wide enough for an elephant to turn around, have a soft, dry bottom, and should include a gradually sloped ramp so that the elephant can easily climb out of the moat or ditch.

Even some relatively recently built elephant houses are very basic in design. The house at Knowsley Safari Park, United Kingdom, was constructed in the 1970s and was essentially a single, unstructured space in which elephants were chained to a concrete floor in a line (Fig. 8.4A). Outside, they had access to a concrete hard standing area (Fig. 8.4B).

It used to be common to find elephants chained in elephant houses for long periods, especially at night — and sometimes during the day, particularly in inclement weather — with a chain around one forefoot attached to a metal loop or ring fixed in a concrete floor and a similar chain attached to the diagonally opposite hindfoot. This arrangement only allowed the elephant to take a small number of steps forward (until the back chain was fully extended) and then a few steps backward (until the front chain was pulled tight) and is likely to be the cause of stereotypic behaviour involving repeatedly pacing backward and forward over a short distance (see Section 7.2.8).

The arrangement for restraining the elephants in this manner in the elephant house at Knowsley Safari Park in the 1970s and 1980s is illustrated in Fig. 8.4A. Modern zoos no longer routinely chain elephants at night and individuals are allowed to associate freely, although bulls may be separated off into a bull pen to prevent them from harassing cows.

Perhaps the most famously inappropriate accommodation for elephants constructed in modern times is the Casson Pavilion in ZSL London Zoo. It was designed by Sir Hugh Casson and constructed between 1962 and 1965 (Fig. 8.5). The rough texture of the concrete walls resembles an elephant's hide — an architectural style that came to be known as brutalist — and the inside accommodation for the elephants was illuminated by natural light entering through the roof via large green lanterns. In 1965 the building was recognised as the best built in London that year by the Royal Institute of British Architects (RIBA). London Zoo no longer keeps elephants.

In the 1930s Lubetkin and the Tecton Group were responsible for the design of a number of iconic buildings in British zoos including the Round House (for gorillas) and the

FIGURE 8.3 The elephant house at the Bronx Zoo, New York, c.1908. Source: *From the New York Public Library, Universal Unique Identifier (UUID): a6da6810-c560-012f-be85-58d385a7bc34. https://digitalcollections.nypl.org/items/510d47e2-8ea3-a3d9-e040-e00a18064a99.*

Penguin Pool at ZSL London Zoo, and bear pits and an elephant house at Dudley Zoo. None of these structures is now used for the species for which they were designed. In fact, most of Lubetkin's zoo buildings are now considered unsuitable for the animals for which they were intended, but have been preserved as important examples of the architecture of their time. At ZSL Whipsnade Zoo, Lubetkin built another elephant house that is no longer used for elephants but is a Grade II* Listed Building (see Box 8.2). An elephant keeper at ZSL London Zoo was killed by one of his charges in 2001. Shortly thereafter the three resident Asian elephants were moved to better facilities at Whipsnade Zoo in Bedfordshire as part of a plan to breed from the animals.

The architecture of some of the best known elephant houses built in Europe has been described by Haywood (2014) (Table 8.3).

In Boston, Massachusetts, an elephant house was built at the Franklin Park Zoo in 1910—12 and demolished in 1978 (Fig. 8.6). Its basic structure has been recorded by the Historic American Buildings Survey (Fig. 8.7). Many zoo buildings have been demolished over time to make room for new exhibits. However, others have been repurposed. In Auckland Zoo, New Zealand, the Old Elephant House is now a stylish restaurant; in London Zoo the Casson Pavilion is used to exhibit tapirs.

In the United States between 1935 and 1937 a new pachyderm house was built at the National Zoo in Washington DC (Fig. 8.8). It was designed by the Chicago architect Edwin H. Clark and constructed largely from poured-in-place concrete, rubble-stone masonry, steel supports and cut limestone. In addition to elephants, the rectangular building was designed to house rhinoceroses (Rhinocerotidae), pygmy and Nile hippopotamuses

FIGURE 8.4 The elephant house at Knowsley Safari Park, United Kingdom, was constructed in the 1970s and later extended and modernised. (A) African elephants chained by one forefoot and the diagonally opposite hindfoot in the elephant house, c.1990. The house was not open to the public. (B) Exterior view of the elephant house and the hard standing area outside bounded by a dry moat and boulders on one side.

(*Choeropsis liberiensis* and *Hippopotamus amphibius*), giraffes (*Giraffa camelopardalis*) and tapirs (Tapiridae). The exhibit was the first in the zoo to adopt the contemporary methods of defining outdoor yards using moats that were popular at that time. The design of the

FIGURE 8.5 The Casson Pavilion, ZSL London Zoo. (A) The exterior of the elephant house and part of the outdoor enclosure. (B) The barrier between the elephant area and the public area. (C) One of the indoor elephant areas illuminated from above by natural sunlight. (D) Outside wall of the building constructed of concrete textured to resemble elephant skin, in a brutalist style. The building was given an award by the Royal Institute of British Architects in 1966. These photographs were taken in 2008 after the elephants had been moved to ZSL Whipsnade Park and the building had been repurposed to hold other species.

building used historicalism (classicism) as the source of its architectural expression, and it is a testament to the increased emphasis on the use of outdoor spaces for healthy recreation designed to mitigate the impoverished circumstances of so many people as a result of the Great Depression. In September 2010 the Smithsonian's National Zoo opened the first phase of its new $52 million immersive elephant exhibit *Elephant Trails*. The finished project includes the renovation of the old elephant house and a new elephant barn.

8.3.3 Enclosure size and substratum

In 1255 King Louis IX of France gave King Henry III of England an African elephant as a gift. On 26 February 1255, Henry ordered the sheriffs of London to have a building of dimensions 40 feet (12 m) long and 20 feet (6 m) wide made at the Tower of London

BOX 8.2 The text from the Historic England entry for the elephant house at ZSL Whipsnade Zoo and part of the entry for the house at ZSL London Zoo

Some zoo buildings are protected for their architectural merit and this may prevent them from being demolished or substantially changed long after they cease to house the kind of animals for which they were originally intended. The elephant houses at ZSL London Zoo and ZSL Whipsnade Zoo in the United Kingdom have been listed under the Planning (Listed Buildings and Conservation Areas) Act 1990 (as amended) for their special architectural or historic interest.

5125 WHIPSNADE WHIPSNADE ZOO
SP 91 NE 19/6 Elephant House
II* Elephant house. 1935. Designed by Lubetkin and Tecton. Reinforced concrete. 1 storey. Long low composition with bowed front and light cantilevered canopy to full length, glazed beneath to centre (glazing altered late C20) and open, supported by pilotis to each side. To rear, projecting above canopy, are 4 cylindrical cubicles to house elephants, each one top-lit through circular opening at apex of low-pitched metal covered roof. Built to house young elephants only.
Lubetkin and Tecton Coe and Reading
Architect and Building News 8-9-1935
Architects' Journal 16-1-1936
Listing NGR: TL0031617319

1900/5/10104 Elephant and Rhinoceros Pavilion London Zoo
GV II*
Animal house. Designed 1961, built 1962—65 by Sir Hugh Casson, Neville Conder and Partners; Jenkins and Potter engineers; landscape architect Peter Shepheard; in consultation with F A P Stengelhofen, architect to the Zoological Society of London. Reinforced concrete, with ribbed walls in three main pours — the rubbed texture designed to prevent the animals rubbing up against them and injuring either party — and inner brick skin; conical copper roofs. The complex is set within walls, ditch and raised paddock enclosure of purple brick, with purple brick paviours and plinth. Internally the animal pens are lined in mosaic, whilst the public spaces are spanned by laminated wood beams set in metal shoes. The tough finishes are carefully considered, tactile and well detailed, and are an important component of the building's exceptional quality.

Complex concave form designed to resemble 'animals drinking at a pool', the pavilion was designed to house four elephants and four rhinoceroses in paired pens, each with access to sick-bay pens and to moated external paddocks. Heavy double doors to entrance, reached via steps for public; slit windows to staff areas only, the others incorporated as skylights. The pens are arranged round a central hall for the public, who circulate through the building at a slightly lower level on an S-Shaped route, with sunken viewing area and integral fixed benches between them and the animals. The pens themselves are circular, reflecting the turning movements of the animals and avoiding sharp corners or spaces difficult to clean. The pens are top-lit, the funnels combining light with extract fans, and the public spaces kept dark; the effect has been likened to a theatrical cyclorama. Elephant bathing area and keeper's mess room — the latter said to be the only area whose form is not the logical consequence of functional requirements.

https://historicengland.org.uk/listing/the-list/list-entry/1235411 and https://historicengland.org.uk/listing/the-list/list-entry/1323694

TABLE 8.3 Some historically and architecturally important European elephant houses.

Zoo	Year of construction	Building	Architect	Architectural style
London, United Kingdom	1831	Elephant stables	Decimus Burton	Hybridised blend of imperial and subaltern indigenous architectures
London, United Kingdom	1867	Elephant and rhinoceros house	Anthony Salvin Jr	English Gothic Revival
Berlin, Germany	1873	Elephant house	Ende und Böckmann	Indian pagoda
Whipsnade, United Kingdom	1935	Elephant house	Berthold Lubetkin/ Tecton	Modernist
Dudley, United Kingdom	1935–37	Elephant house	Berthold Lubetkin/ Tecton	Modernist
London, United Kingdom	1962–65	Elephant and rhino pavilion	Sir Hugh Casson	Brutalist
Copenhagen, Denmark	2008	Elephant house	Sir Norman Foster	Postmodernist

Source: *Based on information in Haywood, M., 2014. Human places for large non-humans: from London's imperial elephant stables to Copenhagen's postmodern glasshouse. In: Pauknerova, K., Stella, M., Gibas, P., et al. (Eds.), Non-Humans in Social Science: Ontologies, Theories and Case Studies. Pavel Mervart, Červený Kostelec, Czech Republic, pp. 201–218.*

FIGURE 8.6 Derelict elephant house in Franklin Park Zoo, Boston, Massachusetts, constructed in 1910–12 and demolished 1978. Source: *US Library of Congress. Reproduction Number: HABS MASS,13-BOST,68B--3.*

menagerie to house his elephant (Calendar of Liberate Rolls 1251–60, 197). The elephant was the first to be exhibited in England and was popular with the public. It had a dedicated keeper but the animal died after a couple of years (Fig. 1.3).

An enclosure of 72 m² is obviously no size for an elephant. However, there is no objective method for determining the appropriate size for an elephant enclosure in a zoo. The size of

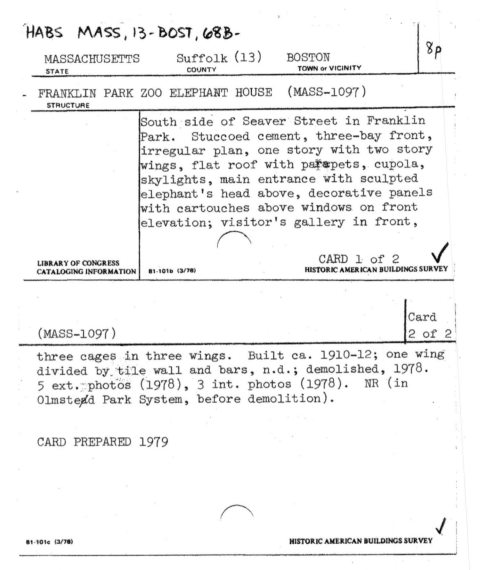

FIGURE 8.7 Record cards of the Historic American Buildings Survey, for Franklin Park Zoo elephant house. Source: *US Library of Congress Cataloging Information, 81-101b and c (3/78)*.

the four circular 'silos' where elephants were kept in Lubetkin's elephant house at ZSL Whipsnade Zoo was apparently based on German research that suggested that, when deprived of space, elephants walked endlessly in circles so there was no point in building them large enclosures. Lubetkin (mistakenly) believed he was moulding his concrete walls around the animal's behaviour in captivity (Baratay and Hardouin-Fugier, 2002). The

FIGURE 8.8 The old elephant house at the National Zoological Park, Washington DC, constructed in 1935—37 and designed by Edwin H. Clark: (A) The east entrance; (B) The view looking northeast from elephant yard number 1; (C) The interior (D) Indoor elephant accommodation. Images created in 2006. Source: *(A) US Library of Congress. Reproduction Number: HABS dc-777-C-7; (B) US Library of Congress. Reproduction Number: HABS dc-777-C-16; (C) US Library of Congress. Reproduction Number: HABS dc-777-C-22; (D) US Library of Congress. Reproduction Number: HABS dc-777-C-24.*

elephants were prevented from standing near the edge of the raised floor of the silo by two rows of metal 'teeth'.

Clearly, a small circular space is inadequate for an elephant. However, Hutchins (2006) has argued against using '"nature" as a yardstick for gauging the adequacy of zoo animal management and care' in relation to elephants because of the flexibility of their ecology and behaviour. It is clearly unrealistic and inappropriate to use the home range size of wild elephants as a basis for this because home range size is flexible and depends upon resource availability. De Beer and van Aarde (2008) have shown that wild elephants in arid savannahs in southern Africa locate their home ranges in areas of high landscape heterogeneity and that home range size decreases with increasing density of water points. Similarly, the quality of the captive environment is likely to be more important than its size. Elephants living in zoos do not need to move long distances to find food or water — a large homogeneous barren enclosure will offer fewer opportunities for elephants to exhibit natural behaviours than will a smaller heterogeneous enclosure.

TABLE 8.4 Examples of minimum standards for the dimensions of indoor and outdoor spaces for elephants.

Organisation	Outdoor	Indoor	Source
AZA	1800 ft^2 (167.4 m^2) for 1 elephant; increased by 50% for each additional animal.	Male or female with calves: not less than 600 ft^2 (56 m^2) Female: not less than 400 ft^2 (37 m^2)	Tuttle et al. (1997). *Minimum Husbandry Guidelines for Mammals: Elephants*. American Zoo and Aquarium Association.
AZA	Not less than 5400 ft^2 (500 m^2)/elephant	Male or female with calves: not less than 600 ft^2 (56 m^2) Female: not less than 400 ft^2 (37 m^2)	AZA *Standards for Elephant Management and Care* (AZA, 2012)
BIAZA	2000 m^2 with an additional 200 m^2 for every additional cow (over 2 years old) over a herd size of 8 females. Area designed for cows and bulls should be not less than 3000 m^2. Bull pen should be not less than 500 m^2.	200 m^2 for 4 animals; 50 m^2 for each additional elephant (over 2 years old). Minimum herd size taken as 4 females (over 2 years old). Bull stall must be at least 50 m^2.	*Management Guidelines for the Management of Zoo Animals: Elephants* 2nd Edition (Stevenson and Walter, 2006)
BIAZA	2000 m^2 with an additional 200 m^2 for every additional cow (over 2 years old) over a herd size of 8 females. Area designed for cows and bulls should be not less than 3000 m^2 and should allow for separation if necessary. Bull pen should be not less than 500 m^2.	200 m^2 for 4 animals; 80 m^2 for each additional elephant (over 2 years old). Minimum herd size taken as 4 females (over 2 years old). Bull stall must be at least 80 m^2.	*Management Guidelines for the Management of Zoo Animals: Elephants* 3rd edition (Walter, 2010)

Zoo managers have published a number of different minimum standards for the dimensions of indoor and outdoor spaces for elephants, in spite of the lack of evidence to justify them (Table 8.4; Fig. 8.9). This is not an uncommon practice for other species kept in zoos and, indeed, for some domestic and farm animals. In 2009, the Welsh Government published a *Code of Practice for the Welfare of the Rabbits (2009, No. 44)*, under the Animal Welfare Act 2006, in which it recommends that the minimum length of a rabbit hutch should allow the animal to make three or four hops from one end to the other. But why is the distance covered in two hops too short and that covered in five hops more than adequate?

In the third edition of its elephant welfare guidelines, BIAZA increased the recommended minimum indoor space allocation per elephant from 50 to 40–80 m^2 citing evidence from Harris et al. (2008) that individuals with this amount of space exhibited less stereotypic behaviour than those with less space (Walter, 2010). However, the relationship between indoor space and stereotypic behaviour was unclear because elephants in enclosures with more than 80 m^2 per animal (defined by Harris et al. as 'large') exhibited on average four times as much stereotypic behaviour as those in the 'medium' sized indoor

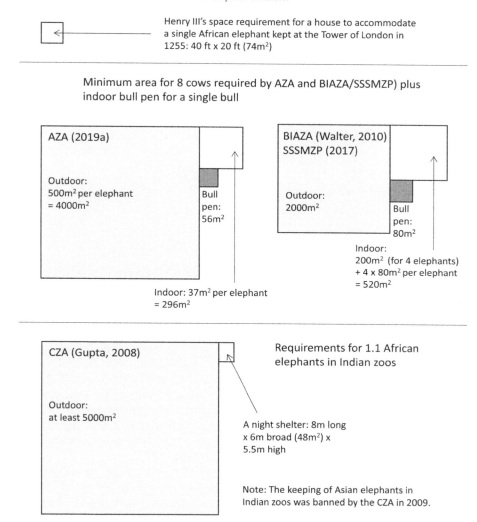

FIGURE 8.9 Examples of standards for enclosure size for elephants living in zoos.

enclosures (40–80 m^2). This suggests that there is no simple relationship between the frequency of stereotypic behaviour and the size of indoor accommodation.

A survey of AZA elephant facilities in North America conducted in 2006 found that the median area available exceeded the recommendations of AZA's *Standards for Elephant Management and Care* published in 2001: 755 ft^2 (230 m^2) per elephant indoors and 10,000 ft^2 (3048 m^2) outdoors (AZA, 2001; Lewis et al., 2010). AZA standards currently specify a minimum outdoor area of 5400 ft^2 (500 m^2) per elephant and indoor area of not less than 600 ft^2 (56 m^2) for males or females with calves, and not less than 400 ft^2 (37 m^2) for females (AZA, 2019a).

Although zoo visitors have no objective means of determining if enclosures for elephants are adequate, Miller et al. (2018) found that visitors' perceptions of the welfare of elephants at nine US zoos were correlated with the exhibit characteristics of being spacious, natural and well maintained.

In the state of Kerala, India, the Kerala Captive Elephant (Management and Maintenance) Rules, 2012 specify the minimum size required for the housing of captive elephants:

3. Housing of elephants.— (1) The owner shall provide a stable (tethering place) in a clean and healthy environment with sufficient shade to keep elephants during its rest period.

(2) Each elephant must be ensured a minimum floor area as specified below:

(i) Weaned Calf (height below 1.50 m)	5 m x 2.5 m
(ii) Sub-adult elephant (height 1.50 m to 2.25 m)	7 m x 3.5 m
(iii) Adult elephant (height above 2.25 m) and cow elephant with unweaned calf	9 m x 6 m

(3) In the case of covered sheds, the height of the structure shall not be less than 5.5 m.

It is interesting to note that the 54 m^2 of space allocated to captive adult elephants in Kerala is 581 ft^2, considerably more space than the minimum indoor space specified for an adult female elephant by the AZA in North American zoos (i.e. 400 ft^2).

Elephant house flooring in older elephant houses is usually concrete or tiles (Fig. 8.10A). Modern houses have rubber flooring or sand (Fig. 8.10B). A number of studies have examined the type of substrates that elephants chose when offered a choice (see Section 6.5.3). Flooring can affect foot health and behaviour in elephants (see Section 7.7.3). Appropriate natural substrates outdoors encourage exercise, especially if landscaped (Fig. 8.11).

AZA standards require outdoor areas to consist of a variety of primarily natural substrates (e.g. soil, sand and grass) that provide good drainage (standard 1.4.5.1; AZA, 2019a). Indoor substrates must be capable of being cleaned daily and must dry quickly (standard 1.4.5.2). The floor must be smooth enough to prevent excessive foot pad wear, but not so smooth that it may become slippery when wet.

In England, standard 8.8.25 of Appendix 8 of the SSSMZP for outdoor substrates is effectively identical to the AZA requirements: primarily natural substrates (sand, soil or grass) with good drainage. For indoor enclosures, sand at least 1 m deep is suggested, but other materials such as mud, clay or bark chip are also considered suitable (SSSMZP, 2017).

Water is an essential feature of elephant exhibits (Figs 8.12 and 8.13). AZA standards require that elephants should have access to pools, waterfalls, misters/sprinklers or wallows for the purposes of enrichment, bathing and to allow them to cool their bodies (standard 1.5.2; AZA, 2019a). Pools should be constructed with round edges and no corners. Artificial pools should have more than one point of entry or exit, each of which should be gently sloping (no more than 30°) and have nonslip surfaces. Vertical pool sides should be avoided where they are accessible to elephants. If steps are present, they should be wide enough for adult elephants to place more than one foot on at a time and small enough for baby elephants to step up and down. The AZA also recommends that one body of water should be deep enough to allow buoyancy (thereby allowing nonweight-bearing exercise) and deep enough for an elephant to immerse itself while lying on its side, or at least 6 ft deep (1.8 m). It also encourages the provision of shallow wading and splashing pools.

FIGURE 8.10 Old and new substrates: (A) The old elephant house at Twycross Zoo, United Kingdom, had a tiled floor. It closed in 2019. (B) The floor of the new elephant house at Dublin Zoo, Ireland, is covered in deep sand.

8.3.4 New enclosures

The 40 ft x 80 ft elephant house built in the Tower of London menagerie by the sheriffs of London in 1255 cost just over £22 to construct, according to the Pipe Rolls kept by the Exchequer (Pipe roll E 372/99, rot. 14). Modern elephant houses are somewhat larger, more elaborate and much more costly.

The past 30 years (1989–2019) have seen many new elephant exhibits built in zoos around the world (Table 8.5). Over the same period many zoos have stopped keeping

FIGURE 8.11 A landscaped outdoor enclosure with a sand substrate encourages exercise.

elephants, either because their last elephant died or because they recognised that their accommodation did not meet modern standards so translocated their elephants to another zoo or elephant sanctuary.

Many modern elephant exhibits are immersive, creating the illusion of a natural habitat (Figs 8.14 and 8.15). Such exhibits are expensive and inevitably include a great deal of cost that does not directly benefit the animals (e.g. elaborate interpretation and elephant-themed children's play areas). Other exhibit designs have concentrated on providing a large amount of mostly unstructured space, such as Blackpool Zoo, United Kingdom (Fig. 8.16).

8.4 The cost of keeping elephants under good welfare conditions

Sach et al. (2019) have estimated the cost of keeping elephants under optimum welfare conditions by analysing the cost of keeping a breeding herd of Asian elephants at ZSL Whipsnade Zoo. They estimated the annual cost to be £593,021–641,863 (US $749,697–811,443) per year. This excluded indirect staffing costs, ground rent and the cost of contributions to field conservation projects and represented up to approximately 1.3% of the funds spent on the ZSL's animal collections (ZSL London Zoo and ZSL Whipsnade Zoo) and conservation (excluding field conservation) in 2016–17 (ZSL, n.d.). Sach et al. also calculated the investment required to comply with improvements to elephant husbandry required by the *Secretary of State's Standards of Modern Zoo Practice* (SSSMZP, 2017).

The demands of recent changes in elephant husbandry are likely to be expensive for some zoos as they adapt or replace their accommodation. In England, changes were made

8.4 The cost of keeping elephants under good welfare conditions

FIGURE 8.12 Waterfall in the *River's Edge* Asian elephant exhibit at St Louis Zoo, Missouri, United States. Source: *Les Rees*.

FIGURE 8.13 Indoor pool in Berlin Zoo's elephant house in Germany (2019).

TABLE 8.5 Examples of new elephant exhibits opened in zoos between 1989 and 2019.

Exhibit/zoo	Taxon[a]	Cost	Date opened	Area
Adventure Africa, Milwaukee County Zoo, United States	*Loxodonta*	$16,600,000	2019	6475 m^2 outside/ 1858 m^2 inside
Project Elephant Base Camp, Blackpool Zoo, United Kingdom	*Elephas*	£2,500,000	2018	12,141 m^2
Elephant exhibit, Artis Zoo, The Netherlands	*Elephas*	—	2017	—
Centre for Elephant Care, ZSL Whipsnade Zoo, United Kingdom	*Elephas*	£2,000,000 (building extension only)	2017	121,406 m^2 outside
Elephant exhibit, Audubon Zoo, New Orleans, United States	*Elephas*	$10,000,000	2016	4047 m^2 outside/ 520 m^2 inside
Kaeng Krachan Elephant Park, Zoo Zürich, Switzerland	*Elephas*	CHF 57,000,000	2014	11,000 m^2
Elephant Eden, Noah's Ark Zoo, United Kingdom	*Loxodonta*	>£2,000,000	2013	81,000 m^2
Elephant Trails, Smithsonian's National Zoo, Washington DC, United States	*Elephas*	$52,000,000	2010	8100 m^2
Realm of the Giants, Dierenpark Amersfoort, The Netherlands	*Elephas*	€2,900,000	2010	3100 m^2 outside/ 841 m^2 inside
Elephant exhibit, Heidelberg Zoo, Germany	*Elephas*	€4,650,000	2010	1600 m^2 outside/ 670 m^2 inside
Elephant Odyssey, San Diego Zoo, United States	*Loxodonta* and *Elephas*	$45,000,000	2009	10,117 m^2 outside only
Watani Grasslands Reserve, North Carolina Zoo, United States	*Loxodonta*	$7,237,169	2008	28,328 m^2 outside/ 1208 m^2 off-exhibit
Elephant exhibit, Copenhagen Zoo, Denmark	*Elephas*	—	2008	—
Giants of the Savanna, Dallas Zoo, United States	*Loxodonta* plus ungulates	$32,000,000	2008	26,466 m^2 outside/ 2416 m^2 inside
Elephants of the Asian Forest, Chester Zoo, United Kingdom	*Elephas*	£3,000,000	2006	6670 m^2 outside/ 1847 m^2 inside
Trail of the Elephants, Melbourne Zoo, Australia	*Elephas*	AUS$13,000,000	2003	3900 m^2
Elephant exhibit, Cologne Zoo, Germany	*Elephas*	—	c.2004	
Wild Asia's Elephants, Taronga Conservation Society, Australia	*Elephas*	AUS$31,800,000	2002–05	7000 m^2
Africa Hall, Dresden Zoo, Germany	*Loxodonta* and *Elephas*	DM12,800,000	1999	550 m^2 inside only

(Continued)

TABLE 8.5 (Continued)

Exhibit/zoo	Taxon[a]	Cost	Date opened	Area
River's Edge, St Louis Zoo, United States	*Elephas*	$3,000,000	1999	1208 m^2 house only
Elephant Park, Vienna Zoo, Austria	*Loxodonta*	ATS80,000.000	1996	4600 m^2 outside/ 2100 m^2 inside
Masai Mara's Elephants, Zoo Atlanta, United States	*Loxodonta*	$1,399,916	1989	1371 m^2 outside/ 370 m^2 inside
Asian Elephant Forest, Woodland Park Zoo, Seattle, United States	*Loxodonta* and *Elephas*	$6,000,000	1989	5076 m^2 outside/ 204 m^2 inside

[a]*Taxa present when the source record was published.*
Source: Zoolex, 2018. Zoolex Zoo Design Organization. <https://www.zoolex.org/> (accessed 07.09.19.).; individual zoo websites; Rees, P. A., 2011. Introduction to Zoo Biology and Management. Wiley-Blackwell, Chichester, West Sussex; Rübel, A., 2014. Kaeng Krachan Elephant Park at Zoo Zürich. Gajah 40, 44–45.

FIGURE 8.14 St Louis Zoo – *River's Edge* Asian elephant exhibit at St Louis Zoo, Missouri. Source: *Les Rees*.

to the SSSMZP. An appendix (Appendix 8.8) specifically relating to elephants was added in June 2017 to bring elephant management standards in line with best practice. The SSSMZP requires facilities that hold elephants to produce long-term management plans spanning a minimum of 30 years, which is longer than most zoo masterplans and reflects elephants' longevity. In reality, provision must be made for the elephants' entire lives (Sach et al., 2019). The elephant facility at ZSL Whipsnade Zoo has been estimated to

FIGURE 8.15 The *Kaziranga Forest Trail* elephant exhibit at Dublin Zoo, Ireland. (A) The elephant-themed children's play area. (B) An impression of a leaf from a tropical plant in the footpath. (C) A waterfall visible from the footpath but inaccessible to the elephants. (D) A viewpoint. (Inset: the elephants are separated from the public by a wet moat.) (E) The entrance path obscures the elephant house and the main enclosure and leads the visitor through lush vegetation alongside an artificial stream and waterfall.

require an investment of at least £7—8 million (US$9—10.3 million) for a breeding herd of Asian elephants to be expanded and maintained, and if a state-of-the art visitor experience is included this cost could rise to up to £28 million (US$36 million).

As elephant welfare standards continue to increase in line with modern best practice zoos in the United Kingdom (and elsewhere) must decide whether to improve their facilities or stop keeping elephants. In 2018—19 Twycross Zoo sent its Asian elephants to a new facility at Blackpool Zoo — to join their single remaining female — and does not intend to keep elephants in the future. The African elephants from Knowsley Safari Park were moved to ZooParc de Beauval, France, in 2017 while the park builds a new elephant house and redevelops the elephant paddock.

FIGURE 8.16 Outdoor enclosure of the new elephant exhibit at Blackpool Zoo, United Kingdom (2019).

In spite of predicted future costs, Sach et al. (2019) claim that at ZSL Whipsnade there is a clear argument for continuing to keep elephants on commercial grounds and because:

> ...*there is certainty* [sic] *that these animals provide substantial opportunities for education, conservation research and fundraising.*

8.5 Elephants and their caretakers

8.5.1 Keeper−elephant bonds

Many elephant keepers form strong bonds with their charges and these bonds contribute significantly to their job satisfaction. A strong bond has welfare benefits for keepers and elephants. A survey that examined 427 keeper−elephant pairings in North American zoos found that dissatisfaction with the job was associated with weaker keeper−elephant bonds and that stronger keeper−elephant bonds were more common with Asian than with African elephants (Carlstead et al., 2018).

A baby elephant who is rejected by its mother and subsequently bottle-fed may imprint on a keeper and follow the keeper around the enclosure instead of its mother. At Chester Zoo such a strong attachment developed between the head elephant keeper and a rejected female calf, called *Karha*, that she would even follow him into his office (Fig. 8.17). The same keeper was regularly greeted by some members of the herd when he entered their enclosure (Fig. 8.18).

FIGURE 8.17 A hand-reared calf, *Karha*, at Chester Zoo, United Kingdom, following keepers around the enclosure.

In some zoos interaction with keepers and trainers may account for a considerable amount of time. An analysis of the activity budgets of Asian elephants at Busch Gardens Tampa Bay, Florida, found that trainer interaction accounted for 4%—23% of the time during the daylight hours (Lukacs et al., 2016). When keepers are treated by Asian elephants as part of the herd the animals experience lower serum cortisol levels (Carlstead et al., 2018). This is also the case when elephants (of either species) interact with the public.

8.5.2 Traditional elephant expertise versus zoo husbandry

In Western zoos new elephant keepers are trained by more experienced keepers. Turnover is generally high, especially among young keepers. Much of the practical knowledge of elephant husbandry resides in a relatively small number of experienced keepers. This knowledge is easily lost because there is no guarantee that it will be transmitted to suitable, less-experienced keepers who will remain in the profession. An AZA survey found that 78 AZA elephant-holding facilities employed the equivalent of 382 elephant care staff with a combined experience of over 3880 years working with elephants (AZA, 2007b).

There is an important repository of husbandry knowledge within some indigenous populations in Southeast Asia. A study of 16 mahouts and 4 assistants working with elephants in Nagarahole National Park, Karnataka, India, found that head mahouts had a

FIGURE 8.18   Two female Asian elephants at Chester Zoo, United Kingdom, performing a greeting ceremony on seeing the head elephant keeper in their enclosure.

mean of 27.2 years of experience with elephants and assistant mahouts had a mean of 13.8 years working with elephants. Most mahouts were second- or third-generation mahouts (Hart and Sundar, 2000) (Fig. 8.19).

In India, mahouts have a long apprenticeship and the occupation traditionally passes from father to son. Family tradition plays a major role for Hindus, Muslims and tribal people (honey-gathering tribes) in the decision to work with elephants. Mahouts' sons regularly assist in the management of female elephants, and expect to become mahouts themselves (Hart and Sundar, 2000). Mahouts attribute their success in managing elephants to the time they invest in caring for, and becoming familiar with, particular elephants (Hart, 1994).

The traditional knowledge of mahouts, passed from generation to generation over thousands of years, may be of greater long-term value to the survival of the Asian elephant than the husbandry knowledge of western zookeepers, which, at best has only been accumulated over perhaps a century. However, traditional mahout knowledge transfer appears to be under threat.

Crawley et al. (2019) studied the handling system of semicaptive elephants in Myanmar by interviewing 210 mahouts. Traditionally mahouts remained with the same elephant for many years. This study found that the median age of mahouts was 22 years and median length of experience was 3 years. Therefore, current mahouts change elephants much more frequently than in the past and the transfer of traditional knowledge is under threat. Crawley et al. concluded that there is a need for formal training of mahouts and that the effect of the changes in age, experience and commitment of mahouts on elephant welfare

FIGURE 8.19 Mahout riding his bull elephant at a Forest Department elephant camp in Karnataka, India.

should be assessed. They emphasise the global importance of the Myanmar population (around 5000 elephants), noting that around a third of all Asian elephants live in semicaptive conditions in range states.

A study conducted in the Lao People's Democratic Republic (Laos) surveyed 113 mahouts and found that mahoutship was an ageing industry and that the family associations within the profession were vanishing (Suter et al., 2013). Logging mahouts relied on tourism to provide future employment, but this will not absorb all of the mahouts formerly employed in the logging industry. The authors of the study concluded that the need for traditional mahouts and a large population of captive elephants in Laos was rapidly disappearing.

In Tamil Nadu, India, Vanitha et al. (2010) found that the retention of mahouts was related to salary. Retention was highest in the forest departments, where mahouts received the highest salary and benefits, and lowest in Hindu temples, where mahouts received the lowest salary and few benefits. The authors suggested that the retention of traditional mahout talent could be improved by improving their income and providing better welfare benefits.

8.5.3 Keeper and mahout deaths

Keepers in western zoos and mahouts in Asia are frequently killed by elephants. In zoos, aggression towards keepers may be the result of the failure to establish long-lasting keeper—elephant relationships due to the high rate of keeper turnover in some facilities.

Elephant keeping is extremely dangerous. The occupation of 'elephant handler' has the highest fatality rate of any occupation documented by the US Department of Labor Statistics. One study documented 15 elephant-related deaths between 1976 and 1991, approximately one death per year (Lehnhardt, 1991). In some years two elephant keepers have been killed. Based on employment figures of about 600 known elephant trainers in the United States, this would produce a fatality rate of 333 per 100,000 workers compared with 97.4 in aircraft pilots and 16.6 in police employees. This high rate is misleading because the actual number of keeper deaths is low and zero in some years (Toscano, 1997).

Benirschke and Roocroft (1992) analysed data on 36 Asian elephants responsible for serious accidents involving keepers and concluded that bulls were significantly more dangerous than cows. The mean age of bulls involved in accidents was 18 years and the mean age of cows was 25.3 years. These ages coincide approximately with the age when bulls first come into musth and cows become matriarchs (Kurt, 1995).

There has been considerable concern about the safety of elephant keepers in recent years. Most attacks occur when a keeper is in direct contact with an elephant and the injuries caused appear to be deliberate rather than accidental. Many recorded attacks are serious, with 28 out of 119 incidents (23.5%) resulting in the death of a keeper (Gore et al., 2006). Increased use of protected contact (PC) management techniques is likely to reduce the incidence and severity of attacks (Roocroft, 2007).

Case reports of deaths of zoo staff caused by elephants occasionally appear in scientific journals. Hejna et al. (2011) reported a fatal attack of a cow elephant on a male elephant keeper. After the keeper tripped over a foot chain while the elephant was being medically treated, she repeatedly attacked him with her tusks. He died in hospital after sustaining multiple penetrating injuries to the groin and abdomen, prolapse of the small intestine, laceration of the abdominal aorta, multiple rib fractures, contusion of both lungs, laceration of the liver and comminuted fractures of the pelvic arch and left femoral body.

In a second case, a female ex-circus African elephant at the Franklin Zoo and Wildlife Sanctuary in Tuaku, New Zealand, became agitated after being shocked by an electric fence in her enclosure and advanced towards a female veterinarian who consequently fell over (Vuletic and Byard, 2013). The elephant picked the woman up by wrapping her trunk around her torso, held her in the air and carried her for approximately 5 m. While rocking her head from side to side the elephant kneeled down and put the veterinarian on the ground. After her release the victim could not be resuscitated.

Between January 1987 and June 2019 People for the Ethical Treatment of Animals (PETA) recorded elephant incidents in North America that resulted in 20 human deaths and over 140 injuries (PETA, 2019a).

In India, wild elephants are responsible for large numbers of human deaths. Dey (1991) reported 476 deaths in North Bengal from 1980 to 1990 and in 1992, 29 persons were killed by elephants in the Western Dooars (Barua, 1995). Datye and Bhagwat (1995) analysed data on 208 deaths caused by elephants in south Bihar and south-west Bengal between 1980 and 1991. Of the 53 persons for which data were available, only 12 (22.6%) were killed by bulls. However, over 80% of more than 150 incidents analysed by Sukumar (1994) involved bulls.

Mahout deaths in Tamil Nadu, India, have been studied by Vanitha et al. (2010) in three elephant management systems – private, Hindu temple and forest department. They found that cases of human injury caused by elephants were low in forest departments (0.02 incidents/elephant per year), where 68% of the elephants were bulls. In contrast, 95% of temple elephants were female, but temple mahouts experienced the highest frequency of manslaughter cases (0.01 deaths/elephant per year). Vanitha et al. attributed

FIGURE 8.20 Protected contact training. The trainer is holding two training poles. The one in his right hand is indicating where the elephant should stand and the one in his left hand is indicating where he should put his right foot. Source: *Rees, P.A., 2011. Introduction to Zoo Biology and Management. Wiley-Blackwell, Chichester, West Sussex.*

this to a lack of 'traditional compassion and kindness'. They also found that the number of mahouts per elephant was lower than required in the temple system.

8.5.4 Free versus protected contact

There are two methods of handling elephants, free contact (FC) and PC. In FC keepers work directly with the elephants and enter their enclosures. In PC keepers are separated from elephants by barriers and do not share the same unrestricted space (Figs 8.20 and 8.21). The advantages and disadvantages of both methods are listed in Table 8.6.

Direct contact with an elephant in restraints allowed within a PC system is referred to as 'restricted contact' and should only be undertaken after a risk assessment has been completed. Roocroft (2007) has discussed the implementation of PC and the necessary training for keepers and elephants.

There appears to be some experimental evidence that elephants managed under PC experience better welfare than those managed under FC. The extent of behavioural compliance by elephants while being bathed under FC and protected content management systems, along with the rates of reinforcement applied, were compared by Wilson et al. (2015a). Positive reinforcement (food) was delivered on average nearly eight times more frequently in the PC condition compared with the FC condition. The mean rate at which the ankus was used (negative reinforcement) in FC was similar to the mean rate of

FIGURE 8.21 Protected contact walls. (A) A barrier designed to appear as if it is constructed from bamboo inside the old elephant house, Blackpool Zoo, United Kingdom. (B) A barrier at the perimeter of the outdoor enclosure at Elephant Eden, Noah's Ark Zoo Farm, United Kingdom.

TABLE 8.6 Some characteristics of free and protected contact systems.

Free contact	Protected contact
Keepers able to intervene in herd conflicts	Safer for keepers
Keepers able to promote exercise	Makes it more difficult for keepers to interact and bond with elephants
Keepers may attend to newborn calves	Gives elephants more choice and control over their environment
Keepers may access elephants for emergency veterinary care	Does not require keepers to dominate the elephants
No requirement for investment in specialist protected contact fences, etc.	Does not use negative reinforcement
Greater compliance with commands	

application of positive reinforcement under FC. Latencies between verbal commands and the required behaviours were generally longer, and 'refusals' (noncompliance) higher in the PC condition than under FC. The authors of the study concluded that these long latencies and higher refusal rates under PC indicated that the elephants were exercising choice — control over their environment — and thus experienced improved well-being compared with that experienced by elephants under FC.

Proctor and Brown (2015) found no difference in mean serum cortisol concentrations in 112 female elephants (58 African and 54 Asian) managed under FC ($n = 58$) and PC ($n = 54$) conditions in 48 facilities in North America. Their analysis indicated that it may be more important to examine specific facility effects on adrenal activity — such as enclosure conditions, enrichment available and social interactions — rather than handling methods. Proctor and Brown suggest that, in determining whether FC or PC is more appropriate for management it may be necessary to consider the coping style and social needs of individual elephants when considering how to address welfare concerns.

8.6 Transportation

Transporting elephants is dangerous for elephants and keepers. In April 2018 one elephant died and two others were injured when a lorry that was transporting five elephants owned by Circo Gottani overturned on a motorway in southeast Spain (Anon., 2018b).

Elephants' movements have always been big news. The arrival of a 'white elephant from Burmah' at Liverpool Docks made the front page of *The Illustrated London News* on 26 January 1884 (Fig. 8.22). Two years earlier attempts to move *Jumbo* from London Zoo generated a great deal of interest from the press. When *Jumbo* was sold by the zoo to Barnum for his circus, at the age of 21 years, the move to the United States was unpopular with the public and attracted considerable newspaper coverage. The 25 February 1882 editions of both *The Illustrated London News* and *The Graphic* carried a series of drawings showing

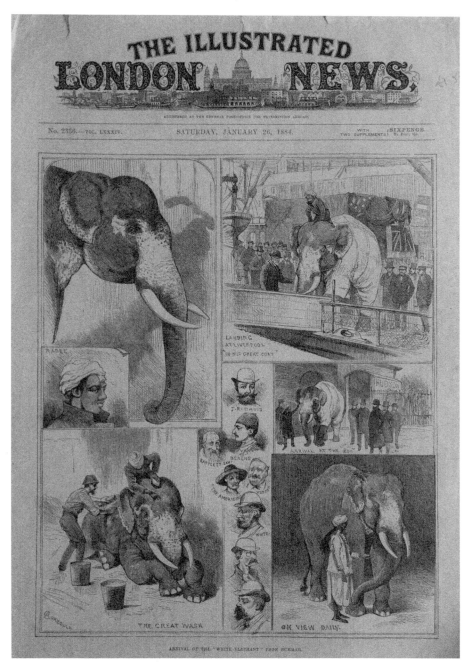

FIGURE 8.22 The arrival of a 'white elephant from Burmah' at Liverpool Docks, United Kingdom, illustrated on the front page of *The Illustrated London News* on 26 January 1884. Source: *Author's collection.*

attempts to persuade the elephant to enter his 'box'. These attempts failed and *The Graphic* (Anon., 1882) reported that:

> *Jumbo will now be left in peace for a fortnight, the large box on wheels being placed at the entrance of his house, so that he will be compelled to pass through it, and in this manner will become accustomed to its appearance.*

Remarkably, until relatively recently, the standard method of transporting elephants was to confine them within a wooden crate (Fig. 8.23).

In some jurisdictions the method of transport of elephants is regulated by legislation, but it does not necessarily follow that this legislation complies with modern standards of elephant welfare. In the state of Kerala, India, the Kerala Captive Elephants (Management and Maintenance) Rules 2012 include rules relating to the nature of the truck that may be used to move elephants:

> *(25) The trucks used for transport of elephants shall have minimum length of 420 cm and tyre shall be of size 900-20. Each vehicle shall have footsteps proportional to the width of the vehicle and each footstep shall have a minimum width of 100 cm. Vehicle shall be strengthened using steel pipes or wooden poles which shall cover at least half the height of the elephant. The horizontal pipes/poles shall be provided on all four sides so that the elephant is secure and is able to hold on to the front bar with its trunk. Elephant should be kept at least 60 cm away from the cabin partition...*

In North America, the AZA's *Standards for Elephant Management and Care* section 1.4.10 deals with the transport of elephants including the type and size of container (AZA, 2019a):

1.4.10.1 Type of transport container

> *... Elephants are typically transported in custom semitrailers, specifically designed for moving elephants. On occasions elephants are moved in crates, most commonly for overseas shipments.*

1.4.10.2 Appropriate size of transport container

> *... The crate or trailer compartment used for shipping should be sized so that the elephant can stand up comfortably, but not turn around. The elephant should not be compressed by the containment front or back. The crate should be equipped with tethering options as needed.*

1.4.10.3 Provision of food and water during transport

> *... Elephants should be provided with food (e.g. hay) and water at regular intervals during transport.*

Standards 1.4.10.4–1.4.10.11 deal with bedding in the transport container, the separation of urine and faeces, temperature range, light and noise levels, group size and the need for separation of animals, transport duration, the timing of release from the container and the size and type of enclosure at the destination. In addition, the standards refer to the obligation to comply with applicable US Federal regulations and International Air Transport Association (IATA) requirements.

Companies now exist that specialise in the transportation of live elephants using specially designed containers. One such company is Stephen Fritz Enterprises, Inc., based in Kingman, Arizona, United States (Box 8.3).

Although elephants have been routinely moved by road and rail for many decades in travelling circuses, by sea and air to supply zoos and circuses with animals and between zoos for breeding purposes, very little work has been conducted on the effects of these movements on their welfare.

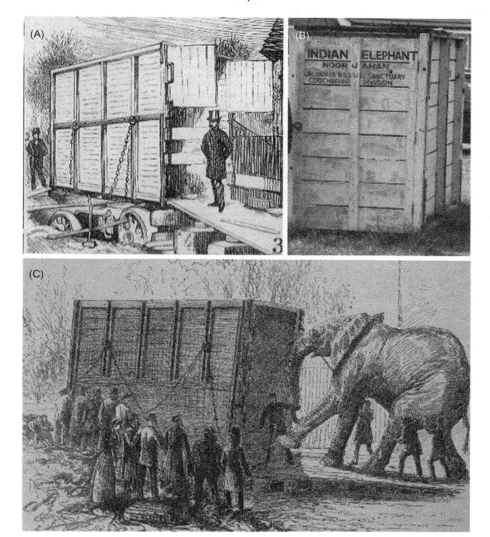

FIGURE 8.23 Until relatively recently moving an elephant involved securing the animal in a small wooden crate. (A) The crate used to transport *Jumbo* from London Zoo to the United States in 1882. (B) The crate used to transport a juvenile Asian elephant, *Noorjahan*, from India to Twycross Zoo, United Kingdom, in 1998. (C) A failed attempt to install *Jumbo* in his crate. Note the men on the far left trying to pull the elephant into the crate with ropes. Source: *(A) The Graphic, 25 February 1882, p. 180. The Illustrated London News, 25 February 1882, p. 200. (A and C) Author's collection.*

The effects of transportation on the welfare of farm animals has been extensively studied (Grandin, 2014). Less is known about the effects on exotic species, although some species have been examined [e.g. tigers (*Panthera tigris*) by Dembiec et al., 2004; dolphins (*Tursiops aduncus*) by Suzuki et al., 2008]. Research has shown that transport can be stressful and have a number of negative effects on animals associated with, for example, loading

BOX 8.3 Elephant transportation: Stephen Fritz Enterprises, Inc.

Transporting elephants from one facility to another is a time-consuming, potentially dangerous and expensive undertaking. It requires detailed planning, expert personnel and specialist equipment that can only be supplied by specialist companies.

Stephen Fritz moved his first elephant in 1985, from San Diego Zoo, California, to Oregon Zoo, Portand, Oregon. Since then, his Arizona-based company — Stephen Fritz Enterprises, Inc. — has transported more than 100 elephants, throughout the United States (including from Alaska to California), from Canada to the United States, from New Zealand to California and from Australia to Florida using specialist vehicles (Fig. 8.24A and B) and lifting equipment (Fig. 8.25A), and where necessary, air cargo transport (Fig. 8.25B and C).

FIGURE 8.24 Elephant transport crates. (A) A specialist vehicle loaded with a single elephant transport crate. The unit immediately behind the cab of the trunk contains an electric heating, ventilation and air-conditioning (HVAC) system that provides climate control inside the crate. (B) A truck carrying two elephant crates. Source: *Stephen Fritz, Stephen Fritz Enterprises, Inc.*

(*Continued*)

BOX 8.3 (Continued)

FIGURE 8.25 Moving elephants: an elephant crate (A) being moved by crane; (B) being lifted into a cargo aircraft; and (C) secured inside an aircraft. (D) Training an elephant to enter an elephant crate. Source: *Stephen Fritz, Stephen Fritz Enterprises, Inc.*

The company takes great care to attend to the welfare needs of the elephants they transport. Prior to transportation, visits are made to the shipping and receiving facilities to assess the elephant or elephants to be moved, meet the keepers and trainers and to gather logistic and other information such as details of access points to the elephant facilities. Following these visits, a transportation plan is made that includes:

- details of all persons involved and their capacities,
- routing for the trip,
- emergency contacts and contingencies, and
- lines of authority over the elephant or elephants and the transport trip itself.

(Continued)

BOX 8.3 (Continued)

Stephen Fritz Enterprises, Inc.

Standards for Elephant Transport.

All Transport Equipment

- Transport crates that are climate controlled, as needed, using all-electric HVAC system for heating and air conditioning.
- Semi-truck and trailers designed specifically for elephant transport.
- Emergency lifting straps that are located inside of crate during transit. Air rescue bags, same as used by Fire Departments, for any needed assistance.
- On board water, inside of crate, for animal to drink during transit as well as additional 65 gallon tank.
- Wireless camera with monitors for observing the animal in transit.
- Three semi-truck team drivers to provide continuous driving. Semi-truck drivers are licensed and experienced in interstate transport and elephant transport.
- Veterinarian experienced in elephant medicine and transport, both land and air.
- Transportation Plan, complete with all necessary information, trip contingencies, routing and emergency details.
- We have permits with USDA/APHIS* and we meet and exceed their requirements.
- Optional crane and/or forklift for loading and unloading.

Stephen Fritz Enterprises, Inc.
101 S. Hayden Ranch Rd.
Kingman, AZ 86409

www.elephanttransport.com

FIGURE 8.26   Stephen Fritz Enterprises, Inc. Standards for Elephant Transport.
*United States Department of Agriculture/Animal and Plant Health Inspection Service. Source: *Stephen Fritz, Stephen Fritz Enterprises, Inc.*

(Continued)

BOX 8.3 (Continued)

The day before departure a pretrip meeting is held to review the transport plan.

Before elephants can be transported they must be trained to enter the elephant transport crate (Fig. 8.25D). A few weeks before the transport, the crate is delivered to the shipping facility and located in a position accessible to the elephant from its enclosure. The goals of the training process are:

- To familiarise the elephant with access to and from the crate.
- To familiarise the elephant with feeding in the crate and drinking from the crate's watering device.
- To familiarise the elephant with connection and disconnection of leg restraints (fitted to both front and at least one rear foot) for the safety of the elephant and the personnel.

On the transport day, the elephants are loaded into the crate at a predetermined time for which they have been specifically trained. Once the elephant is secured in its crate, it is loaded onto the transport vehicle with a crane or forklift. The vehicle has been specifically designed so that the climate in the crate (or crates) can be controlled by an all-electric heating, ventilation and air-conditioning (HVAC) system located behind the cab.

During transportation, the transport vehicle is driven by one of a team of three drivers present to facilitate continuous driving. This vehicle is typically followed by two additional vehicles carrying feed, emergency equipment, keeper/trainers, veterinarians, project managers and other personnel. Staff in these vehicles monitor the elephants via wireless cameras located in the crates. Stops are made approximately every 2 hours to care for the elephants and check their welfare.

Upon arrival at the receiving facility, the elephants are offloaded expeditiously with the assistance of personnel who have been awaiting their arrival with the appropriate equipment.

Stephen Fritz Enterprises, Inc. operates to extremely high welfare standards. These are summarised in the company's own standards document (Fig. 8.26).

---

and unloading, overcrowding, aggression between animals, trip duration, lack of food and water, exposure to noxious gases and temperature extremes (Toscano et al., 2001).

Toscano et al. (2001) examined the environmental conditions experienced by circus elephants during transportation by road and rail in the United States and the effects of these conditions on their body temperature. They measured temperature, relative humidity and radiation within and outside the trailers and rail cars, ammonia and carbon dioxide within these vehicles, and the body temperatures of the elephants — using ingested data loggers — during a total of 12 trips with six different circuses in summer (maximum temperature of 100 °F) and five trips with four different circuses in winter (when exterior temperatures reached below freezing). The study found no evidence of hyperthermia or hypothermia even in extreme climatic conditions and that ammonia and carbon dioxide were always below detectable levels. The authors concluded that the transport of circus elephants does not compromise their well-being when proper care is taken, but emphasised the need to use experienced handlers and monitor conditions in the vehicles.

Toscano et al. did not study the behaviour of the elephants during transportation or attempt to assess the stress they experienced by making physiological measurements. However, Williams and Friend (2003) studied 11 circus elephants during transportation and found that they exhibited stereotypic behaviour for 40% of the time they were observed.

Faecal corticosteroid assays and behaviour measurements were used as indicators of welfare during the relocation of a single Asian elephant by Laws et al. (2007). They measured cortisol metabolites in faecal samples for 10 days before and 10 days after a 24 h transportation of the elephant to a new herd. Following relocation, cortisol metabolite secretion increased by almost 390%. Maximal excretion occurred 2 days after relocation and it remained elevated during the establishment of the new herd. The relocation affected sleep patterns; the elephant spent less time sleeping during the night and slept standing. After relocation, stereotypic behaviour increased by approximately 400%.

Few opportunities arise to study the effect of very long-distance travel on elephants. However, the response to relocation was studied in eight Asian elephants (originating in Thailand) who were flown from the Cocos (Keeling) Islands, in the Indian Ocean, to mainland Australia and then transported by road to Melbourne Zoo and Taronga Zoo (Fanson et al., 2013). Inconsistent signals of adrenocortical activity were provided by analyses of serum, urine and faeces. Faecal glucocorticoid metabolites (FGM) increased following transport, but urinary glucocorticoid metabolites (UGM) decreased and there was no significant change in serum cortisol levels. The authors suggested that the observed differences in FGM and UGM may reflect changes in steroid biosynthesis, with different primary glucocorticoids being produced at different stages of the stress response. The adrenocorticoid responses of the elephants were correlated with personality traits. Those elephants who were described by keepers as 'curious' exhibited a more prolonged increase in FGM posttransfer, and those described as 'reclusive' elephants had a greater increase in FGM values. This work illustrates the importance of understanding elephant personality types in ensuring their well-being.

The behaviour and physiology of African elephants temporarily held captive for the purpose of translocation by the Kenya Wildlife Service from Shimba Hills National Reserve and Mwaluganji Elephant Sanctuary in Kenya to Tsavo National Park (East) were compared with those of the resident Tsavo population by Pinter-Wollman et al. (2009) as proxies for long-term survival and reproductive success. The welfare of the elephants was not scientifically assessed while in captivity or during transportation, but six animals died during the translocation. After release, no difference was found in the stress hormone levels between the resident and the translocated populations as measured by faecal corticosterone levels and the activity budgets of the translocated animals converged with those of the resident elephant population over time.

CHAPTER 9

Ethics, pressure groups and the law

> *Counsel asked if he* [a circus elephant called Pickaninny] *had been hurt, and he* [the elephant] *shook his head vigorously from side to side; when asked if he was generally well treated he nodded and made grunts of assent.*
>
> Richard Carrington (1958)

9.1 Introduction

In a perfect world the law would follow a simple set of moral rules with which everyone agrees. However, in the real word, developments in law generally lag behind changes in our views about morality and what counts as moral or immoral depends on who you ask, especially which philosopher you ask. Ethics is a branch of a philosophy and ethical decisions should be made by considering the views of ethicists. The problem is, specialists in this area may come to diametrically opposite conclusions after considering identical evidence. This situation is complicated by well-meaning interest groups claiming the moral high ground by arguing that, on the one hand, it is acceptable to allow some welfare compromises to keep elephants in a zoo breeding programme, and on the other, it is unacceptable to keep elephants in captivity because the welfare compromises are too great. These groups often selectively quote the results of research that supports their own view and rarely invoke serious ethical arguments. In some cases, parties who support these competing views resort to the law and must argue their cases in court. In this chapter I discuss some of the ethical and legal dilemmas faced by those who care about the conservation and welfare of elephants, and attempts that have been made to improve the lives of elephants using the law.

9.2 Is it ethical to keep elephants in captivity?

Peter Singer published *Animal Liberation* — a book that was to become the 'bible of the animal liberation movement' — in 1975 (Singer, 1975). The work was revised in 1995, yet even this updated version largely focuses on the treatment of animals living in factory farms and laboratories, making little reference to animals in zoos and circuses, and no reference at all to elephants in captivity (Singer, 1995). In 1975 scientists knew little about the complexity of the social and mental lives of elephants, but by 1995 this was most certainly not true. Iain Douglas-Hamilton's work on the Lake Manyara elephants in Tanzania was published in popular form in 1975 (Douglas-Hamilton and Douglas-Hamilton, 1975) and Cynthia Moss's book on the elephants of Amboseli, in Kenya, appeared 13 years later (Moss, 1988).

In his book *Animal Rights*, DeGrazia suggests that in considering whether or not an animal should be kept in a zoo, two tests should be applied (DeGrazia, 2002). First, can a zoo provide for the basic physiological and psychological needs of the animal (the basic needs test)? Second, can a zoo provide a life at least as good as the life the animal could expect in the wild (the comparable life test)? DeGrazia suggests that there should be a strong presumption against keeping great apes and dolphins in zoos, but makes no specific reference to elephants.

In 2003 a symposium entitled *Never Forgetting: Elephants and Ethics* was held at the Conservation and Research Center in Front Royal, Virginia (now the Smithsonian Conservation Biology Institute). This resulted in the publication of *Elephants and Ethics: Towards a Morality of Coexistence* (Wemmer and Christen, 2008). In spite of its title, this book contains little serious philosophical discussion of the ethics of keeping elephants in captivity and of the 37 contributing authors, only one appears to have held an academic post in a university philosophy department at the time the work was published.

Unfortunately it is not uncommon for biologists and others to write about ethics without using the terminology and principles of the discipline of ethics developed by philosophers. In a discussion of the animal rights—conservation debate, Hutchins (2007) suggested that:

> *A conservationist's work is analogous to that of an emergency room doctor's, and emergency room ethics apply. Sometimes desperate acts are going to be necessary to preserve life, even if it means causing the patient short-term harm.*

He acknowledged that the 'patients' are endangered species and the harm is done to a subset of captive individuals, thereby demonstrating that this is not, in fact, an analogy. In an emergency room doctors may sometimes need to cause short-term suffering in their patients in their future long-term interests, but they do not harm one patient for the benefit of others. Elephants do not care if they become extinct, only (some) people do. Any welfare compromises suffered by captive elephants in the name of conservation are ultimately for the benefit of people, not other elephants.

Doyle (2014), an anthrozoologist, has recently argued against keeping elephants in captivity as a contributor to *The Ethics of Captivity* (Gruen, 2014), but, like many of the contributors to *Elephants and Ethics*, she largely reiterated well-rehearsed arguments, out-of-date statistics and her account contained no ethical − philosophical − arguments whatsoever (Fig. 9.1).

Some authors have attempted a serious philosophical analysis of the ethics of keeping elephants. Varner (2008) discussed the extent to which elephants may be considered to be persons and how this should inform our attitude to keeping elephants in captivity. He considered the extent to which elephants can be shown to have a 'robust, conscious sense of their own past and future' and to have the 'biographical life' described by Rachels (1986), which some ethicists believe is necessary if an animal species is to be afforded a special

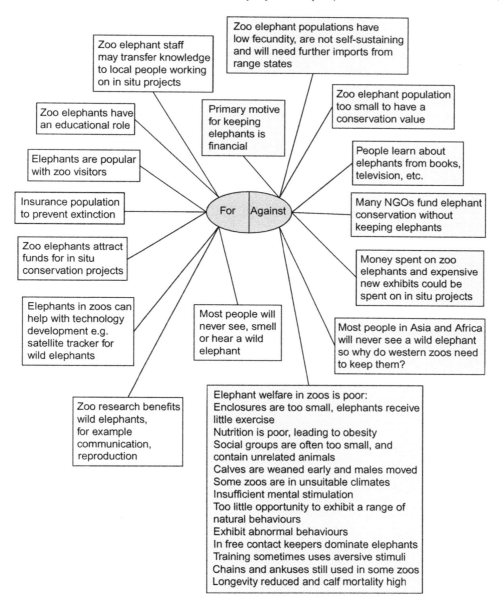

FIGURE 9.1 Arguments for and against keeping elephants in zoos.

moral status similar to that given humans. After reviewing the evidence from studies of elephant memory, mirror self-recognition and theory of mind, Varner concluded that holding elephants in captivity could be justified if they are both born in captivity and their keepers treat them like 'domesticated partners'. He defines such a partner as 'a companion animal who works with humans in ways that emphasize and exercise [their] mental and/or physical faculties in a healthy way'. Varner recognises that adult elephants 'sometimes pose special difficulties', but concludes that keeping captive-born elephants in a circus or zoo environment may be consistent with respecting them as near-persons, provided that their management is 'enlightened'. Furthermore, he suggests that a well-treated working elephant or an elephant performing in an idealised circus could be better off than one in a zoo.

A different approach to analysing the ethics of keeping elephants in captivity has been taken by Alward (2008). The 'capabilities approach' — when applied to animals — analyses ethical dilemmas by asking what an individual animal in a given situation is able to do and to be; whether they are able to live a full life. This is defined by first developing a list of the central functional capabilities of the species. Alward proposes the following as a list of central elephant functional capabilities:

1. **Life.** Not dying prematurely.

2. **Bodily health.** Living in good health — including reproductive health — in an environment conducive to bodily health.

3. **Bodily integrity.** Freedom to move, freedom from assault, opportunities for self-directed sexual satisfaction, freedom to choose food and shelter and other requirements for normal growth and well-being.

4. **Senses, thought and imagination.** Able to use the senses and to imagine in an elephantine manner, communicate over long distances and live in an environment to which elephants are naturally adapted (rather than one to which they have learned to adapt).

5. **Emotions.** Able to form attachments to conspecifics (friends and family), places and things. Freedom from overwhelming fear and anxiety, and abuse and neglect.

6. **Affiliation.** Living with and having normal social relationships with others.

7. **Other species.** Being able to live with concerns for, and in relation to, other elements of nature.

8. **Play.** Freedom to play.

9. **Control of the physical environment.** Being able to seek and acquire food, water and resting places. This control must be shared with other species and humans.

This approach focuses on autonomy and freedom, rather than health, normal lifespan and suffering. Alward concludes that the capabilities approach demonstrates that circuses are unsuitable for elephants — and cannot be made suitable — because these species cannot exercise most of their abilities in a circus environment, but that some zoos ('natural habitat zoos') may be able to provide a suitable environment where elephants are allowed to roam freely and associate with conspecifics and other species.

Alward's analysis concludes that circuses are unsuitable for elephants, but that some zoos may provide a suitable environment; however, Varner's analysis led to the opposite conclusion that working in an 'enlightened' circus may provide elephants a better life than they would have in a zoo. Indeed, Dennis Schmitt, a veterinarian and research scientist who has worked with elephants owned by Ringling Bros. and Barnum & Bailey circus, has argued

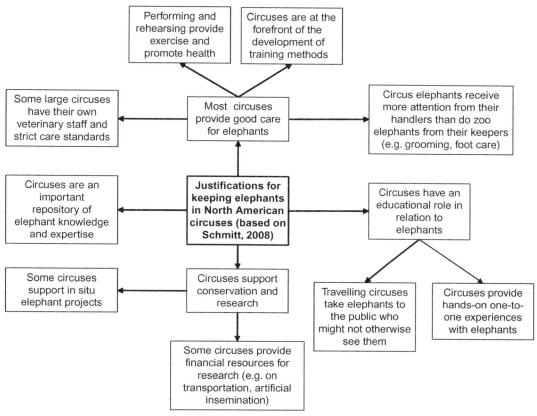

FIGURE 9.2 Justifications for keeping elephants in North American circuses. Source: *Based on the discussion in Schmitt, D., 2008. View from the big top. Why elephants belong in North American circuses. In: Wemmer, C., Christen, C.A. (Eds.), Elephants and Ethics. Towards a Morality of Coexistence. Johns Hopkins University Press, Baltimore, MD. pp. 227–234.*

that elephants belong in North American circuses, where they are well-treated and perform a useful educational purpose (Schmitt, 2008). His arguments are summarised in Fig. 9.2.

Conservationists work to prevent the extinction of species. So, from an ethical perspective, would the extinction of elephants be a harm? Should elephants have a right to not become extinct? If so, this could help to justify keeping elephants in captivity for captive breeding purposes, but only if we can show that these programmes have a realistic prospect of assisting the future survival of elephants as a species.

An individual elephant clearly has interests and may suffer harm; elephant *species* have no interests and thus may not suffer harm. Russow (1994) proposes that we protect animals because of their aesthetic value, rarity, adaptations and for many other reasons, but not because they belong to a particular species. Russow would argue that it is not the species *Elephas maximus* or *Loxodonta africana* that we admire, but individual elephants. We value encounters with rare animals because they are less frequent than encounters with common animals. Russow says we should preserve these animals because we value possible future encounters with other individuals of the same species.

International nature conservation law increasingly recognises the rights of future human generations to exist in a world with a high degree of biodiversity (e.g. the UN Convention on Biological Diversity 1992). However, this is difficult to argue in law. The traditional view of rights is that individuals who have not yet been born cannot have rights because they do not yet exist. Nevertheless, many international treaties clearly recognise the importance of protecting wildlife for the 'good of mankind' and for the 'benefit of future generations'.

9.3 Pressure groups

A number of pressure groups concern themselves with the welfare of elephants under human care. Some function as general advocates for animals at a national or international level, others have been established to focus solely on elephants (particularly in zoos and circuses) and a few have been established just to improve the life of a single elephant. People for the Ethical Treatment of Animals (PETA) believes that zoos cannot provide for the basic needs of elephants and that zoo elephants make no contribution to elephant conservation or educating the public. Some zoo professionals agree with this view.

In 2002 the Royal Society for the Prevention of Cruelty to Animals (RSPCA) campaigned to remove all elephants from zoos in Europe as a result of the findings of research it funded into zoo elephant welfare (Clubb and Mason, 2002). At the time of writing this book, Born Free is campaigning for an elephant-free United Kingdom. In 2013 the Born Free Foundation published *Innocent Prisoner: The Plight of Elephants Living in Solitary Confinement in Europe* that focussed on elephants living alone (Anon., 2013). In Thailand, World Animal Protection (formerly the World Society for the Protection of Animals) has campaigned, with some success, to prevent contact between visitors and elephants at Thai elephant camps (WAP, 2019).

Between 1995 and 2019 elephant numbers in the United Kingdom and Ireland, as well as the number of zoos holding elephants, decreased. In 1995, 21 zoos held elephants (Spooner and Whitear, 1995), but by 2019 this had fallen to 13 (Sach et al., 2019) (Table 9.1). Asian elephant numbers fell by 23% and African elephant numbers fell by 27% over this period of approximately 24 years. The number of zoos holding elephants decreased by 38%.

PETA is campaigning for elephant-free zoos and against the use of elephants in circuses and the tourist trade. *In Defense of Animals* (IDA) — a nonprofit organisation based in California — 'has a proven commitment to ending elephant captivity' and lists over 30 zoos in North America that, since 1991, have either closed their elephant exhibits or plan to close them (see Appendix). The Performing Animal Welfare Society (PAWS) investigates elephant cruelty, assists in prosecutions and provides a sanctuary for elephants rescued from zoos.

At a local level a number of individual campaigns have been organised in the United States and elsewhere against zoos keeping elephants alone or in what some perceive as poor conditions (Table 9.2). On 20 June 2009 IDA organised the *International Day of Action for Elephants in Zoos*. Protests were held at 23 zoos in the United States, 2 in Canada, 1 in Spain and 1 in Thailand (Rees, 2011).

TABLE 9.1 Elephants held by zoos in the United Kingdom and Ireland in 1995 and 2019.

Year	Asian		African	
	Males	Females	Males	Females
1995	4	35	6	27
Species total	39		33	
Holding zoos	13		8	
Elephant total	72[a]			
2019	6	24	8	16
Species totals	30		24	
Holding zoos	7		6	
Elephant total	54[a]			

[a] Plus one female Asian elephant in private ownership at a Hindu temple.
Source: Based on data in Spooner, N.G., Whitear, J.A. (Eds.), 1995. Proceedings of the Eighth UK Elephant Workshop, 21st September 1994, North of England Zoological Society, Chester Zoo, Chester, UK; Sach, F., Fitzpatrick, M., Masters, N., Field, D., 2019. Financial planning required to keep elephants in zoos in the United Kingdom in accordance with the Secretary of State's Standards of Modern Zoo Practice for the next 30 years. International Zoo Yearbook. https://doi.org/10.1111/izy.12213.

TABLE 9.2 Examples of campaigns to remove elephants from zoos.

Elephant/zoo	Campaign/organisation
ZSL London Zoo, England	Elephant-Free London/Born Free Foundation
Jenny, Dallas Zoo, United States	Concerned Citizens for Jenny
Henry Vilas Zoo, United States	Citizens for Human Treatment of Vilas Elephants
Woodland Park Zoo, Seattle, United States	Friends of Woodland Park Zoo Elephants
Billy, Los Angeles Zoo, United States	Save Billy Campaign/Free Lucky Campaign/Voice for the Animals Foundation
Lucky, San Antonio Zoo, United States	
Los Angeles Zoo, United States	Elephant Sanctuaries Not Captivity/Last Chance for Animals
Philadelphia Zoo, United States	Friends of Philly Zoo Elephants
Reid Park Zoo, Arizona, United States	Save Tucson Elephants
Valley Zoo, Edmonton, Canada	Valley Zoo Elephant Campaign
Susi, Barcelona Zoo, Spain	Free Susi Campaign/Libera (Spain)
Taronga Zoo, NSW, Australia	Campaign to Stop Importation of Thai Elephants to New South Wales/World League for Protection of Animals/Animal Liberation NSW

Source: Based on information in Rees, P.A., 2011. Introduction to Zoo Biology and Management. Wiley-Blackwell, Chichester, West Sussex.

Some zoos have moved their elephants to sanctuaries, or to zoos with better facilities, because they no longer believe that they can provide suitable accommodation for them. In the United States, these include Detroit Zoo (Michigan), Frank Buck Zoo (Texas), Mesker Park Zoo (Indiana), Henry Vilas Zoo (Wisconsin), Chehaw Wild Animal Park (Georgia), Alaska Zoo and the Louisiana Purchase Gardens and Zoo. In the United Kingdom, zoos in Bristol, Dudley, Edinburgh, London, Cricket St Thomas, the Welsh Mountain Zoo and Twycross Zoo no longer keep elephants. PETA has produced a list of elephants that have been transferred to better zoo facilities or to sanctuaries, along with zoos that have stopped keeping elephants (see Appendix).

9.4 Law

9.4.1 Introduction

Elephants and objects made from elephant body parts, especially ivory, exist in the custody and care of humans — and institutions created by humans — all over the world. In order to protect elephants, we must use the law to:

1. Protect wild populations and their habitats.
2. Protect the welfare of elephants in human care.
3. Strictly control the trade in, and international movements of, live elephants, their parts and derivatives.

In many jurisdictions elephants are protected by laws, both in the wild and when they are under human care. A number of legal cases have considered incidents of elephant cruelty and neglect. Others have examined whether or not elephants held in zoos and circuses should remain there and whether it is right to import more elephants from their range states to supplement unsustainable zoo populations. Clear-cut cases involving cruelty and neglect have seen some success. However, although there has been considerable public concern about the welfare of elephants kept in zoos and the merits of further importations, this concern nevertheless has not always been reflected in the decisions made by the courts (Rees, 2018b).

The legal protection of wild elephants is beyond the scope of this book. However, objects made of ivory are effectively 'under human care' and their fate directly affects the fate of living elephants if a market exists for them and if they can be traded freely.

9.4.2 Ivory in human ownership

Government attitudes towards elephant protection are beginning to improve. This is exemplified by recent improvements in the control of the ivory trade. The international trade in elephants and their parts is controlled by the Convention on International Trade in Endangered Species of Wild Fauna and Flora 1973 (CITES)(see Section 9.4.9). Some legal trade still exists, but illegal trade has been difficult to control due to the high demand for ivory, particularly in countries in the Middle East and parts of Asia.

When an animal species gets into real difficulty in the wild, humans do things that were previously unthinkable. China is undoubtedly the largest single market for elephant ivory. However, since 1 January 2018, all trade in ivory and ivory products in China has been illegal. All commercial processing and sales of ivory were stopped by 31 March 2017 and thereafter all registered traders were phased out. Until this point in time, China had been the world's largest illegal ivory market.

In July 2019 the authorities in Singapore seized 8 tonnes of elephant ivory worth £10.4m (about US$12.9m) after receiving information from China's customs department. The shipment was found in shipping containers falsely declared as timber en route from the Democratic Republic of Congo to Vietnam, and consisted of almost 300 African elephant tusks along with 11.9 tonnes of pangolin (*Manis* sp.) scales (BBC, 2019a).

In response to increasing concerns about the decline in wild elephant populations, the United Kingdom passed the Ivory Act 2018, one of the strongest bans on ivory trading anywhere in the world. It prohibits dealing in ivory with only very limited exemptions:

s1 Prohibition on dealing in ivory

(1) Dealing in ivory is prohibited.

(2) "Dealing" in ivory means—

(a) buying, selling or hiring it;

(b) offering or arranging to buy, sell or hire it;

(c) keeping it for sale or hire;

(d) exporting it from the United Kingdom for sale or hire;

(e) importing it into the United Kingdom for sale or hire.

The Act allows for exemptions for pre-1918 items of outstandingly high artistic, cultural or historical value (s.2(2)), pre-1918 portrait miniatures (s.6), pre-1947 items with low ivory content (s.7), pre-1975 musical instruments and acquisitions by qualifying (accredited) museums in the United Kingdom (s.9).

No one would expect a zoo to participate in the illegal trade in elephant parts, but investigators found one that did exactly that in Myanmar, at Mong La on the border with China (Nijman and Shepherd, 2014). They found the zoo shop of Mong La Zoo offering around 200 pieces of ivory and 25 pieces of elephant skin for sale in 2009. In 2014 the same shop carried 22 pieces of ivory, including a tusk tip, two elephant molars and the skull of an immature elephant without tusks. A shed nearby contained a near-complete elephant skeleton minus tusks. The tusk tip found at the zoo and another found close to the zoo were said to have come from zoo elephants. During visits made in 2009 and 2014 only a single female Asian elephant was seen at the zoo, but it was reported to have previously kept 10 elephants; 2 had died and the remaining 7 were said to have been sent to one or more elephant tourism camps in Thailand. Nijman and Shepherd identified Mong La as a significant hub for the ivory trade, serving the internal Chinese market.

9.4.3 The UN Convention on Biological Diversity and the IUCN

The ex situ conservation of a species involves measures taken in a captive environment, that is captive breeding. The United Nations Convention on Biological Diversity 1992 (CBD) recognises that ex situ conservation measures may be important for conservation (Box 9.1). However, it should be noted that the emphasis in Article 9 is on the adoption of ex situ measures 'in the country of origin' of the organisms to be conserved.

The International Union for the Conservation of Nature and Natural Resources (IUCN) is not supportive of the captive breeding of African elephants. It explicitly stated this in a statement made by the Species Survival Commission (SSC) African Elephant Specialist Group (AfESG) membership in 1998 (Box 9.2). However, the IUCN/SSC Asian Elephant Specialist Group takes a different view regarding the Asian elephant in its Draft Position Statement on Captive Asian Elephants:

> *The IUCN/SSC Asian Elephant Specialist Group (AsESG) recommends that captive Asian elephants be considered in conservation strategies,...*
>
> IUCN (2017b)

The CBD requires the richer parties to fund conservation work in the territories of the poorer parties (Art. 21) and some in situ elephant populations have benefitted from this (see Section 9.4.11).

BOX 9.1 Ex situ conservation and the UN Convention on Biological Diversity 1992

The UN Convention on Biological Diversity

Article 9. Ex-situ Conservation

Each Contracting Party shall, as far as possible and as appropriate, and predominantly for the purpose of complementing in-situ measures:

(a) Adopt measures for the ex-situ conservation of components of biological diversity, preferably in the country of origin of such components;

(b) Establish and maintain facilities for ex-situ conservation of and research on plants, animals and microorganisms, preferably in the country of origin of genetic resources;

(c) Adopt measures for the recovery and rehabilitation of threatened species and for their reintroduction into their natural habitats under appropriate conditions;

(d) Regulate and manage collection of biological resources from natural habitats for ex-situ conservation purposes so as not to threaten ecosystems and in-situ populations of species, except where special temporary ex-situ measures are required under subparagraph (c) above; and

(e) Cooperate in providing financial and other support for ex-situ conservation outlined in subparagraphs (a) to (d) above and in the establishment and maintenance of ex-situ conservation facilities in developing countries.

BOX 9.2 The position of the International Union on the Conservation of Nature and Natural Resources (IUCN) on the captive breeding of African elephants

The IUCN's Position on the Captive Breeding of African Elephants
STATEMENT AND RESOLUTIONS ON THE ROLE OF CAPTIVE FACILITIES IN IN-SITU AFRICAN ELEPHANT CONSERVATION

The IUCN Species Survival Commission African Elephant Specialist Group (AfESG) membership, at its general meeting in Burkina Faso in January 1998, debated the role played by captive facilities in the conservation of the African elephant.

The Group agreed on the following:

The Statement

The AfESG recognizes there is some role for captive facilities in the conservation of African elephants, through the fields of public education, scientific research, development of technologies, professional training and direct support to the conservation of the species in the field.

The AfESG also recognizes the role that some zoos and zoological societies play in mobilizing public support for funding these activities.

However, the AfESG is concerned by the poor breeding success and low life expectancy of captive African elephants and does not see any contribution to the effective conservation of the species through captive breeding per se.

Where African elephants are held in captivity, the AfESG believes that special care should be accorded to their physical and psychological well-being.

The AfESG encourages captive facilities to maintain and expand field programmes directed to African elephant populations in African range states, but wishes to point out that the holding of African elephants by a captive facility is not a necessary precursor for involvement for in situ African elephant conservation.

The Resolutions

1. Regarding the question of what action the Chair of the AfESG should take when asked by captive facilities in importing countries to issue a letter on behalf of the AfESG to their CITES Management Authorities endorsing the importation of live African elephants from the wild, the Group agreed by consensus that this should not be done.

2. The Group also decided that there should be no formal linkages between the AfESG and the AZA, but there could be continued dialogue regarding prioritization of their input into field research and conservation programmes in the field, as well as improving the content of their public awareness and education programmes, as and where necessary.

IUCN, 2017a. Statement and resolutions on the role of captive facilities in in-situ African elephant conservation. https://www.iucn.org/ssc-groups/mammals/african-elephant-specialist-group/afesg-statements/role-captive-facilities (accessed 28.12.17.).

9.4.4 Elephants, zoos and the law

Where they are kept by zoos in England, elephants are classified as Category 1 animals under the *Secretary of State's Standards of Modern Zoo Practice* (SSSMZP) hazardous animal categories (Appendix 12) along with other dangerous taxa such as big cats and bears. These standards are published under the authority of s9 of the Zoo Licensing Act 1981 and animals listed in Category 1 are considered to be 'greater risk':

Category '1' (greater risk)
1.1 Contact between the public and animals in Category '1' is likely to cause serious injury or be a serious threat to life, on the basis of hazard and risk of injury, toxin or disease, irrespective of the age and vulnerability of the visitor.
1.2 Animals in Category '1' must either be separated from the public by a barrier of suitable design in order to prevent physical contact between the animals and members of the public within their designated areas, or, with the prior approval of the local authority, be provided with adequate supervision to allow the public and the animals to be in the same area without hazard.

SSSMZP (2012)

In June 2017, as a result of concerns for the welfare of elephants in zoos, the Department for the Environment, Food and Rural Affairs (Defra) issued a technical update to the SSSMZP that contained more exacting standards on the keeping of elephants (Defra, 2017). Although not law per se, these standards include a requirement to produce an individual welfare plan (IWP) for each elephant that must contain a behaviour profile (8.8.4); a long-term management plan (LTPM) that must include herd compatibility details; and a bull profile for each male from the age of 4 years to inform his management (8.8.5). Section 8.8.12 of the standards requires that all elephants must have the option to get away from other elephants if so desired, through use of space and/or physical barriers within the enclosure.

Zoos in India are licensed and regulated by the Central Zoo Authority (CZA). In 2009 the Authority instructed all zoos in India to release their elephants in an advisory (Central Zoo Authority Letter No. 7-5/2006-CZA (Vol. II) dated 7 November 2009. Advisory — Banning Elephants from Zoo Collections). It directed that:

Elephants are banned from zoo collections throughout the country with immediate effect. All captive elephants in zoos should be rehabilitated in elephant camps/rehabilitation camps/facilities available with the forest department at National Parks/Wildlife Sanctuaries/Tiger Reserves for departmental use.

The Authority gave a number of reasons for this decision: the stress caused by the inadequate space allocated to elephants and the long hours for which they are chained in most zoos; the risk to visitors posed by bulls in musth; and the financial liability of maintaining elephants in the zoos. The CZA also noted that:

...more often than not, such captive elephants in zoos hardly breed.

There has been considerable concern expressed by welfare organisations about the lack of space available to elephants in most zoos. In 2006 IDA petitioned the US Department of Agriculture (USDA) asking for clarification of space and living requirements for captive elephants. The IDA surveyed the veterinary records from 2000 to 2005 of elephants at 46%

of the Association of Zoos and Aquariums (AZA)-accredited zoos which held elephants. The survey revealed high levels of foot disease (62%), joint disorders (42%) and reproductive problems (Doyle and Roy, 2006).

9.4.5 The law and elephants in private ownership

Elephants living in captivity may be owned by individuals or by zoos, circuses, tourist facilities, forestry departments, logging companies, governments or other organisations that use them for work or entertainment.

In the past, the ownership of animals, including dangerous wild animals such as elephants, was not closely regulated by laws. There was little control of the private ownership of exotic animals in the United Kingdom until the 1970s. However, in the mid-1970s the situation changed. The Endangered Species (Import and Export) Act 1976 Schedule 1 prohibited the importation of many species – including elephants – into the United Kingdom except under licence, thereby implementing the United Kingdom's obligations under CITES. Prior to this it had been possible to purchase a wide range of exotic taxa from animal traders. Knight (1967) discussed the cost of acquiring an elephant from overseas:

> From Calcutta to London, an Indian elephant over five feet in height will cost you £135, and one under five feet about £84. If you choose an African [elephant] the rates will be £135 and £78 respectively for a passage from Mombasa to London.

In the 1960s the London department store *Harrods* was famous for selling exotic animals, including a baby Asian elephant called *Gertie* who was purchased by King Leka of Albania in 1967 for his friend Ronald Reagan – when Reagan was Governor of California – prior to his being elected President of the United States. The elephant is the mascot of the Republican Party.

Since 1976, ownership of dangerous wild animals in Britain has been regulated by local government and requires a licence for those species listed in the schedule to the Dangerous Wild Animals Act 1976, which includes elephants. This legislation does not apply to zoos. A female Asian elephant called *Valli* is kept under licence by a religious group located in Carmarthen in Wales. At the time of writing, she was the only nonzoo elephant in private hands in Britain.

Many elephants in Southeast Asia are privately owned. In Sri Lanka, under the Fauna and Flora Protection Ordinance 1937, elephants owned or used by people must be licensed:

> *s.22A. Registration and licensing of elephants.*
>
> *(1) No person shall own, have in his custody or make use of an elephant unless it is registered and unless a licence in respect of the elephant has been obtained in accordance with the provisions of this section.*
>
> *(2) Every person who owns or has in his custody an elephant shall register such elephant with the prescribed officer.*
>
> *(3) Every person who owns or has in his custody an elephant shall, prior to registration, pay such registration fee as may be prescribed.*
>
> *(4) The Director shall maintain a register of elephants in such form as may be prescribed.*

The term 'elephant' is defined in s.28:

s.28. Interpretation Part II.

In this Part of this Ordinance, unless otherwise expressly provided or unless the context otherwise requires—

"elephant" means a wild elephant and includes a tusker and for the purposes of section 22A, includes a tame or domestic elephant or tusker;

..."tusker" means a wild elephant with a pair of incisors* (teeth) or a single incisor.

[* An elephant's tusks are its upper incisors.]

9.4.6 Elephants, entertainment and the law

Many countries have banned the use of wild animals in circuses. At the time of writing, Animal Defenders International listed 54 countries in which the use of animals in circuses was banned, mostly on a nationwide basis. This included 92 partial or full bans on performing animals in jurisdictions in the United States, in 32 states, and state-wide bans in New Jersey and Hawaii. It also included over 200 local authorities in the United Kingdom that had banned the use of animals in circuses (more than two thirds of which banned all performing animals, the remainder banning just wild animals) (IDA, 2019).

A general ban on the use of wild animals in circuses has the effect of banning the use of elephants. Scotland banned the use of wild animals in travelling circuses in 2018 in the Wild Animals in Travelling Circuses (Scotland) Act 2018, although there were no elephants in any circus in Scotland at the time:

s1 Wild animals in travelling circuses: offence

(1) A person who is a circus operator commits an offence if the person uses, or causes or permits another person to use, a wild animal in a travelling circus...

s2 Meaning of wild animal

(1) In this Act, "wild animal" means an animal other than one of a kind that is commonly domesticated in the British Islands.

In 2017 the State of New York banned the use of elephants in entertainment (Box 9.3). In the same year, the producer of Ringling Bros. and Barnum & Bailey Circus, Feld Entertainment, Inc., announced its intention to close the circus after 146 years of operation, due to falling ticket sales. Although this was an important success for campaigners for the removal of elephants from circuses, the American Society for the Prevention of Cruelty to Animals (ASPCA) paid a high price for its efforts in this regard. On 28 December 2012 the company announced that it had reached a legal settlement amounting to $9.3 million with the ASPCA in connection with two federal court cases. These cases related in part to more than a decade of litigation attempting to outlaw the use of elephants in circuses (Anon., 2012a).

In Germany, Circus Roncalli has replaced its live animal performers with three-dimensional holograms, including one of an African elephant (Katz, 2019).

There is growing opposition to elephant polo in Asia. In August 2011, a district magistrate in Jaipur, Rajasthan, cancelled an elephant polo match following appeals by PETA and a notification from the Animal Welfare Board of India (AWBI) that the organisers had

BOX 9.3 The State of New York ban on the use of elephants in circuses: The Elephant Protection Act 2017

STATE OF NEW YORK
2098--B

Cal. No. 775

2017−2018 Regular Sessions
IN SENATE
January 12, 2017

Introduced by Sens. MURPHY, BOYLE...

AN ACT to amend the agriculture and markets law and the environmental conservation law, in relation to prohibiting the use of elephants in entertainment acts

THE PEOPLE OF THE STATE OF NEW YORK, REPRESENTED IN SENATE AND ASSEMBLY, DO ENACT AS FOLLOWS:

Section 1. Short title. This act shall be known and may be cited as the "elephant protection act".

§ 2. Legislative findings. The legislature hereby finds that:

a. elephant abuse is a matter of worldwide concern, and the state as a recognized environmental leader should help assure the protection and welfare of elephants;

b. it is widely recognized that elephants used for entertainment purposes ("entertainment elephants") suffer physical and psychological harm due to the living conditions and treatment to which they are subjected, resulting in increased mortality with life spans only one-half as long as wild elephants;...

§ 3. The agriculture and markets law is amended by adding a new section 380 to read as follows:

§ 380. USE OF ELEPHANTS IN ENTERTAINMENT ACTS. 1. NO PERSON SHALL USE OR CAUSE TO BE USED ELEPHANTS IN ANY TYPE OF ENTERTAINMENT ACT.

2. ANY PERSON THAT VIOLATES THE PROVISIONS OF THIS SECTION, OR ANY RULE OR REGULATION PROMULGATED PURSUANT THERETO, MAY BE ASSESSED, BY THE COMMISSIONER, A CIVIL PENALTY NOT TO EXCEED ONE THOUSAND DOLLARS FOR EACH SUCH VIOLATION.

3. THE PROVISIONS OF THIS SECTION SHALL NOT APPLY TO:

(A) INSTITUTIONS ACCREDITED BY THE ASSOCIATION OF ZOOS AND AQUARIUMS; AND

(B) WILDLIFE SANCTUARIES AS DEFINED IN SUBDIVISION THIRTY-TWO OF SECTION 11-0103 OF THE ENVIRONMENTAL CONSERVATION LAW.

4. AS USED IN THIS SECTION:

(A) "ELEPHANT" MEANS THE THREE SPECIES OF THE FAMILY ELEPHANTIDAE:

(I) LOXODONTA AFRICANA AND LOXODONTA CYCLOTIS, ALSO KNOWN AS THE AFRICAN ELEPHANT; AND

(II) ELEPHAS MAXIMUS, ALSO KNOWN AS THE ASIAN ELEPHANT OR INDIAN ELEPHANT.

(B) "ENTERTAINMENT ACT" MEANS ANY EXHIBITION, ACT, CIRCUS, TRADE SHOW, CARNIVAL, RIDE, PARADE, RACE, PERFORMANCE OR SIMILAR UNDERTAKING WHICH IS PRIMARILY UNDERTAKEN FOR THE ENTERTAINMENT OR AMUSEMENT OF A LIVE AUDIENCE...

failed to register the elephants with the AWBI as required by the Performing Animal (Registration) Rules 2001 and had not submitted valid elephant ownership certificates (Anon., 2011). In response to pressure from PETA India, Carlsberg India withdrew its sponsorship of the Polo Cup.

9.4.7 Legal personhood and *habeas corpus*

The concepts of legal personhood (legal personality) and *habeas corpus* are pivotal to the operation of an effective legal system. Only a legal person can hold legal rights. A legal person may be a natural person — a human being — or an artificial person (an organisation). The principle of *habeas corpus* gives natural persons the right not to be held against their will unlawfully and the lawfulness or otherwise of their detention to be determined by a court.

A small number of attempts have been made to secure the release of animals by asking a court to issue a writ of *habeas corpus*. This is a judicial mandate requiring that the animal be 'brought to court' by their owner so that it might be determined whether or not it had been lawfully deprived of its liberty. Such a writ would normally only be issued in an attempt to release from custody someone who had the status of a 'natural person' in law, that is a human being, who was, for example, being held unlawfully in prison or in a mental institution.

This approach was used for the first time on behalf of elephants in November 2017 when the Nonhuman Rights Project (NhRP) filed a petition for a writ of *habeas corpus* in the Connecticut Superior Court in an attempt to secure the release of three elephants owned by Commerford Zoo to the care of the Performing Animal Welfare Society's ARK 2000 natural habitat sanctuary. It failed (Box 9.4).

BOX 9.4 Extending the definition of personhood to nonhumans

One of the tasks of the courts is to determine the meaning of words. The definition of legal persons has changed with time. In 1927 five Canadian women asked the Supreme Court of Canada to decide whether the word 'person' in s24 of the British North America Act 1867 included female persons:

24. The Governor General shall from Time to Time, in the Queen's Name, by Instrument under the Great Seal of Canada, summon qualified Persons to the Senate; and, subject to the Provisions of this Act, every Person so summoned shall become and be a Member of the Senate and a Senator.

The court decided it did not. The women appealed to the Judicial Committee of the Privy Council of Great Britain, which was, at that time, Canada's highest court of appeal. In 1929 the Privy Council overturned the Supreme Court's decision and declared that women were persons, paving the way for women to participate equally in all aspects of Canadian life, including serving in the Senate.

(Continued)

BOX 9.4 (Continued)

In 2015 the NhRP petitioned the Supreme Court of New York State, in the United States, for a writ of *habeas corpus* on behalf of two chimpanzees, *Hercules* and *Leo*, in an attempt to secure their release from the State University of New York to a sanctuary in Florida [*The Nonhuman Rights Project, Inc. v. Stanley* (2015)]. The petition was denied as the court did not recognise the chimpanzees as legal persons and only a legal person is entitled to bring a writ of *habeas corpus*. Nevertheless, the judge acknowledged that this may not be the case in the future, and in his judgement concluded:

> The similarities between chimpanzees and humans inspire the empathy felt for a beloved pet. Efforts to extend legal rights to chimpanzees are thus understandable; some day they may even succeed. Courts, however, are slow to embrace change, and occasionally seem reluctant to engage in broader, more inclusive interpretations of the law, if only to the modest extent of affording them greater consideration. As Justice Kennedy aptly observed in Lawrence v Texas, albeit in a different context, "times can blind us to certain truths and later generations can see that laws once thought necessary and proper in fact serve only to oppress." The pace may now be accelerating...[granting right to marry to same-sex couples and acknowledging that institution of marriage has evolved over time notwithstanding its ancient origins].

It is not beyond the bounds of possibility that personhood could one day be extended to the members of a number of taxa, including primates (especially apes), cetaceans and elephants.

Anyone who doubts the ability of governments to extend rights to nonhumans should consider the action of the New Zealand Government when, in 2017, it granted legal personhood to the Whanganui River in the passing of the Te Awa Tupua (Whanganui River Claims Settlement) Act 2017 in recognition of the inseparability of the river and the Māori:

> 14 Te Awa Tupua declared to be legal person
>
> Te Awa Tupua is a legal person and has all the rights, powers, duties, and liabilities of a legal person.

9.4.8 Elephant cruelty and cruel methods

Elephant handlers traditionally use an ankus – also called a goad or bullhook – to control elephants. The practice of directing an elephant with a hook attached to the end of a stick appears to have its origin in India in ancient times (Searle, 2018). A Roman coin dated to CE 248 commemorating the 1000th anniversary of Rome bears a bust of Philippus I on the obverse and, on the reverse, an elephant walking to the left ridden by a mahout carrying a goad and a wand.

A modern ankus is essentially a short stick with a metal tip consisting of a hook and a pointed end used to prod and pull the animal (Fig. 9.3). In Southeast Asia mahouts still use ankuses with long handles (Fig. 7.27).

In the United States the use of bullhooks and chains on elephants has been banned in some jurisdictions. From 1 January 2017 the use of bullhooks or similar devices on elephants in travelling shows was banned in the state of Rhode Island, the first US State to pass such a measure. California banned the use of bullhooks and other devices (including baseball bats and pitch forks!) from 1 January 2018 (Box 9.5).

The Pittsburgh City Council banned bullhooks in December 2017 (Rhode Island General Laws § 4-1-43). This ban affects Pittsburgh Zoo which previously lost its AZA accreditation over a dispute regarding its handling of elephants (AZA, 2015). The AZA requires these

FIGURE 9.3 An elephant keeper using an ankus (or bullhook) during a public demonstration in a British zoo.

BOX 9.5 Senate Bill No. 1062, banning the use of bullhooks and other devices to inflict pain on elephants in the State of California

Senate Bill No. 1062
CHAPTER 234

An act to add Section 2128 to the Fish and Game Code, relating to elephants.

[Approved by Governor August 29, 2016. Filed with Secretary of State August 29, 2016.]

LEGISLATIVE COUNSEL'S DIGEST

SB 1062, Lara. Elephants: prohibited treatment...

THE PEOPLE OF THE STATE OF CALIFORNIA DO ENACT AS FOLLOWS:

SECTION 1.
Section 2128 is added to the Fish and Game Code, to read:

2128.

(a) (1) Any person who houses, possesses, manages, or is in direct contact with an elephant shall not do either of the following:

(A) Use a bullhook, ankus, baseball bat, axe handle, pitchfork, or other device designed to inflict pain for the purpose of training or controlling the behavior of an elephant.

(B) Authorize or allow an employee, agent, or contractor to use a bullhook, ankus, baseball bat, axe handle, pitchfork, or other device designed to inflict pain for the purpose of training or controlling the behavior of an elephant.

(2) Use prohibited by this subdivision includes brandishing, exhibiting, or displaying the devices in the presence of an elephant.

(b) Any person who violates this section is subject to the civil penalty set forth in Section 2125 for each violation, and the restricted species permit for the elephant is subject to immediate suspension or revocation by the department. A person whose restricted species permit is suspended or revoked pursuant to this section may appeal the suspension or revocation to the commission by filing a written request for an appeal with the commission within 30 days of the suspension or revocation. A person who violates this section is not subject to the criminal penalties set forth in this code.

(c) The provisions of this section are in addition to, and not in lieu of, any other laws protecting animal welfare. This section shall not be construed to limit any state law or rules protecting the welfare of animals or to prevent a local governing body from adopting and enforcing its own animal welfare laws and regulations.

(d) This section shall become operative on January 1, 2018.

animals to be handled using protected contact whereas the zoo was using Australian cattle dogs to herd elephants (PETA, 2014). The zoo was ordered to stop this practice by officials from USDA (Gambino, 2015).

In England the SSSMZP requires that the use of an ankus be restricted (Defra, 2017). Standard 8.8.52 requires that:

> *The ankus must not hit, injure, damage or break the skin or be used in any other way that could cause physical or mental injury.*

All injuries caused by an ankus must be recorded, reviewed by management, and the records made available to zoo inspectors. The use of an electric goad (cattle prod) is restricted to trained staff in extreme situations – such as to protect human life – not for routine control (standard 8.8.53).

Concern about the use of an ankus is not confined to western societies. In 2006 the Rajasthan High Court in India passed an order banning the use of the ankus as a result of public interest litigation moved by Naresh Kadyan, founder of People for Animals, Haryana [DB Civil Writ petition no. 8987/2006 (PIL)].

We could reasonably expect that individuals who spend their lives working with large animals in zoos would treat them with respect and provide them with the best possible living conditions. Unfortunately, this is not always the case and cruelty cases concerning the treatment of elephants in zoos are occasionally brought to the courts.

When the Radford Report on *Wild Animals in Travelling Circuses* was published in the United Kingdom in 2007 there were just 47 nondomesticated animals in four British circuses, including one Asian elephant – *Anne (Annie)* – kept by Bobby Roberts' Super Circus – who was still touring, but had been retired from performing (Radford, 2007). She was subsequently the subject of a legal case and rehomed at a safari park. *Anne* was the last elephant to work in a circus in Britain.

In September 2005 the *Sunday Mirror* – a national newspaper – reported that it and the charity Born Free were campaigning to free *Anne*. At the time she was 52 years old, suffering from arthritis and had been seen trailing one of her hind legs. More than £20,000 (US $34,000) was raised to pay to allow *Anne* to retire, but the circus refused to part with her. Instead, they continued to bring her into the ring during the interval so that members of the audience could pay to be photographed with her.

In 2011 Animal Defenders International commenced a private prosecution at Northampton Magistrates' Court in England against Mr and Mrs Bobby Roberts – owners of the circus – after their investigation into the cruel treatment of their Asian elephant *Anne*, who was now 59 years old (Anon., 2012b). As a result of public concern, the Crown Prosecution Service (CPS) took over the case and the defendants were jointly charged with the following offences under the Animal Welfare Act 2006:

1. Causing the elephant to suffer unnecessarily, by requiring the elephant to be chained to the ground at all times, contrary to section 4(1).

2. Failing to take reasonable steps to prevent their employee from causing unnecessary suffering to the elephant, by repeatedly beating her, contrary to section 4(2).

3. Failing to take reasonable steps to ensure that the needs of the elephant were met to the extent required by good practice, contrary to section 9.

As a result of this case, *Anne* was retired to Longleat Safari Park, United Kingdom, where a new £1.2 million facility (*Anne's Haven*) was built for her with the help of £350,000 (US$500,000) donated by *Daily Mail* readers (Ellicott, 2016).

In 2012 three keepers were recorded on CCTV beating two Asian elephants with canes at Twycross Zoo in Leicestershire, United Kingdom (BBC, 2012). The zoo acted quickly by dismissing the keepers and reporting the incident to the police. The three men were arrested, but not charged. Although a file was sent to the CPS no further action was taken against the keepers because the CPS believed that there was insufficient evidence for a realistic prospect of conviction. The mistreatment of the elephants could potentially have been a case for a prosecution under the Animal Welfare Act 2006.

In 2013 USDA charged a circus supplier, Lance Ramos, with violating the Animal Welfare Act of 1966. Five years earlier, in 2008, USDA had confiscated a starving Asian elephant called *Ned* from Ramos for failing to provide adequate veterinary care. The emaciated elephant was taken to The Elephant Sanctuary in Tennessee following a complaint filed with USDA by PETA, but died 6 months later. Ramos' exhibitor's licence was permanently revoked by USDA in 2009 (PETA, 2015).

9.4.9 Zoos and the wildlife trade

Zoos cannot survive without animals, and if captive populations are not self-sustaining, they turn to wild populations to supplement their stock. Since the coming into force of CITES in 1975 international movements of protected animals have been strictly regulated by a system of import and export permits. This system applies as much to live animals being moved between zoos for breeding programmes as it does to animal trophies.

Occasionally, zoos import wild elephants, but there has been considerable opposition to this from animal welfare NGOs and the general public (Box 9.6).

In March 2016, 18 wild African elephants (3.15) were flown from Swaziland to the United States to be distributed among three zoos, the Dallas Zoo, Texas, Sedgwick County Zoo, Kansas, and Henry Doorly Zoo, Nebraska. This move took *Friends of Animals* – a non-profit international animal advocacy organisation – by surprise as they were preparing to petition a US federal court in an attempt to prevent the transfer. The organisation filed for an emergency injunction to stop the transfer, but this was not granted. The elephants were captured in Hlane Royal National Park apparently in an attempt to reduce pressure on other wildlife during a severe drought, rather than culling them. The US Fish and Wildlife Service (USFWS) granted a permit for the elephants to be imported and the transfer was approved by the AZA (Milman, 2016). Elephant experts have argued that this was a purely commercial transaction for the family-run organisation – *Big Game Parks* – that manages wildlife in the park. At the time there were thought to have been fewer than 35 elephants in Swaziland (Doyle et al., n.d.).

CITES allows trade in live wild elephants from Appendix I populations (i.e. all Asian elephants and African elephants from countries other than Botswana, Zimbabwe, Namibia and South Africa) when an export licence has been issued by the exporting country once a number of conditions have been fulfilled (Table 9.3; Box 9.7).

A total of 24 elephant calves were removed from Hwange National Park in Zimbabwe and taken to Chimelong Wildlife Safari Park in Guangzho, China, in 2015 (IFAW, 2015). In

BOX 9.6 *Born Free USA v. Norton*, 278 F. Supp 2d 5 (D.D.C. 2003)

In 2003 Born Free USA applied to the United States District Court for the District of Columbia for a preliminary injunction to prevent the importation of 11 elephants from Swaziland after the US Fish and Wildlife Service (USFWS) issued permits to the San Diego Zoo and the Lowry Park Zoo for their importation (*Born Free USA v. Norton* 278 F. Supp 2d 5 (D.D.C. 2003)). The injunction was not granted.

In considering the harms to the plaintiffs, in this case the court had to consider the risk that the elephants would be culled if they were not allowed into the United States:

> ...if an injunction is granted the elephants will be culled. This might appear to mean, somewhat ironically, that plaintiffs would be irreparably injured as the result of the very injunction that they request; however, at the August 6 hearing in this matter, counsel for plaintiffs explained that, from plaintiffs' perspective, the elephants will be better off — and thus plaintiffs' interests will be more fully advanced — if the elephants are killed rather than imported and placed in the zoos. Taking the plaintiffs at their word, the Court concludes, on balance, that plaintiffs' interests — interests about which the Court has some concerns in terms of standing — will be harmed if an injunction is not granted, yet somewhat advanced if an injunction is granted.

Some 3 years after this decision, Hutchins and Keele (2006) warned the zoo community that it would face political, legal, ethical, practical and public relations challenges if it imported further elephants from range states.

In 2010 Kane noted that no further attempts had been made to import elephants into the United States following the decision in *Born Free USA v. Norton*, but she predicted that they would eventually come in the face of dwindling zoo populations (Kane, 2010). She was right.

TABLE 9.3 Listing of elephant species and populations under CITES.

Appendix I	Appendix II
Elephas maximus	
Loxodonta africana (Except the populations of Botswana, Namibia, South Africa and Zimbabwe, which are included in Appendix II subject to annotation 2)	*Loxodonta africana*[2] (Only the populations of Botswana, Namibia, South Africa and Zimbabwe; all other populations are included in Appendix I)

Annotation 2.

Populations of Botswana, Namibia, South Africa and Zimbabwe (listed in Appendix II):
For the exclusive purpose of allowing:
a) trade in hunting trophies for non-commercial purposes;
b) trade in live animals to appropriate and acceptable destinations, as defined in Resolution Conf. 11.20 (Rev. CoP17), for Botswana and Zimbabwe and for in situ conservation programmes for Namibia and South Africa;...

2016 it was reported that Grace Mugabe (the wife of former President of Zimbabwe, Robert Mugabe) sent 35 elephant calves and other wildlife species to China in payment for boots and uniforms purchased for the Congolese army (Graham, 2016). In February 2019, *The Times* reported that 35 young elephants — some as young as 2 years — were being held in pens in Hwange National Park in preparation for export to China, and claimed that since 2012 about 100 elephant calves had been sold to Chinese zoos and safari parks (Flanagan, 2019a). Although this operation was stressful for adults and young, since it

BOX 9.7 Conditions that must be fulfilled for the export of wild African elephants from CITES Appendix I and Appendix II populations

The Convention on International Trade in Endangered Species of Wild Fauna and Flora 1973 (CITES) lays down conditions for the export of African elephant populations listed in Appendix I and Appendix II.

Conditions that must be fulfilled for the export of wild African elephants from Appendix I populations:

The importing Management Authority has issued an import permit, having been satisfied that:

- *the animal(s) will "not to be used for primarily commercial purposes"*
- *the Scientific Authority in the importing State is satisfied that the proposed recipient of a living specimen is "suitably equipped to house and care for it" and that the import will be "for purposes which are not detrimental to the survival of the species"*

CITES (2017a)

African elephants from Appendix II populations may be traded to 'appropriate and acceptable destinations'. Until 2019 these were been defined by CITES in Conf. 11.20 (Rev. CoP17) as:

THE CONFERENCE OF THE PARTIES TO THE CONVENTION

1. AGREES that, where the term 'appropriate and acceptable destinations' appears in an annotation to the listing of a species in Appendix II of the Convention with reference to the trade in live animals, this term shall be defined to mean destinations where:

a) the Scientific Authority of the State of import is satisfied that the proposed recipient of a living specimen is suitably equipped to house and care for it; and

b) the Scientific Authorities of the State of import and the State of export are satisfied that the trade would promote in situ conservation; ...

CITES (2017b)

Ho and Lindsay (2017) prepared a background paper on the definition of the term 'appropriate and acceptable destinations' as it relates to the trade in live African elephants for the 69th meeting of the Standing Committee of CITES that met in Geneva (Switzerland) from 27 November to 1 December 2017. They concluded that:

CITES has not established guidance or standards for determining whether a facility that is to receive live African elephants is suitably equipped to house and care for them. Our findings concur with the view of elephant biologists Joyce Poole and Petter Granli, who warned in 2009 that zoos and other captive facilities are "woefully inadequate" to house elephants; we consider that there is no captive facility suitably equipped to house and care for live, wild-caught African elephants forcefully removed from their family groups. In light of this, along with the African Elephant Specialist Group's statement and the views of many respected elephant biologists, we conclude that there should be no trade in live wild-caught African elephants for captive use.

Lindsay et al. (2017) have noted that the split listing of the African elephant is ecologically unsound because 76% of African elephants are found in transboundary populations.

In the summer of 2019 the 18th Conference of the Parties to CITES (CoP18) was held in Geneva, Switzerland 17–28 August 2019 (Fig. 9.4). At this conference the measures for the export of live African elephants to 'appropriate and acceptable destinations' were reviewed. It was decided that future exports outside the natural range of African elephants would only be permitted in 'exceptional circumstances' or 'emergency situations', in consultation with CITES and the IUCN, and only if they provide 'in-situ conservation benefits' (CITES, 2019a,b).

(Continued)

BOX 9.7 (Continued)

FIGURE 9.4 Delegates at CITES CoP18, Geneva, August 2019. Source: *Courtesy Katherine Clark. Photo by IISD/Kiara Worth (enb.iisd.org/cites/cop18/17aug.html).*

Populations of Botswana, Namibia, South Africa and Zimbabwe (listed in Appendix II):

...For the exclusive purpose of allowing:
...b) trade in live animals to appropriate and acceptable destinations, as defined in Resolution Conf.11.20 (Rev. CoP18), for Botswana and Zimbabwe and for in situ conservation programmes for Namibia and South Africa;...
CoP 18, Resolution 11.20

1. AGREES that where the term 'appropriate and acceptable destinations' appears in an annotation to the listing of Loxodonta africana *in Appendix II of the Convention with reference to the trade in live elephants[1] taken from the wild, this term shall be defined to mean in situ conservation programmes or secure areas in the wild, within the species' natural and historical range in Africa, except in exceptional circumstances where, in consultation with the Animals Committee, through its Chair with the support of the Secretariat, and in consultation with the IUCN elephant specialist group, it is considered that a transfer to ex-situ locations will provide demonstrable in-situ conservation benefits for African elephants, or in the case of temporary transfers in emergency situations;*
2. FURTHER AGREES that, where the term 'appropriate and acceptable destinations' appears in an annotation to the listing of a species in Appendix II of the Convention with reference to the trade in all live animals, this term shall be defined to mean destinations where:
(a) the Management and Scientific Authority of the State of import is satisfied that the proposed recipient of a living specimen is suitably equipped to house and care for it sustainably; and
(b) the Management and Scientific Authorities of the State of import and the State of export are satisfied that the trade would promote in situ conservation;...

[1]*Excluding elephants that were in ex-situ locations at the time of the adoption of this Resolution at the 18th meeting of the Conference of the Parties.*
Resolution Conf. 11.20 (Rev. CoP18)

This resolution is now in force and therefore makes illegal any further exports of African elephant calves to zoos outside the range states of the species that do not meet the new standard under international law. Following the conclusion of CoP18, the future of 37 young elephants captured in Hwange National Park and sold to zoos in Asia, but at that time confined in holding pens in Zimbabwe awaiting export, was reported as uncertain (Flanagan, 2019b). However, export of some of these elephants to China began shortly thereafter as they were not covered by the resolution (see footnote 1 above).

required the calves to be removed from the care of their mothers by separating them using a low-flying helicopter, the exportation of elephants from Zimbabwe was at the time permitted under CITES. Zimbabwe's elephants are listed in Appendix II of CITES which allows for trade in live animals to 'appropriate and acceptable destinations' (Table 9.3).

The fact that some zoos are becoming consumers of wild elephants makes the argument that zoos have a conservation role more difficult to sustain than if none was doing so. In August 2019 CITES strengthened its rules regarding the export of African elephants to zoos (CITES 2019a,b) (see Box 9.7).

9.4.10 A right to companionship and retirement

Scientists have been aware of the complex social lives of elephants in the wild for over half a century (Douglas-Hamilton and Douglas-Hamilton, 1975; Moss, 1988). Nevertheless, some zoos continue to keep solitary and very small groups of elephants (Rees, 2009a). In recent years a number of zoos have decided to transfer their elephants to zoos with better facilities where they would have more companions, or to elephant sanctuaries.

In the European Union, Article 3 of the Council Directive 1999/22/EC of 29 March 1999 on the keeping of wild animals in zoos (the Zoos Directive) requires that zoos must accommodate their animals:

...under conditions which aim to satisfy the biological and conservation requirements of the individual species...

For highly social animals this should mean that access to conspecifics is mandatory, since a high proportion of their normal behaviour is social. In Britain, the law governing the operation and licensing of zoos — the Zoo Licensing Act 1981 — requires them to meet the needs of social species. In England this is reflected in the *Secretary of States Standards of Modern Zoo Practice* (Box 9.8).

In January 2009 the Government of New South Wales approved a *Policy on the Management of Solitary Elephants in New South Wales* pursuant to clause 8(1) of the Exhibited Animals Protection Regulation, 2005. It concludes that:

All reasonable efforts should be made to integrate solitary elephants into other groups unless compelling reasons can be provided that warrant the retention of a solitary elephant. Only in the event that all avenues for integration have been exhausted should the maintenance of a solitary elephant be contemplated.

It is difficult to see what legal or moral justification there could be to conclude otherwise.

Working elephants cannot be expected to continue working indefinitely. The government of the Indian state of Kerala requires working elephants to retire at the age of 65 years, under the Kerala Captive Elephant (Management and Maintenance) Rules, 2003 made under powers conferred under s.64(2) of the Wildlife (Protection) Act, 1972:

Rule 9. Retirement of Elephant. (1) An elephant shall normally be allowed to retire from its work on attaining an age of 65 years:

...(2) Healthy elephants above 65 years of age shall be allowed to be put to light work under [a] proper health certificate from the veterinary doctor.

These rules have been superseded by the Captive Elephant (Management and Maintenance) Rules, 2012. Retired animals are cared for at a facility funded by the state.

> **BOX 9.8** The requirements of the *Secretary of States Standards of Modern Zoo Practice* in relation to elephant social groups in zoos
>
> In England the SSSMZP (2017) Section 4.5 requires that:
>
> *Animals of social species should normally be maintained in compatible social groups.*
>
> This reflects the requirements of the Zoo Licensing Act 1981 s1A:
>
> *…(c) accommodating their animals under conditions which aim to satisfy the biological and conservation requirements of the species to which they belong, including—*
>
> *(i) providing each animal with an environment well adapted to meet the physical, psychological and social needs of the species to which it belongs; and…*
>
> Appendix 8 of the standards (Specialist Exhibits) has provided detailed guidance relating to the appropriate social groupings for male and female elephants since June 2017:
>
> *Cows*
>
> *8.8.12 Female elephants must have social contact with other elephants at all times. If herds are kept, groups should contain at least four compatible females over 2 years old. All elephants must have the option to get away from other elephants if so desired, through use of space and visual or physical barriers in the enclosure. Where evidenced compatibility problems arise, the zoo must keep records of the steps taken to try to resolve these issues, and plans (with time frame) should these steps prove unsuccessful. Such records must be made available to zoo inspectors upon request.*
>
> *Bulls*
>
> *8.8.13 Bull elephants can be difficult to manage (particularly in musth) and are not always compatible with cows. Provision must be made for them to be separated from cows and other bulls when necessary.*
>
> *8.8.14 Bulls must be given the option to be in social contact with other elephants if they choose. Acceptable social situations for bulls are:*
>
> *a) housing bulls so they can mix regularly with the family herd, ideally with another bull present, to facilitate social learning (1 older bull, 1 younger); or*
>
> *b) bachelor herd with other bulls of varying ages.*
>
> *Bachelor herd size should reflect best practice in current guidelines. All elephants must have the option to get away from other elephants if so desired, through use of space and visual or physical barriers in the enclosure.*
>
> Defra, 2017. Secretary of State's Standards of Modern Zoo Practice. Appendix 8 – Specialist Exhibits, Elephants. Department for Environment, Food and Rural Affairs, London. https://www.gov.uk/government/publications/secretary-of-state-s-standards-of-modern-zoo-practice#history (accessed 24.10.18.); SSSMZP, 2017. Secretary of State's Standards of Modern Zoo Practice. Defra, London.

Some zoos and circuses have also recognised that elephants should be retired to sanctuaries (see Section 11.6). In spite of the existence of legislation regarding the treatment of captive elephants in Kerala the state Forests and Wildlife Department issued Circular No.01/2019 in January 2019 in response to concerns about the 'alarming rate of deaths of captive elephants in Kerala'. The department attributed this to poor nutrition, overwork and stress causing an increase in susceptibility to disease. The circular established a requirement for the training of mahouts and owners of elephants, inspections, reporting, monitoring of diets and other welfare measures.

9.4.11 Financial support for in situ conservation

In the United States funds have been established under federal law to support conservation projects for African and Asian elephants, but these funds are focussed on animals living in the wild in range states, not on captive populations in zoos (Box 9.9). However, some of these funds are used for projects undertaken in partnership with foreign zoological societies or the Wildlife Conservation Society (formerly the New York Zoological Society) (Box 9.10).

BOX 9.9 The Asian Elephant Conservation Act of 1997 and African Elephant Conservation Act of 1989

African Elephant Conservation Act of 1989

...**4201.** *Statement of purpose* The purpose of this title is to perpetuate healthy populations of African elephants.

...**4211.** *Provision of assistance*

(a) In general. The Secretary may provide financial assistance under this part [16 USCS 4211 et seq.] from the African Elephant Conservation Fund for approved projects for research, conservation, management, or protection of African elephants.

(b) Project proposal. Any African Government agency responsible for African elephant conservation and protection, the CITES Secretariat, and any organization or individual with experience in African elephant conservation may submit to the Secretary a project proposal under this section...

Asian Elephant Conservation Act of 1997

...16 USC 4262. SEC. 3. PURPOSES.

The purposes of this Act are the following:

(1) To perpetuate healthy populations of Asian elephants.

(2) To assist in the conservation and protection of Asian elephants by supporting the conservation programs of Asian elephant range states and the CITES Secretariat.

(3) To provide financial resources for those programs.

...16 USC 4264. SEC. 5. ASIAN ELEPHANT CONSERVATION ASSISTANCE

(a) In General. The Secretary, subject to the availability of funds and in consultation with the Administrator, shall use amounts in the Fund to provide financial assistance for projects for the conservation of Asian elephants for which final project proposals are approved by the Secretary in accordance with section.

(b) Project Proposal. Any relevant wildlife management authority of a nation within the range of Asian elephants whose activities directly or indirectly affect Asian elephant populations, the CITES Secretariat, or any person with demonstrated expertise in the conservation of Asian elephants, may submit to the Secretary a project proposal under this section...

BOX 9.10 In situ conservation projects funded by the Asian Elephant Conservation Fund and the African Elephant Conservation Fund and partnered with zoological societies

In the fiscal year 2018, the US Fish and Wildlife Service (USFWS) awarded $1,804,897 to projects through the Asian Elephant Conservation Fund. This leveraged an additional $2,977,379 in matching funds. In total, the US Congress appropriated $1,557,000 to the Asian Elephant Conservation Fund (USFWS, 2018a). These funds supported 30 projects in nine countries and included two in partnership with zoological societies outside the United States:

ASE1847 Award # F18AP00379 Living with elephants: Improving human–elephant coexistence in Thailand's Western Forest Complex Corridor. In partnership with the Zoological Society of London.

ASE1604 Award # F16AP00320 Conservation and monitoring of Sumatran elephants in Bukit Tigapuluh, Indonesia. In partnership with the Frankfurt Zoological Society.

In addition, 10 awards were made for work to be undertaken in partnership with the Wildlife Conservation Society (formerly the New York Zoological Society):

ASE1861 Award # F18AP00382 Elephant conservation in Keo Seima Wildlife Sanctuary, Cambodia: Phase 10, Years 1–3.

ASE1828 Award # F18AP00370 Prioritizing conservation landscapes for the Asian elephant in Assam, Northeast India.

ASE1829 Award # F18AP00371 Securing Asian elephant habitats and populations in Kerala, South India through a government-endorsed, comprehensive, landscape-scale conservation plan.

ASE1862 Award # F18AP00383 Saving the critically endangered Sumatran elephant in Way Kambas National Park, Sumatra, Indonesia, by preventing human–elephant conflict and poaching.

ASE1654 Award # F16AP00341 Saving the Sumatran elephant through supporting resort-based management in Bukit Barisan Selatan National Park, Indonesia to reduce encroachment and prevent poaching.

ASE1754 Award # F17AP00334 Saving the Sumatran elephant through supporting resort-based management in Gunung Leuser National Park, Indonesia to reduce encroachment and prevent poaching.

ASE1743 Award # F17AP00328 Assessing elephant habitat connectivity and usage in corridor sites in Pahang and Johor State in Peninsular Malaysia and reducing threats to the Asian elephant population in the Endau Rompin Landscape. In partnership with the Wildlife Conservation Society (WCS) – Malaysia

ASE1750 Award # F17AP00331 Long-term protection and monitoring of the Rakhine Yoma Elephant Range, Myanmar.

(Continued)

BOX 9.10 (Continued)

ASE1650 Award # F16AP00339 Elephant conservation and education campaigns in the Western Forest Complex and capacity strengthening of future protected area managers, Thailand, Year 8–10.

ASE1652 Award # F16AP00340 Law enforcement monitoring and human-elephant conflict mitigation in Kaeng Krachan National Park, Thailand, Year 12–14.

Also, in fiscal year 2018, the USFWS awarded $3,408,886 to projects through the African Elephant Conservation Fund, which leveraged $2,450,350 in additional matching funds. In total, the US Congress appropriated $2,582,000 to the African Elephant Conservation Fund (USFWS, 2018b). These funds supported 18 projects in 13 countries including five awards made for work to be undertaken in partnership with the Wildlife Conservation Society:

AFE1828 Award # F18AP00818 Ensuring the long-term protection of the elephants of Mbam and Djerem National Park, Cameroon.

AFE1829 Award # F18AP00928 Ensuring more effective protection of DRC's largest remaining forest elephant population: Enhancing law enforcement and management of the Okapi Faunal Reserve.

AFE1827 Award # F18AP00826 Disrupting African elephant ivory markets in Lao PDR.

AFE1820 Award # F18AP00676 Further strengthen the protection of elephants at Yankari Game Reserve, Nigeria, through enhanced law enforcement action and monitoring.

AFE1824 Award # F18AP00798 Using innovative ivory anti-trafficking methods to protect elephants of South Sudan: Phase 2.

USFWS, 2018a. U.S. Fish and Wildlife Service Division of International Conservation Asian Elephant Conservation Program FY 2018 Summary of Projects. https://www.fws.gov/international/pdf/project-summaries-asian-elephant-2018.pdf (accessed 18.06.19.); USFWS, 2018b. U.S. Fish and Wildlife Service Division of International Conservation African Elephant Conservation Program FY 2018 Summary of Projects. https://www.fws.gov/international/pdf/project-summaries-african-elephant-2018.pdf (accessed 18.06.19.).

The Government of the United Kingdom funds conservation projects in developing countries through its Darwin Initiative in part fulfilment of its international legal obligations under the UN Convention on Biological Diversity 1992. Some of these projects involve UK zoos.

Chester Zoo was awarded £285,350 (US$460,000) in 2009–12 to work in Indonesia on a project whose aim was to secure human–elephant coexistence in Sumatra. The zoo was also awarded £179,750 (US$360,000) to help build capacities for mitigating human–elephant conflict in Assam in India in 2007–10. In 2005–08 the Darwin Initiative provided £188,188 (US$320,000) to the Zoological Society of London for a human–elephant conflict mitigation project in Thailand (Darwin Initiative, 2019).

In spite of attempts to protect wild elephants and elephants under human care using national, European and international laws, wild populations are still being depleted and many elephants in captivity are experiencing less than optimum welfare.

CHAPTER 10

The conservation value of captive elephants

Elephants have been good for the bottom line; they have been an engine that powers attendance at zoos and circuses.

John Seidensticker, Smithsonian's National Zoological Park, Washington, DC
(Seidensticker, 2008)

10.1 Introduction

Most zoos that keep elephants acquired them — or at least their predecessors — long before conservationists, animal welfare groups and the public in general questioned the purpose of keeping elephants in zoos, and when elephants were commonplace in circuses. Zoos, and those who support keeping elephants in zoos, now find themselves in the position of having retrospectively to find a justification for continuing this practice, especially in the light of the increasing interest in elephant welfare being expressed by animal welfare organisations, governments and the general public.

In North America, the Association of Zoos and Aquariums (AZA) represents more than 230 institutions. Zoos in the United States hold more elephants than those in any other country and the AZA is under constant pressure to justify this (Rees, 2009a). Smith and Hutchins (2000) — employees of the AZA at the time — considered the value of elephant captive breeding programmes to the in situ conservation of elephants. They cited their contribution to public education, scientific research, development of technologies relevant to field conservation, professional training of local conservationists and associated technology transfer, ecotourism, political action taken by zoos, direct support for field conservation and cooperation between zoos and the International Union for the Conservation of Nature and Natural Resources (IUCN). The authors distanced themselves from claims that elephants in zoos represent an insurance population, but considered the value of elephant captive breeding programmes to the in situ conservation of elephants.

Supporters of keeping elephants in zoos offer a range of justifications. According to Gore et al. (2006):

> Elephants are important animals for zoos, particularly for visitor interest, education and, more recently, conservation.

Hildebrandt et al. (2006) claim that zoo populations:

> ...have become paramount for the conservation of all elephant species and subspecies in terms of research, education and fund-raising to support habitat protection and fieldwork.

Some authors offer rather more comprehensive justifications. Hutchins and Keele (2006) have argued that ex situ elephant programmes can:

> ...assist field conservation, directly and indirectly, through public education, scientific research, technology development, training and technology transfer and fund-raising to support field conservation initiatives.

This mirrors the earlier claims of Smith and Hutchins (2000). Wiese and Willis (2006) claimed that zoo elephants can:

> ...provide significant benefits to their conspecifics in the wild by increasing awareness, encouraging financial giving and political action, and providing a connection between zoo and field colleagues for a variety of conservation efforts...

The AZA's website discusses what its zoos do to contribute to the conservation of elephants:

> How do AZA-accredited zoos contribute to elephant conservation?
>
> The elephant conservation missions of AZA-accredited zoos encompass a wide range of activities, including conservation education, research, development of relevant technologies, professional training, habitat restoration, ecotourism, community-based conservation, the direct support of national parks and equivalent reserves, and fundraising to support these initiatives.
>
> AZA (2019b)

> While supporting conservation programs in the wild, AZA facilities are also caring for approximately 160 African and 140 Asian elephants in over 60 AZA-accredited facilities.
>
> AZA (2019c)

Although zoo mission statements have previously suggested an important role for zoos in captive breeding of animals and education (Rees, 2011), many now focus on more general aims:

> Preventing extinction [Chester Zoo (NEZS, 2019)].

> A world rich in wildlife and wild places [Newquay Zoo (Anon., 2019c)].

> To connect people with nature and safeguard threatened species [Royal Zoological Society of Scotland (RZSS, 2019)].

In recent studies published as part of the *Using Science to Understand Zoo Elephant Welfare* project in North America — funded largely by the Institute of Museum and Library

Services, in the amount of US$845,720 (IMLS, 2018) — no mention is made of the conservation justification for keeping elephants in zoos. In fact, most do not use the term 'conservation' at all (Brown et al., 2016; Greco et al., 2016a; Holdgate et al., 2016a, b; Meehan et al., 2016a; Miller et al., 2016; Morfeld et al., 2016; Prado-Oviedo et al., 2016) and only one makes just a passing reference to elephants' role as 'conservation ambassadors' (Meehan et al., 2016b). The only justification for the existence of ex situ populations of elephants linked to this project is provided by a short comment paper by Cameron and Ryan (2016) that introduces the collection of papers. Here, they:

> ...call attention to the role of zoo elephant research to in-situ conservation efforts, and stress the importance of understanding welfare, as a component of the scales of integral research into elephant conservation.

Although these papers are primarily concerned with improving elephant welfare, it is remarkable that the authors do not acknowledge that the compromised welfare of elephants living in zoos is a product of human actions and desires and these require justification.

Not surprisingly, most of the authors who believe that elephants living in zoos have a conservation function have a professional connection with zoos. That does not necessarily make them biased. If they did not genuinely support the work of zoos it seems unlikely that they would want to spend their careers working for, or with, them. However, a number of scientists who conduct research on elephants in the field have expressed the view that elephants should not be kept in zoos. In October 2015, 80 scientists, conservationists, elephant care, animal welfare and policy experts and former zoo directors signed a statement opposing the importation of 18 elephants from Swaziland into the United States by Dallas Zoo in Texas, Henry Doorly Zoo in Omaha, Nebraska, and Sedgwick County Zoo in Wichita, Kansas (Anon., 2015). The scientists included Joyce Poole, Cynthia Moss and Phyllis Lee, all of whom have published extensively on the ecology and behaviour of wild elephants (e.g. Lee, 1986, 1987, 1989; Moss, 1983, 1988, 2001; Poole, 1989, 1999, 2010). The list of reasons given included:

> 1. The capture and removal of wild elephants for display in zoos is detrimental to elephants...
>
> ...5. Zoos are capturing and importing wild elephants to restock a dwindling zoo elephant inventory, not to conserve the species.
>
> Despite proof of the systemic failures of zoo practices and policies affecting elephants, the zoo industry has become more resolute in seeking out nations abroad from which to plunder elephants to restock zoo exhibits. This practice is an unacceptable consequence of the unnatural conditions provided by zoos, and should not be allowed to continue.
>
> This importation serves no credible conservation purpose. None of the elephants or their offspring will be returned to the wild, the gold standard of conservation. Instead, it is intended to replenish the zoo industry's dwindling African elephant population in the U.S.
>
> Anon. (2015)

There is some evidence that, in spite of the views of some scientists, visitors to zoos believe that there is a place for elephants in zoos. When 1294 visitors to elephant exhibits in US zoos were asked to respond to the statement, 'I believe it is important to have elephants in zoos', 96% responded at least 'somewhat' on a scale ranging from 1 to 7 (where 1 = not at all, 4 = somewhat, 7 very much) (Miller et al., 2018). However, these responses are meaningless because the question asked had no context — important for what? — and the respondents voluntarily attended a zoo that kept elephants. The same question asked of randomly selected members of the public would probably have produced a different conclusion.

10.2 The popularity of elephants in zoos

Scientists have noted that: 'Elephants are important animals for zoos, particularly for visitor interest' (Gore et al., 2006). There is nothing quite like the smell of an elephant house. Those who advocate the keeping of elephants in zoos as ambassadors for conservation often argue that seeing, smelling and getting close to — or even touching — an elephant does not compare with simply watching elephants in a television documentary. This is undoubtedly true, regardless of whether or not such experiences with living animals affect the attitude of zoo visitors to wildlife in general or elephants in particular.

Modern zoos are putting up more and more physical barriers between visitors and elephants. It was once common to be able to touch the trunk of an elephant by reaching out across a dry moat, stroke a calf on the other side of a low steel fence or meet an elephant on a zoo lawn (Fig. 10.1). Many new elephant houses have increased the separation between the animals and the visitors. The elephants in the elephant house at Dublin Zoo, Ireland, are exhibited behind a thick window, masking both the sound and smell of the animals (Fig. 10.2). When the elephant house was modernised at Chester Zoo, United Kingdom, the animals were moved further back from the public viewing area (Fig. 10.3). These changes improve safety — for elephants and people — but make it harder for visitors to make a connection with an individual animal than had previously been the case.

Elephants are a huge attraction in zoos. Studies have found that elephants were the most popular mammal in the United States (Kellert, 1989) and among the top 10 favourite animals of British children (Morris and Morris, 1966). Dr John Seidensticker, formerly Curator of Mammals at the Smithsonian's National Zoological Park, Washington, DC, acknowledged that:

> *Elephants have been good for the bottom line; they have been an engine that powers attendance at zoos and circuses.*
>
> Seidensticker (2008)

Maryland Zoo knew why it wanted to improve its elephant exhibit. The zoo's application for state funding explained that:

> *The proposed expansion and renovation of the current elephant facilities is intended to address the zoo's declining attendance.*
>
> Anon. (2006)

No mention was made of any conservation-related objectives.

Sach et al. (2019), in considering the cost of keeping elephants at ZSL Whispnade Zoo, United Kingdom, have acknowledged the financial benefits of elephants to the zoo:

> *Commercially, the benefits of keeping elephants are substantial, although very difficult to quantify accurately within a large, multi-exhibit zoo.*

FIGURE 10.1 An Asian elephant calf (*Wendy*) and an African elephant calf (*Christine*) being introduced to visitors on the lawn at Bristol Zoo, United Kingdom (c.1960).

Nevertheless, the authors estimated that the presence of elephants contributes to visitor numbers by up to 24%, increases dwell time by 13% and contributes around 6% of total visitor spend per head. They estimated that the cumulative commercial cost of removing elephants from the zoo could exceed £3.5 million (US$4.5 million) in the first year (based on a 24% reduction in visitor numbers multiplied by a reduced spend per head).

Clubb and Mason (2002) claim that there is some evidence that when a zoo stops keeping elephants it has no effect on visitor numbers. However, a zoo elephant birth generates a great deal of public interest and undoubtedly causes at least a temporary increase in visitors. In 2000 Chester Zoo, after a period of declining visitor numbers, saw annual attendance exceeding 1 million for the first time since 1973 (Rees, 2011). The two Asian elephants born in this year undoubtedly contributed to this achievement. In May 1977 an Asian elephant named *Jubilee* was born at Chester: the first time an elephant had been conceived and bred in the British Isles. The zoo's Annual Report for 1977 noted that:

Largely as a result [of the birth], *we are pleased to report that our attendances showed an increase of 46,612 on those of 1976.*

NEZS (1977)

FIGURE 10.2 Visitors to the elephant house at Dublin Zoo, Ireland, have a somewhat sterile experience because the elephants are viewed from behind glass.

This was in a year that a number of zoos had closed down in Britain.

New technology has allowed the public to engage directly with elephant births. When the birth of an Asian elephant was imminent at Zürich Zoo, Switzerland, in 2000, over 40,000 people joined an Internet mailing list in advance of the birth so that they could be notified of the event, which was then broadcast live on the zoo's website (Rees, 2001b).

A number of studies have shown that visitors, especially children, want to see large animals in zoos, including elephants (e.g. Carr, 2016; Moss and Esson, 2010). Ward et al. (1998) found that exhibits of larger animals were preferred by adults and children. They concluded that, although smaller zoo animals could potentially contribute more to conservation — due to faster breeding rates and lower costs — zoo exhibits could become less popular if they concentrated on smaller species. However, Balmford (2000) has shown that the return per unit investment — in terms of visitor time budgets — is no greater for exhibits of larger animals than for those of smaller animals.

Other studies suggest that we enjoy visiting zoos because they invoke happy childhood memories of time spent with one's family (e.g. Holzer et al., 1998). What could be more memorable than seeing an elephant at close quarters? But fewer children are now seeing elephants in zoos than in the recent past simply because so many zoos that once had elephants no longer keep them. Children were able to see elephants at ZSL London Zoo for 170 years. The last elephants were moved to ZSL Whipsnade Zoo in 2001 after a continuous presence in London since 1831. The childhood association between elephants and zoos will never be made for many of today's children. Although over the past two decades many zoos have closed their elephant exhibits, those zoos have survived.

In a study of the ideal traits of zoo animals as perceived by the general public ($n = 246$) the top three traits were that they should be 'endangered', 'active' and should 'display intelligence' (Carr, 2016). When asked which animals they most wanted to see (anywhere, not just in a zoo), elephants were most frequently selected — by 55 (22.4%) of 246

FIGURE 10.3 Modern elephant houses are moving the public further away from the animals. (A) The rails in the old elephant house at Chester Zoo, United Kingdom, allowed very young calves to come into contact with members of the public. (B) The barrier in the new elephant house at Chester prevents contact with visitors by moving the elephants further back from the public area.

respondents. Over 98% of respondents stated that zoo visits were made more attractive by the presence of baby animals. Monkeys were selected as having the most attractive babies (41% of respondents), followed by orangutans (20.3%) and gorillas (19.9%); surprisingly, only 6 respondents (2.4%) said elephants had the most attractive babies. This study was conducted on the Island of Jersey, which may have influenced the results since its only zoo is Durrell (Durrell Wildlife Park, formerly Jersey Zoo) and almost all the respondents had visited it. Durrell had two baby orangutans and one baby gorilla at the time of the study, and has never kept elephants.

FIGURE 10.4 Members of the press are introduced to a newborn elephant calf at Chester Zoo, United Kingdom.

Public attitudes to the keeping of elephants in zoos vary between cultures and educational level. Gurusamy et al. (2015) found that Australians were more concerned about elephant husbandry conditions in zoos and sanctuaries than were Indians, and that this concern increased with higher education levels. More Australians than Indians were prepared to pay extra to visit a zoo with elephants, but more Indians believed that it was important for zoos to display elephants and to interact with them by touching, feeding and riding them. Indian respondents considered elephants to be primarily of religious, cultural and historical significance while Australians acknowledged their scientific value.

There is considerable evidence that large animals, and especially elephants, are very popular with the public and an elephant birth attracts considerable public and press attention (Fig. 10.4). However, we must acknowledge that knowing that 'Elephants are important animals for zoos, particularly for visitor interest' (Gore et al., 2006) is not of itself a justification for keeping elephants in zoos.

10.3 Zoo elephants as insurance populations

10.3.1 Introduction

Zoos once portrayed themselves as modern 'arks' that held 'insurance populations' of rare animal species against the possibility that they would one day become extinct in the wild and anticipated that these populations could be used to reintroduce animals back into their former ranges. More recently zoos have largely abandoned this justification for their existence — because so few species have been 'rescued' in this way — and turned their attention to their role in education, research and the support of field conservation (Smith and Hutchins, 2000; Wehnelt and Wilkinson, 2005).

Scientists who work with elephants in zoos are beginning to be more cautious about claiming a role for these elephants in supplementing wild populations. In a review of the contribution of zoo elephant research to the conservation of captive and free-ranging elephants, Bechert et al. (2019) state that:

> There is potential [sic] for captive populations to serve as a source of elephants for reintroduction if wild populations decrease to critical numbers (and suitable habitat is available), although many captive populations are not self-sustaining.

Furthermore, they acknowledge that there are 'many populations' in Africa and Asia that are thriving and some of these need to be controlled. They recognise the irony of controlling some populations of elephants despite their global decline. However, they use the fact of the existence of thriving populations to justify using zoo elephants for research on 'population control through reproductive technology' without acknowledging that surplus elephants in one area could be used to increase numbers elsewhere.

10.3.2 The Species Survival Plan in North America

In 1981 the AZA established Species Survival Plans (SSPs) as a cooperative population management programme for selected species at North American zoos and aquariums. The Asian elephant SSP was established in 1985, almost 200 years after the first Asian elephant was taken to the continent (Haufellner et al., 1993). In 1990 the African elephant was included in this SSP as many of the management issues are the same for both species.

The Asian elephant studbook is maintained by Oregon Zoo (formerly Metro Washington Park Zoo). This zoo achieved remarkable early success in breeding Asian elephants. Between 1962 and 1996, the zoo produced 25 calves. Of these, 20 were born between 1962 and 1991 – at the time, the best zoo breeding record in the world (Gröning and Saller, 1999). Schulte (2000) reported that there were 46 males (eight castrated) and 239 females in the North American population in 73 facilities in 1999. Males and females were present together in 23 facilities (31.5%). At November 1999, there were approximately 22 females and 14 males of 10 years and younger. In 2010 the Asian Elephant North American Regional Studbook recorded 725 elephants, of which 269 (53.216) were living and 145 (31.114) were housed in AZA facilities (Keele et al., 2010). As I write, AZA facilities hold approximately 160 African and 140 Asian elephants in over 60 AZA-accredited facilities (AZA, 2019d).

The United States has the largest ex situ captive population of Asian elephants of any single country. However, the US Government has specifically directed federal funds towards in situ (but not ex situ) elephant conservation programmes. The Asian Elephant Conservation Act of 1997, while providing financial assistance for conservation within range states and to support the Convention on International Trade in Endangered Species of Wild Fauna and Flora 1973 (CITES) Secretariat, specifically excludes the use of funds for captive breeding other than for release in the wild [Sec. 5 (h)] (see Box 8.9).

10.3.3 The European Endangered species Programme

The first Asian elephant calf produced in Europe was stillborn in London in 1902, some 500 years after the species was first regularly seen in European menageries (Carrington, 1958). A female born at Schönbrunn Zoo in Vienna, Austria, in July 1906 survived until 1944 and the following December a female calf was born at Berlin Zoo, Germany, but died after 24 days (Blaszkiewitz, 1999). In these early days, there was no coordinated breeding programme for elephants.

The European Endangered species Programme (EEP) — now the European Association of Zoos and Aquaria Ex-situ Programme — for Asian elephants was established in 1991 and the studbook is held by Rotterdam Zoo in The Netherlands. The African elephant EEP is managed separately, unlike the joint species SSP in North America. The first report of the Asian elephant EEP listed 41 bull and 171 cow elephants living in 76 zoos at 31 December 1991 (Anon., 1992). At this time, the largest European herd (1 bull, 9 cows) was held at Emmen Zoo, Netherlands. By 31 December 1998, the EEP included 79 zoos holding 50 bulls and 199 cows (Anon., 2000). At the beginning of 1993, the ratio of bulls to cows was 1:5 but by 1998, it had increased to 1:4.

Schmid (1998) surveyed the status and reproductive capacity of the Asian elephants in European zoos and circuses. She found that in 1992 a total of 503 animals were distributed between 110 zoos and 51 circuses with a male—female ratio of 1:9. Although there had been some breeding success by the end of the 20th century, Kurt (1994) predicted that the European elephant population would not survive more than 30 years. A review by Schmidt and Kappelhof (2019) found that at 1 January 2018 the Asian elephant EEP contained 307 elephants which produced, on average, 15 calves per year.

10.3.4 Are zoo elephant populations sustainable?

Zoo populations of endangered species are often referred to as 'insurance populations'. If there is some prospect of returning individuals to the wild this concept has some merit; however, if there is not, then a captive breeding programme simply becomes a means by which animals are produced to supply the needs of zoos.

Robert Hoage, retired public affairs director of the Smithsonian's National Zoo in Washington, DC, has claimed that

If zoos don't get involved elephants might just exist in museums.

Cohn (2006)

But is it true? Some authorities have considered zoo elephant populations as a 'reservoir for threatened in situ populations' (Faust et al., 2006). However, it is not at all clear that this can be justified in the context of their small size — compared with the relatively large number of elephants still remaining in the wild — low fecundity, the zoo community's inability to produce sustainable populations and continued importations from the wild (Hutchins and Keele, 2006; Rees, 2003b, 2009a; Wiese, 2000).

Zoo professionals themselves have reported that zoo populations of Asian elephants in America are not self-sustaining and that the future for African elephants is uncertain (Olson and Wiese, 2000; Wiese, 2000). More than a decade ago, Hutchins and Keele (2006) suggested that zoos needed to make preparations for an increased number of importations from the wild to prevent further decline. These studies suggest that urgent action is necessary, but none of the authors attempts to explain in detail the purpose of the animals whose numbers they seek to increase.

In his demographic analysis of captive Asian elephants in North American zoos, Wiese (2000) demonstrated that the population was not self-sustaining. He made no mention of the failure of zoos to establish a breeding programme with the potential to eventually return captive-bred elephants to the wild, but appeared more concerned with the potential disappearance of the species from North America. Wiese estimated that, at that time, four elephants per year needed to be imported into North America simply to maintain the population at its current level, but this is based on a model that makes extremely optimistic assumptions. He predicted that the population would drop to approximately 10 elephants within 50 years (i.e. around 2050) and become demographically extinct.

Wiese (2000) suggested that the future need for the importation of Asian elephants into North America may make this species a perfect candidate for the establishment of an extractive zoo reserve (Conway, 1998). This concept proposes to manage habitat reserves intensively such that wild populations can sustain an extractive harvest for use by zoos. Some forest timber camps in India have been reported to be self-sustaining or growing (Sukumar et al., 1997) and Wiese has suggested that these camps may be a logical source for trial extractions.

The coordinators of the African and Asian EEPs have previously stated that it would be unacceptable for elephants from captive situations in Southeast Asia or from culling operations in Africa to 'be the key factor for maintaining a genetically healthy and thriving elephant population within our zoos' (Dorresteyn and Terkel, 2000b).

If wild or in situ captive populations become sustainable to the point where they are able to supply zoos with excess animals, this must put into question the conservation need for an ex situ zoo population with no immediate prospect of returning captive-bred elephants to the wild. The translocation of excess animals to areas where elephants are absent or in decline would arguably make a greater contribution to elephant conservation than would their exportation to foreign zoos. Such translocations have already been used in Asia. For example between 1974 and 1995 the Malaysian Wildlife Department translocated 392 crop-raiding elephants to three protected areas (Abdul et al., 1996). The population ecology of elephants in zoos is discussed in more detail in Section 6.6.

10.4 Scientific research

Tudge (1991) has claimed that zoos — and by implication individual zoo exhibits like those containing elephants — can only be justified as sources of knowledge if the knowledge is used for conservation. Some legislators have taken the same view. In the European Union the Zoos Directive (Council Directive 1999/22/EC of 29 March 1999 on the keeping of wild animals in zoos) requires zoos in Member States to have a conservation function. One way this may be achieved is by undertaking research from which conservation benefits accrue. However, most of the research conducted on animals living in zoos is concerned with some aspect of their biology that is affected by their presence in a zoo, such as their behaviour, welfare, nutrition or reproduction (Rees, 2005a). This is true for elephants.

Of the 115 papers on captive elephants published to December 2017 in the journal *Zoo Biology* ($n = 108$) and the *Journal of Zoo and Aquarium Research* ($n = 7$), 50 (43%) were concerned with some aspect of reproduction (such as calf development, population biology, sex hormones, monitoring of oestrus and ultrasonography of the reproductive organs). These papers were principally concerned with improving the success of captive breeding programmes. Many of the remaining papers were chiefly concerned with problems associated with the husbandry of elephants in zoos (e.g. foot health, disease, stereotypic behaviour, social behaviour, nutrition, well-being, enclosure use). Papers on captive elephants are also published in other academic journals, but the two considered here were the only publications dedicated to zoo research.

Some papers claimed that field studies of elephants benefit from technical support from zoo-based specialists in nutrition, physiology, pathology, reproduction and veterinary science. Some authors also refer to the contributions research on elephants in zoos has made to our understanding of the animals' reproductive biology, the nutrition of captive elephants and elephant communication (Hutchins et al., 1996; Smith and Hutchins, 2000).

After a further two decades of work, Bechert et al. (2019) reviewed the contributions made by zoo elephant research to the conservation of captive and free-ranging elephants. By this time, an impressive body of work had accumulated. They concentrated on studies concerned with pharmacology, nutrition, sensory biology, reproduction and disease (Fig. 10.5), but acknowledge that most research was concerned with welfare and husbandry:

> Research relevant to African and Asian elephants in human care focuses primarily on husbandry and animal-welfare issues, although many of these research areas also benefit free-ranging or semi-free-ranging populations.

While all this work adds to our knowledge of elephants, most of it does not help us to conserve wild elephants and its conservation value will only be realised if we really do need to keep insurance populations in zoos. Wild elephants breed perfectly well left to themselves. We cannot justify keeping elephants in zoos merely to conduct research on the behavioural, health and reproductive problems created by captivity.

10.5 The development of technologies relevant to field conservation

Some wild elephant populations have declined because of poaching and competition with humans for space. A few populations are routinely managed by translocations, the use of contraception or culling. Left alone, and given sufficient space, elephant populations thrive without human intervention. There is no evidence that wild elephants have any significant reproductive problems that would prevent their recovery after a significant decline in numbers. The lowest mean calving intervals recorded for wild elephants are around 3.5 years (Sukumar, 2003). The Addo elephant population in South Africa grew at an annual rate of approximately 7% between 1953 and 2000 (from under 20 individuals to 325) once the park was fenced (Sukumar, 2003). Spinage (1994) claimed that this was the fastest rate recorded for an elephant population where there was zero immigration. In Tarangire National Park, Tanzania, following a reduction in poaching (after the ivory trade ban in 1989) and several consecutive wet years, the population of female elephants and infants increased at an annual rate of 10% (Foley, 2002). Furthermore, elephants are perfectly capable of recolonising their former habitats after an extended period of absence.

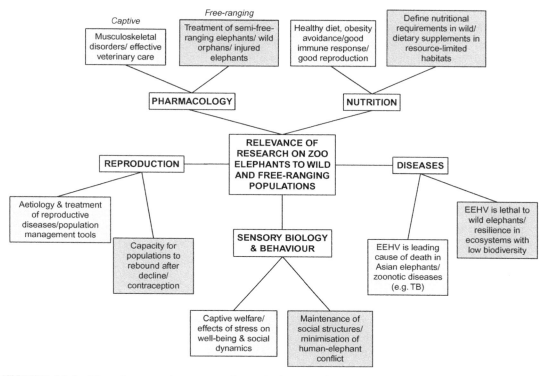

FIGURE 10.5 The relevance of zoo research on elephants to free-ranging (grey boxes) and captive (white boxes) elephant populations. Source: *Based on information in Bechert, U.S., Brown, J.L., Dierenfeld, E.S., Ling, P.D., Molter, C.M., Schulte, B.A., 2019. Zoo elephant research: contributions to conservation of captive and free-ranging species. International Zoo Yearbook 53, 1–27.*

The recolonisation of Kruger National Park in South Africa (partly by immigrants from Mozambique) took almost 80 years, although it took 150 years for elephants to reach the extreme southwest corner, in 1982. In the 1950s elephants repopulated Etosha National Park from the surrounding area after an absence of 70 years (Spinage, 1994).

Smith and Hutchins (2000) argue that research on elephant reproduction has led to the development of contraceptives that have been used to control expanding populations in Kruger National Park and elsewhere (Fayrer-Hosken et al., 1997). But this development, like advances in artificial insemination (AI), is attempting to solve a problem that is entirely man-made. Protecting African elephants from hunting and providing them with water-holes during droughts that would otherwise have controlled their numbers has led to populations being able to double within a decade (Pimm and van Aarde, 2001). Some ecologists oppose the use of contraception on ecological grounds and because it uses up financial resources allocated for conservation. There is also evidence that contraception affects the social behaviour of bulls (Doughty et al., 2014).

Some of the knowledge gained from zoo research on elephants is clearly intended to be applied to wild populations [e.g. feeding ecology (Rees, 1982); satellite tracking

(Nobbe, 1992); noninvasive collection of DNA (Ryder, 1990)] or increases our scientific knowledge about the general biology of elephants [e.g. dust bathing (Rees, 2002a)]. But we cannot pretend that all research on zoo elephants is directly related to the conservation of wild elephant populations. However, that does not mean it is not useful and interesting. But if there were no elephants in zoos it seems unlikely that we would establish facilities containing elephants purely for zoological research. Scientists conduct research on zoo elephants because they are there; elephants are not kept in zoos for the purpose of carrying out research. If we accept this argument, then we cannot justify the keeping of elephants in zoos simply because they allow us to increase our knowledge. Scientists have conducted research on circus elephants (e.g. Benedict, 1936; Gruber et al., 2000), but would we accept this as a justification for the presence of elephants in circuses?

Would we know less about elephants if they had never been kept in zoos? Undoubtedly yes. Would the lack of this knowledge compromise efforts to conserve elephants in situ? If it would, we would expect there to be unanimous support for the work of zoos from elephant scientists working in the field, and yet some of the world's best-known elephant scientists have spoken out against keeping elephants in zoos (Anon., 2015).

10.6 Educational function

Zoos generally claim to have an educational function, and many function as educational charities. Indeed, in some parts of the world — especially the European Union (Directive 1999/22/EC) — they are required by law to demonstrate such a function (Rees, 2005a, b). Zoo professionals claim specifically that elephants in zoos have an educational function (Gore et al., 2006; Hildebrandt et al., 2006; Wiese and Willis, 2006). More recently, in their review of zoo elephant research, Bechert et al. (2019) state that:

> The purpose of zoos has changed over time from primarily providing entertainment for visitors to creating opportunities for public education and research (Barongi et al., 2015).

The citation in this quotation is, however, not an academic source based on any evidence of an educational value, but a reference to the World Zoo and Aquarium Association's conservation strategy. This publication mentions education 29 times and elephants just three times, but never in the same sentence.

Most zoos that keep elephants have signage explaining various aspects of their biology and some provide demonstrations to the public to illustrate their intelligence and physical abilities, showcase the zoo's training methods or convey a conservation message (Fig. 10.6). In the past at Blackpool Zoo, the Asian elephants demonstrated cooperation by carrying a log, collecting litter in their enclosure and pulling a sled using a harness (Fig. 10.7). The elephants at Blackpool no longer do this, but their new enclosure contains a great deal of useful elephant facts on its signage and there is a static display of equipment used to store bees kept to deter elephants from damaging crops (Fig. 10.8). The Asian elephants at ZSL Whipsnade Zoo have been taught to 'demolish' a specially built hinged hut to demonstrate human—elephant conflict in Asia (Fig. 10.9).

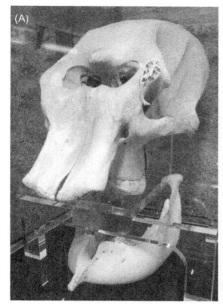
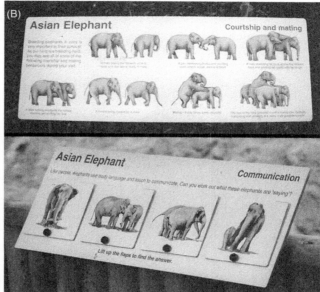
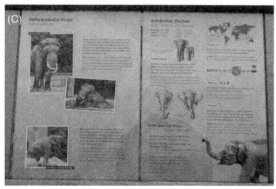

FIGURE 10.6 Interpretation and signage vary between elephant exhibits. (A) The skull of an Asian elephant that forms part of an educational exhibit in the elephant house at Blackpool Zoo, United Kingdom. (B) Signage at Chester Zoo, United Kingdom, showing courtship and mating in Asian elephants (top) and communication (bottom) (based on author's photographs). (C) Signage providing information about the Asian elephants at Berlin Zoo, Germany. (D) A child-friendly sign designed to convey a single fact about elephants at Blackpool Zoo, United Kingdom.

In their paper justifying the keeping of elephants in zoos, Smith and Hutchins (2000) presented no scientific evidence for an educational role and merely quoted a small number of studies that appeared to show the value of live animals – not specifically elephants – in stimulating curiosity (e.g. Saunders and Young, 1985).

If zoos claim an educational function for 'their' elephants they cannot merely say it is so, but should be held to the same standards as others who provide education; they need

FIGURE 10.7 Elephants performing in their old enclosure at Blackpool Zoo, United Kingdom: (A) collecting litter; (B) two elephants cooperating to carry a log (sled in foreground); (C) fitting a harness; (D) pulling a sled (out of sight) with a harness.

to demonstrate an increase in knowledge and understanding, or a change in long-term behaviour following exposure to elephants. This is particularly important in the current climate because there has been much recent concern about the welfare of elephants in zoos, notably in North America (e.g. Carlstead et al., 2013) and the United Kingdom (e.g. Chadwick et al., 2017).

The educational value of zoo visits in general is equivocal. Most studies are small-scale and many are concerned with a single exhibit in one zoo. Some studies appear to demonstrate an educational function, while others do not (Rees, 2011). Indeed, some zoo educators have acknowledged that:

> *Zoos exude a certain self-confidence regarding their roles as education providers*

FIGURE 10.8 A model of a beehive fence used to deter elephants from raiding crops is part of the Asian elephant exhibit at Blackpool Zoo, United Kingdom: (A) model of two hives suspended from a rope fence; (B) explanatory sign.

FIGURE 10.9 A demonstration of the effects of human–elephant conflict in Asia: (A) Asian elephants at ZSL Whipsnade Zoo, United Kingdom, 'destroy' a village hut; (B) the simulated building is hinged so that it can be returned to its original shape after the demonstration.

and that,

> ...[educational] *outputs do not necessarily lead to outcomes.*
>
> Moss and Esson (2013)

Balmford et al. (2007) examined the effects of a single informal visit by 1340 respondents to one of seven zoos in the United Kingdom. They found little evidence of any effect on adults' conservation knowledge, concern or ability to do something useful. Broad (1996) found that knowledge about threatened species gained from a zoo visit had not influenced visitors in any way in 80% of cases when they were contacted 7–15 months later. Some zoo visits have even apparently led to a decrease in wildlife knowledge and an increase in 'dominionistic' attitudes to conservation in visitors (Kellert and Dunlap, 1989). A recent study of the impact of a global biodiversity education campaign – relating to the Aichi Biodiversity Targets of the UN Convention on Biological Diversity – examined almost 5000 visitors to 20 zoos and aquariums located in 14 countries and concluded that visitors to these institutions showed an increased understanding of biodiversity and actions to protect it (Moss et al., 2017). However, this study measured only the effects of exposure to campaign materials, not the effects of seeing captive animals.

Educational claims made by zoos are often vague and unsupported by good evidence. A survey of 50 member zoos of the European Association of Zoos and Aquaria (EAZA) with elephants found that 90% had signs that taught about general biology, 40% included an explanation of elephant training and 60% included information on conservation (Terkel, 2004). However, Terkel conceded that:

> ...the message on conservation is somewhat confusing as there are surplus populations in major areas of Africa and many instances of human–animal conflict for both species.

According to the AZA, 95% of American adults surveyed in 2005 agreed that seeing elephants in real life fosters a greater appreciation of the animals (AZA, 2007a). On its own, this claim means very little. What does 'greater appreciation' mean in terms of changes in behaviour? Are these people more likely to contribute to elephant conservation (financially or in some other way) than those who have not seen an elephant in a zoo?

Zoos have used the argument that elephants have a role in conservation education in an American court to justify further importations of elephants from a range state. In 2003, Born Free USA attempted to obtain an injunction to prevent the importation of African elephants from Swaziland into the United States. In considering the public interest in allowing the importation, the court considered the zoos' claim that viewing elephants in zoos promotes and encourages conservation (*Born Free USA v Norton*, 2003):

> ...D. Public Interest
>
> It is not clear where the public interest lies in this dispute. The zoos aver that sustaining a viable population of African elephants in North America will serve the public interest because the public will be able to view the elephants and because such viewing will promote and encourage conservation of the world's resources. But however sensible that may seem, it may be that the public (or a substantial portion thereof) sympathizes with plaintiffs' view that zoos are improper places to keep wild animals and that the elephants are even better off culled than at these zoos. [The zoos claimed that culling was the likely fate of the elephants if they remained in Swaziland].

The court refused to grant an injunction to prevent the importation of the elephants because the plaintiffs failed to show that the interests of the elephants would be irreparably harmed or that an injunction would be in the public interest. Although this case did not settle the argument regarding whether or not elephants living in zoos have an educational function – as that was not its purpose – nevertheless it is a further example of zoos making unsubstantiated claims for their educational value.

A very small number of studies have looked specifically at elephant exhibits in zoos. When Smith and Hutchins (2000) claimed an educational function for elephants living in zoos the only study that appears to have been conducted was that of Swanagan (2000). He studied 350 visitors to Atlanta Zoo, Georgia (where he was deputy director at the time), and found that those who had an interactive experience with the zoo's elephant demonstration and 'bio-fact program' were more likely to support elephant conservation — by signing a petition or returning a solicitation card lobbying against the importation of elephant parts — than those who simply viewed the animals in their exhibit and read the associated signage. Only 18.3% of participants returned the solicitation cards — just 64 visitors. Smith and Hutchins did not mention this study or, indeed, any other study that was specifically concerned with the educational role of elephants in zoos.

Smith and Broad (2008) studied visitor behaviour at the *Trail of the Elephants* exhibit at Melbourne Zoo, Australia, but were concerned with determining dwell time — the time spent by visitors in the exhibit — rather than measuring the extent to which visitors improved their knowledge or changed their behaviour following their visit. Moss et al. (2008) studied the relationship between the size of the viewing area and visitor behaviour in the Asian elephant exhibit at Chester Zoo, United Kingdom. Neither of these studies addressed the question of whether the presence of living elephants affected attitudes towards their conservation or had any other specific educational function. Subsequently, Moss and Esson (2010) studied the popularity of different animal taxa — including Asian elephants — at Chester Zoo, and showed — not surprisingly — that zoo visitors are particularly interested in seeing large mammals.

Two recent studies have attempted to link close-up encounters with elephants in zoos and 'conservation intent' in visitors. Hacker and Miller (2016) reported that experiencing close-up encounters with elephants and observing active elephant behaviours resulted in the greatest changes in 'conservation intent' reported by visitors to the San Diego Zoo Safari Park in California. A similar study conducted by Miller et al. (2018) analysed 1294 questionnaires completed by visitors following observations of elephants kept at nine zoos in the United States accredited by the AZA. They concluded that 'up-close experiences' watching elephants engage in active species-typical behaviours correlated with a positive emotional experience and visitors' interest in getting involved in conservation. However, they also noted that most visitors arrived at elephant exhibits with very receptive predispositions towards wanting to learn more and get more involved in conservation. In response to the statement, 'I now have a better understanding of what actions I can take that will help protect and preserve elephants and their habitats', the mean visitor response was 4.55 (s.d. = 1.61) on a scale from 1 (strongly disagree) to 7 (strong agree). Overall, the respondents, before they visited the elephant exhibits, reported that they already supported conservation organisations (mean = 4.29, s.d. = 1.74; where 1 = not at all, 4 = somewhat, 7 = very much), but their visit did not appear to have encouraged them to want to donate to an elephant conservation organisation and the question asking about this received the lowest score of the 29 'questions' posed (mean = 3.93, s.d. = 1.69).

Asking the public about their future intentions is problematic. Recording a future *intention* to become involved in conservation on a questionnaire is not the same as recording an actual *change* in future behaviour. There is evidence that 'intention' in relation to the protection of the environment fails to predict future behaviour (Davies et al., 2002). Indeed, the very act of measuring intention may inflate the association between intention and behaviour — a phenomenon known as 'self-generated validity' (Chandon et al., 2005). Put simply, we cannot determine what someone will do in the future simply by asking their intentions and research that implies that we can is misleading.

ElephantVoices, an organisation run by Joyce Poole, believes that elephants do not belong in zoos and they are not necessary for zoos to educate the public about elephants (see Section 11.2).

Elephants living in zoos may have an educational value. But the scant evidence available for this is unconvincing and, of the five studies concerned specifically with elephant exhibits discussed above, only Smith and Broad (2008) was written by authors who did not have a vested interest in zoos. Unless independent scientists are able to demonstrate conclusively that visitors exposed to elephants living in zoos are more likely to support elephant conservation as a result — and more likely to do so than if only exposed to other information about elephant conservation in the media and elsewhere, or to museum specimens — then it is difficult to see how zoos can continue to justify an educational function in keeping them.

In spite of the lack of evidence that elephants living in zoos have an educational role, zoo professionals continue to refer to its importance when discussing elephants:

> *Positive animal welfare is essential for a zoo to fulfil the underlying mission to raise awareness of conservation issues, public education and (re)-connecting people with animals.*
>
> Sach et al. (2019)

10.7 Professional training of local conservationists and associated technology transfer

Human—elephant conflict may cause fatalities in humans and elephants, and may result in elephants being taken into captivity or being moved away from the areas where they are causing problems (Fig. 10.10).

Some zoos have established strong links with in situ conservation projects for elephants. In 2004 the North of England Zoological Society (which operates Chester Zoo) began cooperating with the Assam-based NGO EcoSystems-India and created the 'Assam Haathi Project for human—elephant conflict mitigation'. The project received funding from the Darwin Initiative operated by the UK Government as part of its obligation under the United Nations Convention on Biological Diversity to assist developing countries with conservation projects. The project integrates research on, and monitoring of, local elephants using a community-based approach to mitigate human—elephant conflict and protect the livelihoods of the local people. This has resulted in a number of research papers on the in situ work (Chartier et al., 2011; Davies et al., 2011; Wilson, 2015b; Zimmermann et al., 2009) and an educational project at a school close to Chester Zoo (Esson et al., 2014).

Two decades of work undertaken by North Carolina Zoo, United States, to support field-based conservation initiatives for wild African elephants have been described by Wilson et al. (2019). This work includes the use of novel anaesthesia techniques to facilitate the fitting of satellite tracking collars in West and Central Africa, to monitor elephant movements and reduce human—elephant conflict, and the implementation of Spatial Monitoring and Reporting

FIGURE 10.10 Asian elephants that come into conflict with local communities are sometimes captured and moved by forestry and wildlife authorities. These two bulls were being held by the Forestry Department in a logging camp in Karnataka, southern India. They were part of a group that had been raiding crops and damaging property. At the time this photograph was taken, the bulls had been in captivity for about 6 months and their front legs were chained to restrict their movements.

Tool (SMART) to collect and analyse antipoaching patrol data in 14 protected areas in five countries.

While some zoos have provided useful expertise and equipment for in situ elephant conservation projects it would be wrong to think that there is no indigenous elephant expertise. People working in elephant range states have considerable expertise in the capture, transportation and veterinary treatment of elephants. For example, studies of immobilisation of the Asian elephant using etorphine (M-99) have taken place in the wild (Gray and Nettashinghe, 1970). India has a long history of providing veterinary care for elephants (Somvanshi, 2006). This has developed from traditional knowledge and European practice introduced by the British during the colonial period (Krishnamurthy and Wemmer, 1995).

10.8 Fundraising for in situ conservation

Zoos claim to be important in fundraising for elephant conservation in the field and habitat protection (Hutchins and Conway, 1995). There is some evidence to support this. In North America, the AZA Conservation Grants Fund (CGF) supports the cooperative conservation-related scientific and educational initiatives of the association, its members and their collaborators. Some of the past elephant-related CGF awards include:

Average Daily Walking Distance of Captive African Elephants *(Loxodonta africana)* (Zoological Society of San Diego, 2007)

Developing Rapid Hormone Kits for Monitoring Reproductive Activity in Elephants (Smithsonian's National Zoological Park (CRC), 2006)

Development of Sex Selection Protocols for African and Asian Elephant Sperm (Indianapolis Zoo, 2005)

Sumatran Elephant Healthcare and Conservation Program (Oregon Zoo, 2001)

Development and Validation of Simple and Rapid LH and P4 Field Kits for Prediction of Ovulation in the Elephant (Indianapolis Zoo, Southwest Missouri State University, Six Flags Marine World, 2000).

AZA (2020)

Between 2014 and 2018, over 90 AZA-accredited facilities reported providing almost $16 million in support of in situ elephant conservation projects (AZA, 2020).

In the United Kingdom, ZSL Whipsnade Zoo keeps a herd of Asian elephants and held an annual elephant-specific fundraising weekend event each year between 2010 and 2017, raising a total of over £100,000 (US$129,000) through sponsorship and visitor donations. This has been used to fund elephant disease research and field conservation (Sach et al., 2019).

However, many organisations raise money for elephant conservation without keeping elephants. In its 2013 Christmas Appeal, *The Independent* — a national newspaper published in the United Kingdom — raised approximately half a million pounds (US$ 825,000) which was then used by its partner charity, *Space for Giants*, on elephant conservation projects in East Africa (Independent, 2014). The funding provided by governments for elephant conservation is discussed in Section 9.4.11.

10.9 Do zoo and conservation authorities support captive breeding in zoos?

The United Nations Convention on Biological Diversity 1992 (Article 9) recognises that ex situ conservation measures may be important for the conservation of some species, especially within their country of origin, requiring Parties to:

> ...(b) Establish and maintain facilities for ex-situ conservation of and research on plants, animals and microorganisms, preferably in the country of origin of genetic resources;...

However, the extent to which conservation organisations accept the need for captive breeding varies between organisations and between species.

Most of the world's Asian elephants live wild in India. Until recently, Asian elephants were kept by many Indian zoos. These zoos are regulated by the Central Zoo Authority (CZA), which was established under s38A of the Wildlife (Protection) Act 1972. In November 2009 the CZA issued a notice requiring the 26 zoos and 16 circuses holding elephants (approximately 140 animals in total) to send them to national parks, sanctuaries and tiger reserves (Lalchandani and Chopra, 2009). The CZA clearly attached no conservation importance to elephants living in Indian zoos and required the animals to be released due to welfare concerns.

In North America and Europe, some zoos have been giving up their elephants and sending them to sanctuaries or other zoos with better facilities (Rees, 2009a). People for the Ethical Treatment of Animals (PETA) lists 33 zoos that have already given up their elephants, or intend to stop keeping elephants (PETA, 2019b; see Appendix). This is largely due to increasing concerns about the welfare of elephants in zoos, but some zoos have decided to invest in new, bigger, better and more expensive facilities — for example Chester Zoo [£3,000,000 (building only)]; Heidelberg Zoo, Germany (€4,650,000); Seneca Park Zoo, New York (US$4,400,000); Reid Park Zoo, Arizona (US$8,500,000); North Carolina Zoo (US$8,000,000) (see Table 8.5).

The IUCN is not supportive of the captive breeding of African elephants in zoos but believes that captive Asian elephants should be considered in conservation strategies (IUCN, 2017a,b) (see Section 9.4.3).

10.10 Captive breeding in range states

There is anecdotal evidence that the successful breeding of captive Asian elephants at elephant stations has been known for around a century. Hundley (1923), writing while in Burma (Myanmar), referred to the fact that Asian elephants in logging camps bred successfully and supplied the camps with new stock:

> Numbers of calves born amongst the herds of elephants employed and owned by the large timber working firms are to be seen nowadays, born and bred, trained, worked and growing aged in captivity.

It was perfectly clear to Williams (1951), while working as an elephant manager in Burma, that a population of captive elephants could be self-sustaining:

> When...the Bombay Burma Corporation had built up considerable herds of elephants, it realised the importance of the elephant calves born in captivity...when the Corporation's herds had reached a strength of two thousand animals, it was found that births balance deaths, and that new supplies of elephants were required only on rare occasions.

In many camps, elephants are allowed into the forest at night to feed and rest. They often consort with wild elephants and this may result in calves being born in the camps.

Thitaram (2012) has argued that maintaining the captive populations in tourist and timber elephant camps, circuses, zoos and other elephant facilities is essential to prevent the depletion of wild populations and to reduce the illegal capture of wild elephants, and discusses steps being taken to establish breeding programmes in countries in Southeast Asia. The Pinnawala Elephant Orphanage in Sri Lanka holds one of the largest captive Asian elephant (*Elephas maximus maximus*) populations in the world, with 79 animals reported in 2016 — 45 females and 34 males (Pushpakumara et al., 2016). The orphanage has recorded 65 live births and its data indicate that calf mortality (7.6%) and stillbirth rates (4.4%) were low. However, Bechert et al. (2019) claimed that many working elephant populations are not sustainable, based on modelling the captive population of Myanmar (Burma), due to high death rates and low birth rates (Leimgruber et al., 2008).

10.11 'Domestication' of African elephants

At the end of the beginning of the 20th century the attention of the conservation community was drawn to the possibility of protecting African elephants by taming them. In 1908 the *Journal of the Society for the Preservation of the Wild Fauna of the Empire* — the forerunner of the journal *Oryx* — carried a short article by P. L. Sclater entitled 'The domestication of the African elephant' (Sclater, 1908):

> In my article 'On the Best Mode of Preserving the Existence of the Larger Mammals of Africa for Future Ages'...
> I suggested that a 'kheddah' should be moved across from British India to British East Africa and established on the slopes of Mount Kenia [sic] for the purpose of capturing and taming the African elephant. Much to my surprise, however, I have lately ascertained that the authorities of the much-abused Congo Free State have been before us in this matter and at their station on the River Welle have already established a mission expressly for this object.

Sclater included in his paper a translation of an article published in *La Tribune Congolaise* on 5 August 1906 that provided further details of this project. It had been begun in 1899 under the direction of Commandant Jules Laplume — a Belgian cavalry officer — and at the time the newspaper article was published 132 elephants had been captured, of which 22 were released due to 'their tender age', 86 died of various causes (some of which were probably being too young to be separated from their mothers), thereby leaving 24 elephants who were subsequently 'the subject of attempts at domestication'.

When the wildlife filmmaker Armand Denis visited the elephant station located in Gangala-na-Bodio — around 1935 — there were 35 elephants (Denis, 1963). The professional hunter T. Murray Smith wrote of his visit to an 'elephant farm' at Aba in the Belgian Congo in 1947 (Smith, 1963). He described the farm as being of 'immense area' and having 36 elephants ranging in age from 7 to 40 years. According to Smith, the adults 'pulled ploughs, lifted and hauled timber and fetched their own fodder'. Each elephant had its own mahout. The elephants — forest elephants (*Loxodonta cyclotis*) — were usually caught when aged between 5 and 10 years, by men on horseback, on foot and riding elephants (*moniteurs*). Once a calf was captured, it was roped and taken back to the farm escorted by the *moniteurs* (a full-grown bull and a cow) ridden by their mahouts. Smith noted that, 'the elephants breed in captivity, a sure sign that they are contented and find the life normal'.

In spite of these optimistic historical accounts, there is no captive breeding programme for forest elephants.

10.12 Conclusion

It is naïve and disingenuous for zoos to suggest that elephants living in zoos have any current, or will have any future, impact on elephant conservation. There is no substantive evidence that the presence of elephants in zoos contributes to the education of the public on matters concerning elephant conservation. A great deal of interesting research has been conducted on elephants in zoos, but most of it has been concerned with understanding reproduction, the development of AI and alleviating the various harmful effects of zoo-living on the animals. Some impressive technological advances have been made, but this work is seeking solutions to the human-made problems of captive breeding in nonrange states and the ecologically unnecessary contraception of wild populations whose natural controls have been removed by people. Very little of what has been discovered about the biology of zoo elephants has had a direct impact on the animals' survival in the wild and the keeping of elephants in zoos receives little support from elephant field biologists.

Some zoos raise money for in situ elephant conservation projects, but the amount of money raised is infinitesimal when compared with the cost of building and operating zoo elephant facilities. In any event, money is raised for many species, including elephants, without the help of animal 'ambassadors' held in zoos.

Keeping and breeding elephants in new, expensive, state-of-the-art zoo exhibits are the most expensive way of conserving these animals (San Diego Zoo's *Elephant Odyssey* cost US$ 45,000,000 and Los Angeles Zoo's *Elephants of Asia* cost US$ 42,000,000). The type of 'elephant barn' constructed by elephant sanctuaries is cheaper — two 20,000 ft^2 (1858 m^2) barns and a 100-acre enclosure built in California by the Performing Animal Welfare Society (PAWS) cost a total of US$3,250,000 (PAWS, 2019). When Sach et al. (2019) calculated the cost of keeping Asian elephants at ZSL Whipsnade Zoo, they estimated the annual cost of elephant keeper salaries at the zoo to be between approximately £235,000 (US$ 303,150) and £284,000 (US$ 366,360).

In situ conservation measures can be very cost effective compared with ex situ management. Leader-Williams (1990) estimated that the annual cost of keeping one black rhinoceros (*Diceros bicornis*) and one African elephant in a zoo was 50 times more than protecting them in the wild, along with all of the other species in the same area (based on comparing the costs of caring for these two animals at ZSL London Zoo and in a protected area in Africa).

In India, solar-powered electric fences are a relatively inexpensive method of reducing human—elephant conflict. Appayya (1995) quoted costs of just US$300 per kilometre. However, more recently PAWS (2019) quoted US$1,000,000 to fence a 100-acre elephant enclosure. Some conservationists are advocating the construction of fences from palmyra trees (*Borassus*) because they are much cheaper than electric fences, provide fruit for the elephants and require minimal long-term maintenance, unlike electric fences which, in any event, have a lifespan of perhaps merely a decade or so.

Common sense would suggest that when animal populations go into decline they should be helped to recover in situ; establishing ex situ captive breeding programmes should be a last resort and only used when all other efforts have failed. Where ex situ efforts are essential, they should take place in the range states of the species and not in distant developed, nonrange states. Walls (2018) discussed the economic and conservation benefits of taking proactive conservation measures — as opposed to waiting for a species to reach crisis point and implementing reactive conservation measures — and advocates a 'conservation triage' approach. She notes that for long-lived species with slow intrinsic growth rates, status assessments need to be conducted over long periods. Turkalo et al. (2017) have suggested that such assessments need to be made for 60-year time periods in the forest elephant (*L. cyclotis*) — approximately twice the generation time.

One way of thinking about the relevance of elephants living in zoos to elephant conservation is to consider what we would do now if they did not exist. If there were no elephants in zoos anywhere in the world, and each species became rare in the wild, would individuals be captured and taken from their natural habitats in Africa and Asia to zoos, predominantly in Europe and North America? Would these zoos then use these elephants to generate interest in, and funding for, in situ elephant conservation, carry out research on them and establish captive breeding programmes with the intention of returning their progeny to the range states at some time in the future? I suggest that this would be unthinkable. It would cost an extraordinary amount of money, it would require building large specialist facilities for housing the animals and there would be no guarantee of success. We can attempt retrospectively to justify keeping elephants in captivity as aids to elephant conservation, but the reality is that the hundreds of elephants kept in zoos around the world are little more than an historical relic population.

CHAPTER

11

The future of elephants in captivity

...managed African and/or Asian elephants in North America could potentially enhance biodiversity and economically benefit ranchers through grassland maintenance and ecotourism.

Donlan et al. (2006)

11.1 Introduction

Although humans have been blamed for previous mass extinctions of animals (Sandom et al., 2014) Faith et al. (2018) have challenged this view and have shown that megaherbivores in eastern Africa (e.g. elephants, hippopotamuses, rhinoceroses) began to decline about 4.6 million years ago, preceding evidence for hominin consumption of animal tissues by more than 1 million years. They suggest the decline may have been triggered by a decrease in atmospheric carbon dioxide that led to an expansion of grasslands and the disappearance of the woody vegetation that many megaherbivores used for food.

Whether or not humans have been responsible for previous extinctions of proboscidians it is clear that human activity is responsible for the decline in elephant populations in recent times, especially due to poaching and human—elephant conflict. Unlike most wildlife species, in addition to living in the wild and in zoos, elephants also live in a semi-domesticated state in some countries. This affords a number of possibilities for securing their future survival.

11.2 Elephant ranching

In 1981 Dr Keith Eltringham — who at that time was lecturing at the University of Cambridge — delivered a paper to the British Association for the Advancement of Science warning of the possible extinction of the African elephant. Three years later he suggested that there should be a dual attitude towards elephants in captivity (Eltringham, 1984). First, traditional zoos would continue to keep small numbers of elephants. He suggested that elephants should be treated as domestic animals and trained for riding or given other jobs, provided with trees for shade and sand-pits for dust bathing, and that contact with the keeper should act as a substitute for the presence of other elephants. Second, Eltringham suggested that if zoos are to achieve self-sustaining populations of elephants, large breeding groups must be established and that this would require 'a specialized approach more like ranching than zoo-keeping'. Eltringham clearly considered the possibility that it might one day be 'necessary to conserve elephants in captivity with the aim of eventual reintroduction into the wild'.

Eltringham did not predict that most zoos in the 21st century would not be giving elephant rides and that zoo visitors would be kept at a greater distance from elephants than was the case in the 1980s, when it was not uncommon for zoos to walk elephants through their grounds in close proximity to members of the public, while many safari parks allowed visitors to drive their own cars through elephant enclosures where there was no barrier to protect them from contact with the animals. Furthermore, he did not foresee the possibility that, such would be the concern for the welfare of elephants, some zoos would voluntarily stop keeping them and animal welfare organisations would create elephant sanctuaries where the animals would not be allowed to breed. Neither did he predict that the free contact system of handling elephants in operation by keepers in the 1980s would be replaced by a preference for protected contact.

In 2020, 36 years after the publication of Eltringham's article, the zoo community is nowhere near achieving the level of commitment to the creation of a ranching system for elephants that he envisaged. Many zoos have given up keeping elephants, while others have built better facilities and are building larger herds, but nothing on the scale that Eltringham imagined. These zoos need animals and the zoo community will inevitably put pressure on governments to allow more animals to be imported into North America and Europe.

The concept of elephant ranching is not as outlandish as it might at first appear. In South Africa, white rhinoceroses (*Ceratotherium simum*) are bred on private farms. Buffalo Dream Ranch is owned by John Hume, who was removing rhino horns and selling them legally until the South African Government imposed a moratorium on the sale of rhino horns in 2009. Thereafter he stockpiled the horns in the hope that the moratorium would come to an end. In 2017 he successfully applied to the South African High Court to allow him to sell his horn by auction to raise money to fund his project.

Eltringham (1984) was right. If we need to breed elephants in captivity, this could be done in a method akin to ranching, preferably in the range states of the animals. There would certainly be challenges with this approach, but they would not be related to reproduction. We know how to construct elephant-proof fences – this has already been done in many locations in the wild (Kioko et al., 2008).

The organisation *ElephantVoices* (2019) advocates an approach similar to that of Eltringham, or replacing elephants in zoos with virtual exhibits:

> *ElephantVoices' standpoint is that elephants require complex social and environmental settings to thrive; to meet the interests of elephants zoos need to start thinking on the order of square kilometers.*
>
> *Or, as an even better alternative, zoos can offer high-end virtual educational exhibits that through animatronics and multimedia connect the public to the capabilities and lives of wild elephants, while stimulating the interest in their conservation.*

11.3 Rewilding — shades of Jurassic Park

Although Indian scientists and the Central Zoo Authority do not see a conservation role for Asian elephants living in Indian zoos — let alone a need to captive-breed them in other countries — some American scientists think they know better. In 2006 a group of 12 American scientists suggested that, rather than allow large mammals to be squeezed into smaller and smaller refuges in poor countries that cannot afford to protect them properly, they should be 'reintroduced' into the United States as substitutes for the large mammals that previously roamed the Great Plains in the Pleistocene. They proposed that elephants (*Elephas maximus* and *Loxodonta africana*) could be included in a 'Pleistocene re-wilding' of the Great Plains — where they might suppress the woody plant species that threaten the western grasslands — using animals from American zoos and domesticated elephants from Asia to replace the five species of proboscidians that once roamed North America (Donlan et al., 2005, 2006). They identify the construction of elephant fences to prevent conflict with humans as the major economic cost, but suggest the benefits of such a project would be ecological and economic, providing employment and income through tourism. Donlan et al. patronisingly suggested that many of the large mammal species of Africa and Asia are likely to be lost within this century and would have a better chance of long-term survival in America.

Donlan et al. were not the first to suggest that the warmer parts of the United States would be suitable for elephants. Eltringham (1984) suggested this some 20 years earlier:

> *There are parts of the southern United States where winter cold is not a problem and perhaps it is there that one should look for a site for the elephant ark should one ever become necessary.*

Prof. David Bowman, a fire ecologist at the University of Tasmania, has suggested that elephants — and other large herbivores such as rhinoceroses — could be introduced into parts of Australia to help reset the ecological balance. In particular, he suggests that they could help to control gamba grass (*Andropogon gayanus*), an African species that has invaded savannahs in north Australia, is a restricted invasive plant under the Biosecurity Act 2014 and a major source of fuel for wildfires (Bowman, 2012). This 'proposal' — he does not believe this is a realistic possibility — would not primarily be an elephant conservation measure, but rather a vegetation and fire control measure, and he acknowledges that any large herbivores that were introduced would have to be managed to prevent further ecological damage in fragile ecosystems.

These proposals to create novel ecosystems — involving elephants and other large mammals — have generated considerable academic discussion. Some conservation biologists have warned of the possibility of the unintended consequence of creating new and undesirable interactions between species, while also recognising that 'ecological replacement' may be a useful response to the challenges of climate change (Seddon et al., 2011).

Donlan et al. and Bowman appear to have overlooked the fact that the United States and Australia have legal restrictions on introducing exotic species, for good ecological reasons such as the protection of indigenous biodiversity and preventing the spread of disease.

11.4 Release to the wild

Translocation is increasingly being used as a strategy to reduce human−animal conflict. The Kenya Wildlife Service (KWS) has used translocation to address human−elephant conflict and restock elephant ranges since 1996. In 2002 the KWS relocated 56 elephants from Sweetwaters Rhino Sanctuary − which is completely surrounded by an electric fence − to Meru National Park (Omondi et al., 2002). This was the first time that the KWS had moved elephants in family groups. Five elephants died during the operations (8.9%). Prior to this, 141 elephants had been moved by the KWS in eight translocations, recording a mortality of 9.2%.

There are few carefully documented cases of the release of captive elephants to the wild, but there is some scientific evidence that such releases may succeed and that animals may not only thrive after release, but may also contribute to the growth of wild elephant populations.

Evans et al. (2013) reported the release of a captive-raised female African elephant in the Okavango Delta, Botswana. *Nandipa* was orphaned in 1989 (when she was 1 year old) by the elephant cull in Kruger National Park, South Africa, and taken − along with five other youngsters − to the Okavango Delta where she was trained to carry tourists on a saddle on her back at the age of 7 years. These animals joined a group of three adult elephants who had previously been held by zoos and circuses in Canada and the United States. *Nandipa* was fitted with a satellite collar and released in an area with which she was familiar in September 2003. The collar collected spatial and behavioural data for 17 months and the elephant was monitored less intensively for a further 5½ years. Although *Nandipa* did not fully integrate with a wild herd, she produced three calves and formed a social group with a female and her calf that were subsequently released from the same herd. *Nandipa* did not appear to experience aggression from the wild elephants with which she socialised and utilised an area similar to that used by wild adult females. Evans et al. suggested that releasing elephants in the wild should be considered as a viable option for some captive elephants.

The success of elephant reintroductions depends upon the establishment of stable social groups. Social groups in wild elephants consist of individuals who are genetically related, but this is not always the case when elephants are reintroduced. The genetic relatedness and behavioural relationships among elephants who had been reintroduced into forested areas in central and northern Thailand were studied by Thitram et al. (2015). They collected blood samples − for DNA analysis − from 53 elephants prior to their release in Sublanka and Doi Phamuang Wildlife Sanctuaries, made direct observations of social bonding behaviours for the following 12 months and calculated an association index (AI) for each animal. Of the total of 53, 33 elephants formed 11 groups across both locations, ranging in size from two to six individuals per group. The average AI for elephants within groups was 0.517 ± 0.039. Many individuals preferred social isolation − 20 of the 53 elephants released were not associated with particular groups and had an average AI of 0.002 ± 0.001. Pairwise genetic relatedness in the groups was 0.078 ± 0.019 and 0.047 ± 0.013 in those animals not associated with groups. There was no correlation between pairwise genetic relatedness and AI for either the group-living elephants or those that lived alone. Thitram et al. found that social group bonding was not influenced by genetic relatedness, but groups formed in association with the presence of a calf. They suggested that the chance of stable group formation in reintroduced elephants would be increased by including adults with calves.

11.5 Welfare concerns

Attitudes towards captive elephants have changed significantly over the past 40 years and there is no reason to think that this will not continue. Although some reports have suggested that animals − including elephants − do not suffer in circuses any more than in other captive environments (Kiley-Worthington, 1990b; Radford, 2007), nevertheless animal welfare organisations will continue to campaign against the use of 'wild animals' in circuses and keeping elephants in zoos.

In the United Kingdom, the Zoos Expert Committee − formerly the Zoo Forum − provides independent technical advice to the Government on zoo policy matters. In 2010 the Zoos Forum review of issues in elephant husbandry in UK zoos (Zoos Forum, 2010) considered the report of Harris et al. (2008) on elephant welfare in UK zoos and concluded that:

> 10. ...if solutions to welfare problems and threats cannot be found, if no or negligible evidence of improved health and welfare can be observed, and if there is no compelling reason to breed elephants in the UK, then, in our opinion zoos should take steps to stop keeping elephants.

In effect the zoo community was put on notice that if it could not improve the welfare of zoo elephants, the UK Government would consider banning keeping elephants in zoos. At the time of the Zoo Forum's report (2010) there were approximately 70 elephants in 13 UK zoos. The total number of elephants was unclear because the data presented for the individual zoos did not add up to the totals given for all zoos. It is not unknown for governments to intervene to protect elephants. In 2019 the Danish Government purchased the last four circus elephants in Denmark with the intention of retiring them to a sanctuary (BBC, 2019b).

11.6 Sanctuaries

It seems highly likely that the number of sanctuaries for captive elephants will increase as abused and elderly animals are 'retired' from logging camps, circuses and other captive situations and zoos continue to decide that a sanctuary is a better option for their elephants than a solitary or socially inadequate existence in a zoo enclosure.

The first elephant sanctuary to be created in Latin America was Santuário de Elefantes Brasil, which received its first two (Asian) elephants in 2016 (Anon., 2019d). Elephant Aid International is creating a new sanctuary in Georgia in the United States, Elephant Refuge North America (Anon., 2019e). In Europe there are ambitious plans to create the Elephant Haven European Elephant Sanctuary on land in the (Limousin) Nouvelle-Aquitaine Region, in the Department of Haute Vienne, France (Anon., 2019f). The site has an area of

28 hectares (over 70 acres) and there is the possibility of purchasing adjoining land in the future. The sanctuary plans to offer homes to elephants from circuses and zoos.

Professionally run sanctuaries in nonrange states are likely to provide a better home for most elephants than circuses or zoos with poor facilities. Ultimately they cannot have a role to play in the survival of elephant populations, but rather provide a sink for unwanted elephants.

Four Paws is an international animal charity that was founded in Austria in 1988 to campaign against fur farming and battery-farmed eggs. It has since extended its remit and is now involved with elephant welfare and conservation (Anon., 2018c). In Myanmar, logging restrictions and export bans have reduced the need for elephants to work in the teak industry, leaving around 1000 elephants 'jobless'. Four Paws has constructed Elephant's Lake, one of the largest elephant sanctuaries in Southeast Asia, to care for former logging elephants — along with orphaned and injured elephants — on 6880 acres (2784 hectares) of land in the Bagon region in Myanmar. The Ministry of Environmental Conservation and Forestry provided the land and Myanmar Timber Enterprise — a state-owned forestry organisation — is responsible for placing the elephants. The ultimate aim of this project is — where possible — to release rehabilitated elephants into the adjacent North Zar Ma Yi Forest Reserve.

11.7 A repository of useful genes

Elephants use their tusks for fighting, to dig for water and to remove bark from trees. The loss of tusks could alter the behaviour and distribution of wild African elephants. It is conceivable that captive populations of elephants could act as a repository of genes that are being lost from poached populations and that at some time in the future captive-bred elephants could help to redress the evolutionary changes caused to wild populations by poaching and other human pressures (Figs 11.1 and 11.2).

In some parts of Africa there has been an increase in the frequency of tuskless female elephants apparently as a result of increased poaching pressure. In poached populations in Tanzania more than 6% of adult females were found to be tuskless (Jones et al., 2018).

In a study of historical records (1989—96) of elephants randomly culled in Kruger National Park, South Africa, only 15 out of 735 females (2%) were tuskless and no males without tusks were found in the 605 bulls examined (Whyte and Hall-Martin, 2018). They suggested that the absence of one or both tusks was cause by accidental injury rather than genetic inheritance. Jachmann et al. (1995) found that the percentage of tuskless females in parts of Zambia's Eastern Province increased from 10.5% in 1969 to 38.2% in 1989, apparently as a result of poaching. They noted that being tuskless appears to run in families and is sex-linked.

Whitehouse (2002) reported that 98% of the 174 females in South Africa's Addo Elephant National Park were tuskless, but she suggested that this was largely the result of genetic drift caused by the population's small size, isolation and genetic bottlenecks that occurred in the 1800s and 1920s.

FIGURE 11.1 A tusker in the forest around the rim of Ngorongoro Crater, Tanzania, c.2000. Bulls with tusks of this length are becoming increasingly rare in Africa due to poaching.

FIGURE 11.2 Mr Wilfred Foya from the College of African Wildlife Management (CAWM) in Moshi, Tanzania, holding a large tusk stored at the ranger station in Ngorongoro Crater, Tanzania.

In Asian elephants, tuskless bulls are rare, but in Sri Lanka 93% of adult and subadult bulls have been reported to be tuskless due to the selective hunting and capture of large bulls for the logging industry from an isolated population (Kurt et al., 1995).

11.8 Cloning

Can you imagine a world where large mammals are cloned for the tourist industry? Daniel Wright of the University of Central Lancashire, in the United Kingdom, can. He has suggested that cloning could provide animals for sport hunting and for 'future safari zoos' that would focus on education and conservation (Wright, 2018). Whether or not elephants could be part of this imagined future for zoos depends, of course, partly on suitable advances being made in cloning technology, but also on developments in elephant husbandry sufficient to satisfy the public, zoo professionals, animal welfare organisations and other stakeholders that their welfare in zoos could be assured.

Could Asian elephants one day be used to resurrect the woolly mammoth? The genome of the woolly mammoth (*Mammuthus primigenius*) was sequenced in 2008 (Miller et al., 2008). The Asian elephant is the ideal surrogate and egg donor for cloning a mammoth (Nicholls, 2008). It has been argued that it would be unethical to use Asian elephants for this purpose because of their endangered status (Pina-Aguilar et al., 2009). However, Cottrell et al. (2014) have noted that, once in existence, a cloned mammoth would be of even greater conservation concern:

> Once a mammoth is cloned, it will be the only one of its kind to become an endangered species immediately at birth, and legally will qualify for those protections allocated to other endangered species. We would be creating an endangered species from one that is not currently endangered (nor currently living), and we would need to protect and preserve it.

Realistically, cloning a mammoth would be a technically challenging and extremely expensive undertaking with no guarantee of success, so it seems highly unlikely that we will face the dilemma of having to justify keeping one proboscidean species in captivity to provide surrogates for its extinct relatives any time soon.

11.9 Elephants as therapy

Animal-assisted therapies have been used to improve the well-being of humans with a number of conditions, such as anxiety, depression, cerebral palsy and autism, by providing them with close physical contact with horses, dogs, dolphins and other species (Morrison, 2007). Improving human well-being is a relatively recent role for captive elephants.

The Thai Elephant-Assisted Therapy Project (TETP) was founded in 2007 and has the dual missions of assisting young people with autistic spectrum disorder and supporting the welfare and conservation of elephants in Thailand. The project is a collaboration between conservationists, occupational therapists and researchers at Chiang Mai University in northern Thailand.

In one study at the TETP, staff investigated the use of a captive Asian elephant as a therapy for a person with autism (Satiansukpong et al., 2008). A comparison of the participant's performance before and after a treatment programme of 3 weeks showed an improvement in adaptive behaviour, sensory processing, postural control and balance.

In a second experiment, 16 children with Down's syndrome from Kawila Anukul School, Chiangmai, participated in a study of the effects of elephant-assisted therapy in which the eight children in the control group continued to attend school while the experimental group of eight children attended school and therapy sessions twice a week for 2 months. These sessions consisted of feeding elephants, cleaning up food debris and dung, and mounting and dismounting an elephant after practising this on an artificial elephant. The children who received the therapy showed no significant improvement in balance or postural control, but a significant improvement in visual motor integration (VMI) was shown in this group (Satiansukpong et al., 2016). VMI is concerned with the ability to make sense of visual information and then use it appropriately for a motor task such as using a tool, writing, or playing sports.

11.10 Climate change

The global ecosystem is undergoing unprecedented change as a consequence of global warming and it is difficult to predict how this will affect the distribution and dynamics of elephant populations. An international multidisciplinary team of 26 experts in biodiversity conservation, climate change, macroecology and spatial modelling have predicted that increased climatic variability resulting from global warming will cause a range shift in elephants in South Asia and increase human—elephant conflict (Kanagaraj et al., 2019). They have predicted that around 42% of the approximately 256,500 km^2 of the habitat available at the time of their analysis will be lost by the end of the 21st century due to the combined effects of climate change and human pressure. They believe that changes in climatic water balance in monsoon areas will cause elephants to shift their range upwards along gradients of water availability and seasonal drought, and they emphasise the necessity to create wildlife corridors to facilitate this.

If climate change does cause distributional changes in elephant populations, this could cause far greater pressure on their numbers than they currently experience, especially if they move out of protected areas. Clearly, any large-scale changes could be used as a further justification for ex situ captive breeding projects.

11.11 A role for zoos?

Zoos face a dilemma. On the one hand some zoo populations cannot be sustained into the future without importing animals from the wild (Hutchins and Keele, 2006). On the other hand, some breeding programmes have achieved sufficient success that they have outgrown existing elephant accommodation and now face significant challenges in managing their elephants. Importation of wild elephants to zoos in nonrange states has almost ceased, preventing zoos from selecting the sex of their animals. Natural breeding in zoos combined with artificial insemination has resulted in a more or less 1:1 ratio of male-to-female births. Taken together, these facts give rise to a predictable demographic certainty — the proportion of bulls in the population will increase with time. This will inevitably create significant financial and logistical challenges for ex situ breeding programmes because, historically, there has been no need for zoos to accommodate large numbers of bulls. Indeed, as discussed in Chapter 3, Elephant social structure, behaviour and complexity, in 1952 there were only six bull elephants in the United States among the 264 elephants of both genera kept in zoos and circuses (Lewis and Fish, 1978).

The management of Asian elephants in the European Association of Zoos and Aquaria Ex-situ Programme (EEP) has been reviewed by Schmidt and Kappelhof (2019) and their analysis exemplifies the challenges faced by successful elephant breeding programmes. Improvements in breeding within the EEP resulted in an increase in annual births to 15 in the 5 years between 2013 and 2017. In 1999 some 19.4% of Asian elephants in zoos were captive-born (Clubb and Mason, 2002), but at the time of their review, Schmidt and Kappelhof calculated that 60.4% of the EEP population had been born in captivity. Although not explicitly stated, we must assume they meant born in zoos, as opposed to logging camps, since they claimed this increase was a measure of the success of the EEP.

This increase in birth rate has produced a number of challenges for the EEP, the most urgent of which is the need to develop a strategy to deal with the growing number of bulls in the population. Schmidt and Kappelhof suggest a number of possible approaches (Fig. 11.3). Studbook data show that some female elephants in zoos have given birth at the age of 6 years. They suggest this should be extended to 8 years and that the interbirth (calving) interval should be 7 years. The combined effect of these two measures would be to reduce the reproductive rate and consequently reduce the rate at which bulls are added to the population. However, some zoos keep bulls more-or-less continuously with the rest of their herd and so cannot control mating. At Chester Zoo, *Thi* produced 9 calves between September 1993 and May 2018 — a calving interval of just 2.75 years, and 2.5 times the rate now being recommended (Fig. 6.18).

Schmidt and Kappelhof suggest that zoos need to develop a complex fission—fusion housing system similar to that used for Bornean orangutans (*Pongo pygmaeus*) to allow group composition to be easily changed (Amrein et al., 2014). This would allow bulls to remain in their natal herd until sexual maturity as they do in the wild. It would also facilitate the association of bull calves with older bulls, which may be important in the development of normal sexual behaviour (Rees, 2004b). Even if zoos are able to develop the type of elephant accommodation envisaged here there would still be a need for additional sites that could accommodate around 30 bulls based on current predictions for population growth.

In addition to the expensive option of adapting existing accommodation and building new bull facilities, Schmidt and Kappelhof have also discussed the option of using contraception to reduce population growth — although this could increase acyclicity in the population if reversibility proved to be problematic — and even posited 'management euthanasia' as a possibility, while acknowledging that the latter would be unacceptable to the public. While zoos have a 'surplus' of bull Asian elephants, some wild populations

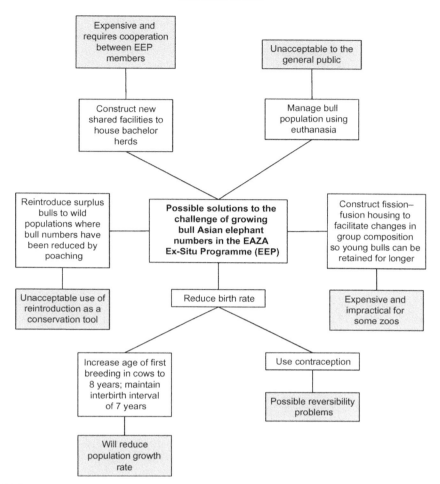

FIGURE 11.3 Possible solutions to the challenge of managing the growing number of bull Asian elephants within the European Association of Zoos and Aquaria Ex-situ Programme (EEP). Source: *Based on information in Schmidt, H., Kappelhof, J., 2019. Review of the management of the Asian elephant* Elephas maximus *EEP: current challenges and future solutions. International Zoo Yearbook 53, 1–14.*

have a sex ratio of bulls to cows as low as 1:100 as a result of poaching, because most cows bear no tusks (Sukumar, 2003). Captive-born bulls could be translocated to these populations, but this would effectively be using wild populations as a sink for surplus zoo animals and an unethical use of translocation as a conservation tool.

As we have already seen, some zoos have decided that keeping elephants is no longer an option and it is clear that even zoos that have a long history of keeping elephants may decide that these animals are no longer part of their future.

Twycross Zoo in the United Kingdom illustrates the transient nature of the role of some zoos in the captive breeding of elephants. The zoo opened on 26 May 1963 and kept Asian elephants from 1964. It imported *Noorjahan* from Jaldapara Wildlife Sanctuary in India in 1998 at the age of around 2.5 years. In 2014, at 18 years old, she gave birth to a calf conceived by artificial insemination. Another artificial insemination conception in a different cow resulted in a stillbirth. In 2010 and 2014 the elephant enclosure was improved. In 2018 the zoo decided to transfer its four elephants to a new facility at Blackpool Zoo, United Kingdom, permanently citing the following explanation (Anon., 2018d):

> *Twycross Zoo's all-female elephant herd are leaving for a new home. Their new home will be at Blackpool Zoo and will allow the herd to breed naturally with a bull elephant. This will ensure the long-term survival of this beautiful, endangered species.*
>
> *Although our facilities are amongst the best in the world, and are suited to house an all-female herd, we cannot accommodate an adult male elephant. Elephants are highly social creatures and the latest research has shown us that having a bull in their herd mix is vital for their well-being.*

The statement that this move will 'ensure the long-term survival of this beautiful, endangered species' is disingenuous as there is no scientific evidence that the future survival of elephants depends upon zoos. Furthermore, wild Asian elephants live in groups consisting of adult cows and their calves with no permanent adult bull members. There is no evidence that the absence of an adult bull in a zoo elephant group seriously compromises the animals' welfare. It does, of course, prevent natural breeding and may inhibit the development of natural sexual behaviour in juveniles (Rees, 2004b). Also, there is some concern that keeping adult bulls permanently within a herd is producing — in some zoos — the short calving intervals that are helping to increase the bull population at a rate that the Asian elephant EEP cannot currently accommodate (Schmidt and Kappelhof, 2019).

In the long term, zoos may take a more pragmatic view of their potential to return endangered species to the wild. In May 2019, Barcelona Zoo, in Spain, announced that it would no longer breed most of the species it was keeping at the time, including elephants (Fig. 11.4). Against a backdrop of increasing public concern for animal welfare, the city council amended its animal protection ordinance, prohibiting breeding programmes for species that cannot later be released back into their natural habitat (Congostrina, 2019). If other zoos take this view, or if governments accede to the wishes of the animal welfare lobbies, we may see an end to the keeping of elephants in zoos, and a consequent end to the associated breeding programmes, within a relatively short time scale.

11.12 Consumptive use or intensive protection zones?

The future for elephants is uncertain, but the evidence of history is that states make extraordinary efforts to protect their wildlife when its future is seriously threatened. Some countries are using paramilitary personnel and tactics to protect elephants and other big game species. Indeed, in the 1990s the Director of the Kenya Wildlife Service, Dr Richard Leakey equipped and trained his field staff in a paramilitary fashion to facilitate the protection of elephants and other wildlife (Leakey and Morell, 2001).

The alternative to captive breeding within elephants' range states is to establish intensive protection zones (IPZs) similar to those created in Zimbabwe (Kock et al., 1996) and Kruger National Park in South Africa for black rhinoceroses (*Diceros bicornis*). IPZs may be equipped with high-tech antipoaching equipment such as gun-fire detection systems and movement detectors that warn wildlife authorities about illegal activities.

The protection of elephants and other wildlife must be dramatically improved by professionalising protected area management. *African Parks* is a nonprofit organisation that works with governments and local communities to provide rehabilitation and long-term management of protected areas in several African countries. It takes direct responsibility for these areas and has had some remarkable successes. In 2017 *African Parks* completed the largest elephant translocation in history when it moved 520 elephants from Liwonde National Park and Majete Wildlife Reserve in Malawi to restock the depleted population in Nkhotakota Wildlife Reserve after improving the reserve's security (Anon., 2018e).

Elsewhere in Africa, for example Zimbabwe, sport hunting of elephants gives them an economic value and can help provide an income to rural communities, while giving the indigenous population a reason to protect them from poachers (Bond, 1994; Frost and Bond, 2008).

In elephant range states it is common to find that rich people in urban areas value elephants while poor farmers see them as a threat to their livelihoods and treat them as pests. Exploiting the willingness of urban dwellers to pay for elephant conservation can help to fund measures to mitigate human—elephant conflict (Bandara and Tisdell, 2004; Neupane et al., 2017). In addition, a permanent and comprehensive ban on the sale of ivory would be beneficial; for example China and the United States have recently put an end to their domestic ivory trade (Harvey et al., 2017).

FIGURE 11.4 African elephants at Barcelona Zoo, Spain, in 2019.

11.13 The court of public opinion

In 2016 the tour operator Thomas Cook stopped promoting holidays that involved elephant rides and shows as a result of receiving a petition signed by almost 175,000 people organised by World Animal Protection (Anon., 2016d). In July 2018 the company decided to stop selling tickets to SeaWorld in the United States and Loro Parque in Tenerife because of concerns about the welfare of the captive killer whales (*Orcinus orca*) held by these organisations (Haslam, 2018). In 2017 the company sold 10,000 SeaWorld tickets at £100 (US$130) each; a total of £1million (US$1.3 million).

It is not inconceivable that in the near future zoos of all types will have to adjust their operations to accommodate a greater public understanding of, and concern for, animal welfare and this could mean the end of elephants in zoos just as it has meant the end of elephants (and other nondomesticated animals) in circuses in many countries. Animal welfare is no longer the sole concern of activists dressed in balaclavas wielding baseball bats. It is an academic subject taught in universities based on a set of principles acknowledged by legislators, and a legal specialism practised by lawyers — it is a concern of ordinary people when they buy their food and clothing, and book their vacations.

If the public stops visiting zoos with elephants and tourists stop travelling to resorts offering elephant rides the fate of many captive elephants may be sealed by the court of public opinion. Scientists lack the tools to measure welfare precisely but that may not matter. Perception is everything and if the public perceives the keeping of elephants in captivity to be wrong that will be enough.

There is no convincing evidence that captive elephants play a significant role in elephant conservation. Wildlife biologists who spend their time studying wild elephants rarely if ever refer to any beneficial role of captive elephants in conservation in their scientific papers or textbooks; indeed, many have actively spoken out against zoos and the recent importation of elephants from the wild to zoos. They include Joyce Poole, Iain Douglas-Hamilton, Beth Archie, Harvey Croze, Phyllis Lee and Cynthia Moss (Poole, 2010).

11.14 Predictions

The future of captive elephants is uncertain. Scientists are developing methods of improving captive breeding in zoos as politicians ban, or threaten to ban, zoos from keeping them. Zoos — or at least some zoos — claim a conservation benefit from keeping elephants while other zoos release them to sanctuaries that will not breed them. Working elephants in Asia are redeployed to the tourist industry — amid ethical and welfare concerns — while others are used to provide therapy for autistic children. Many of these developments were not predicted 25 years ago and, no doubt, 25 years from now there will be other unpredicted developments in the fate of elephants under human care.

My predictions are as follows:

1. Little evidence will be published that elephants living in zoos have an educational role that benefits elephant conservation.

2. Elephants in Europe and North America will be kept in fewer, but larger, groups compared with today and bachelor herds will be maintained in some facilities.

3. Elephant ranching on the scale envisaged by Eltringham (1984) will not happen.
4. The range states of the Asian elephant will establish captive breeding herds from the redundant elephants formerly employed in logging and use these as a source of animals to release into the wild.
5. The range states of elephants will create special protection areas for wild herds and translocate animals to them in social groups.
6. Animal welfare organisations will continue to use the courts to apply pressure on zoos and circuses to release elephants to sanctuaries and prevent the importation of elephants from their range states to zoos in nonrange states.
7. Circus elephants will not exist or will be very rare as a result of legal bans on the keeping of exotic animals by circuses.
8. The courts will not acknowledge an elephant as having a legal personality and thereby deprive elephants of basic rights beyond those available under animal welfare laws.
9. Fewer small children will be able to see, smell and hear an elephant at close quarters in a zoo.

Although this book was intended to consider only those studies made of elephants under human care in zoos, circuses, logging camps and other situations where humans exert direct control over them, it is clear that all of the Earth's wild elephant populations are very much 'under human care' by virtue of the fact that, to a very large extent, humans determine their ultimate fate.

In a perfect world, where elephants are not hunted for their ivory or captured for entertainment, exploited for our amusement, our religious ceremonies, as forestry workers or confined in inappropriately small and barren enclosures, we should need only to exert the minimum of human influence on them. But because human attitudes towards elephants vary between and within cultures and because experts disagree about how best to secure the future of the Earth's elephant populations, it is impossible to predict the ultimate fate of elephants living under human care with any confidence.

Postscript

As this book goes to print zoos around the world are reopening after many weeks of closure following a novel coronavirus pandemic (COVID-19) in the human population. This has brought many zoos to the brink of financial failure and most have launched public appeals for donations including Berlin Zoo, Chester Zoo, ZSL Whipsnade Zoo, Taronga Zoo, San Diego Zoo Safari Park and the Smithsonian's National Zoo, all of which keep elephants.

These events should act as a stark reminder that the financial future of zoos cannot be guaranteed. If they disappear, so too will most, if not all, of the captive breeding programmes for elephants. It is clear that the future survival of these species, and many others, should not be left to institutions as financially fragile as zoos and if breeding programmes are considered essential for their survival they should be funded by governments.

Appendix

Examples of zoos that have stopped keeping elephants.

Zoo of origin	Destination	Year	Species	m.f.u.	Notes
Twycross Zoo United Kingdom	Blackpool Zoo, United Kingdom	2018–19	Asian	0.4.0	Four females transferred to Blackpool to join a single female at a new facility in Blackpool.
Buffalo Zoo (New York)	Audubon Zoo (Louisiana)	2018	Asian	0.2.0	In August 2018, Buffalo Zoo announced that it would be moving Asian elephants *Jothi* and *Surapa* to the Audubon Zoo.
Mendoza Zoological Park, Argentina	Elephant Sanctuary, Brazil	2016	African Asian	1.0.0 1.2.0	One African bull (*Kenya*), two Asian females (*Pocha* and *Guillermia*) and one Asian bull (*Tamy*).
Virginia Zoo (Virginia)	Zoo Miami (Florida)	2016	African	0.2.0	In November 2015, Virginia Zoo announced that it would be moving African elephants *Lisa* and *Cita* to Zoo Miami.
Woodland Park Zoo (Washington)	Oklahoma City Zoo (Oklahoma)	2015	Asian	0.2.0	It transferred elephants *Bamboo* and *Chai* to the Oklahoma City Zoo.
Greenville Zoo (South Carolina)	N/A	2014	African	0.1.0 died during move	Greenville Zoo sent its remaining elephant, *Joy*, to another facility but she died during transportation.
BREC's Baton Rouge Zoo (Louisiana)	National Zoo (Washington DC)	2013	Asian	0.1.0	The zoo transferred its last elephant (*Bozie*) to the National Zoo in Washington, DC.
Niabi Zoo (Illinois)	Little Rock Zoo (Arkansas)	2013	Asian	0.2.0	It transferred the last two elephants (*Sophie* and *Babe*) to the Little Rock Zoo in Arkansas.
Toronto Zoo (Canada)	PAWS, United States	2013	African	0.3.0	Transferred three African elephants – *Thika*, *Iringa*, and *Toka* – to the Performing Animal Welfare Society in San Andreas, California.
Central Florida Zoo and Botanical Gardens (Florida)	Zoo Miami (Florida)	2011	Asian	0.1.0	It transferred the last elephant (*Maude*) to Zoo Miami.

(*Continued*)

(Continued)

Zoo of origin	Destination	Year	Species	m.f.u.	Notes
Brookfield Zoo (Illinois)	Six Flags Discovery Kingdom (California)	2010	African	0.1.0	The last remaining elephant, *Joyce*, who was on loan from Six Flags Discovery Kingdom, was returned there.
Lion Country Safari (Florida)	Dallas Zoo (Texas)	2010	African	0.2.0	It transferred African elephants *Stumpy* and *Mama* to the Dallas Zoo in March 2010.
The Jackson Zoo (Mississippi)	Nashville Zoo (Tennessee)	2010	African	0.2.0	It transferred two African elephants (*Juno* and *Rosie*) to the Nashville Zoo in Tennessee.
All zoos in India	Wildlife parks and sanctuaries	2009	Asian	0.0.140	The Central Zoo Authority (CZA) announced the transfer of all 140 elephants living in 26 Indian zoos to wildlife parks and sanctuaries.
Philadelphia Zoo (Pennsylvania)	Pittsburgh Zoo (Pennsylvania)	2009	African	0.2.0	It transferred two African elephants (*Bette* and *Kallie*) to the Pittsburgh Zoo's elephant-breeding centre in June 2009.
Abilene Zoo (Texas)	Cameron Park Zoo, Waco, Texas	2007	African	0.1.0	It transferred the remaining elephant, 29-year-old *Tanya*, to the Cameron Park Zoo in Waco, Texas.
Alaska Zoo (Alaska)	PAWS, San Andreas, California	2007	African	0.0.1	It transferred an African elephant to the Performing Animal Welfare Society in San Andreas, California.
Gladys Porter Zoo (Texas)	Milwaukee County Zoo (Wisconsin)	2006	African	0.1.0	Citing its inability to increase the size of its elephant exhibit, it sent its only elephant, *Ruth*, a 28-year-old African, to another facility.
Detroit Zoo (Michigan)	PAWS Sanctuary, California	2005	Asian	0.2.0	In 2004 Detroit Zoo announced its decision to close its elephant exhibit and send the two female Asian elephants — *Winky*, age 51, and *Wanda*, age 46 — to a sanctuary.
Lincoln Park Zoo (Chicago)	Three elephants died	2005	African	0.0.3	After all three of its elephants died within a 6-month period, the zoo announced that camels would be moved into the empty elephant exhibit.
Chehaw Wild Animal Park (Georgia)	The Elephant Sanctuary	2004	African	0.2.0	It retired *Tange* and *Zula*, both 30-year-old African elephants, to The Elephant Sanctuary.
San Francisco Zoo (California)	PAWS Sanctuary	2004	Asian	0.1.0	It announced its decision to close its elephant exhibit and send *Tinkerbelle*, a 37-year-old Asian elephant, and *Lulu*, a 38-year-old African elephant, to a sanctuary.
			African	0.1.0	
Dudley Zoo (United Kingdom)	Planète Sauvage, France	2003	African	0.2.0	Transferred African elephants *Flossie* and *Flora* to Planet Sauvage in Nantes, France

(Continued)

(Continued)

Zoo of origin	Destination	Year	Species	m.f.u.	Notes
Longleat Safari Park (United Kingdom)	Zoo Parc de Beauval, France	2003	African	1.4.0	Five elephants permanently moved to France.
Bristol Zoo (United Kingdom)	Last elephant died	2002	Asian	0.1.0 Euthanised	Euthanised the lone 42-year-old female elephant, *Wendy*, after years of suffering from arthritis. (She had been kept alone in a tiny enclosure since 1986. Bristol Zoo elected not to replace her.)
London Zoo (United Kingdom)	Whipsnade Wild Animal Park, United Kingdom	2001	Asian	0.3.0	Elephants were transferred out of the city to Whipsnade following the death of a keeper.
Henry Vilas Zoo (Wisconsin)	The Elephant Sanctuary, Riverbanks Zoo, N. Carolina	2000	Asian African	0.1.0 0.1.0	It retired *Winkie*, a 34-year-old Asian elephant, to *The Elephant Sanctuary*, and transferred *Penny*, a 21-year-old African elephant, to Riverbanks Zoo in North Carolina.
Louisiana Purchase Gardens and Zoo (Louisiana)	The Elephant Sanctuary	1999	Asian	0.1.0	One Asian female sent to *The Elephant Sanctuary*, Tennessee.
Mesker Park Zoo (Indiana)	The Elephant Sanctuary	1999	Asian	0.1.0	*Bunny*, a 46-year-old Asian elephant, was retired to *The Elephant Sanctuary*.
Frank Buck Zoo (Texas)	The Elephant Sanctuary	1998	Asian	0.1.0	It transferred *Sissy*, a 20-year-old Asian elephant, to the Houston Zoo, then to El Paso Zoo, and finally to *The Elephant Sanctuary*.
Welsh Mountain Zoo (United Kingdom)	Knowsley Safari Park, United Kingdom	1995	African	0.1.0	*Myfanwy* was sent to Knowsley Safari Park and the only elephant exhibit in Wales was closed.
Sacramento Zoo (California)	Detroit Zoo (Michigan)	1991	Asian	0.1.0	It sent lone elephant *Winky* to the Detroit Zoo because the zoo's elephant enclosure was considered totally inadequate.
Edinburgh Zoo (United Kingdom)	–	1988	–	–	No longer keeps elephants.

m.f.u = males.females.unknown/unspecified sex.
Source: Based on PETA, 2019b. Elephant-Free Zoos. People for the Ethical Treatment of Animals (PETA), Norfolk, VA. https://www.peta.org/issues/animals-in-entertainment/zoos/elephant-free-zoos/ (accessed 03.09.19.). and the author's records.

Glossary

The purpose of this glossary is to help readers who are not scientists understand some of the terminology used in this book. The definitions given here are not intended to be definitive.

Sex ratios of animals living in zoos or listed in studbooks are commonly expressed as numbers using the following format:

2.5

males: females

or

1.4.0

males: females: unknown sex

α-Tocopherol A form of vitamin E.
A–B–A design A type of experimental design used in behavioural research in which a baseline condition is established for a behaviour (phase A); a treatment or intervention is introduced (phase B); and then the treatment or intervention is removed to see if the behaviour returns to the baseline condition.
Accelerometer A device that measures the rate of change of velocity.
Acid detergent fibre (ADF) The percentage of highly indigestible or slowly digestible fibre in food. It contains cellulose, lignin and silica. Low ADF feeds are more digestible than those with high ADF and they have higher energy values.
Activity budget An analysis of the amount of time an animal spends on various behaviours during some specified period of time.
Acyclicity The condition of not exhibiting an oestrous cycle.
Adrenal glands A pair of endocrine glands that produce adrenaline, cortisol, aldosterone and other hormones.
Adrenaline A hormone secreted in response to fear, excitement and anger. Also called epinephrine.
Affiliative behaviour Social behaviour that serves to reinforce social bonds.
Agonistic behaviour Behaviour related to aggression and fighting.
Allele A variant of a particular gene.
Alloparenting Parental care provided to individuals who are not direct descendants but may be related.
Androgen Any hormone that regulates the development and maintenance of male characteristics.
Ankus A tool used in the handling and training of elephants consisting of a pole ending in a metal hook and pointed tip. Also known as a bullhook or goad.
Anoestrus A period of infertility in females when they are not sexually receptive.
Anovulatory Relating to an individual female who does not ovulate.
Antibody A protective blood protein produced by the immune system in response to the presence of a pathogen or foreign substance (antigen).
Anticipatory behaviour Behaviour exhibited by an animal based on predictions, expectations or beliefs about future states or events, for example the imminent arrival of food.
Antigen A toxin or other foreign substance that is capable of inducing an immune response in the body, especially the production of antibodies that bind to and neutralise it.
Antiphonal calling An exchange of calls (rumbles) in alternating sequences: call–response–call–response, etc.

Appeasement behaviour A type of social behaviour whereby one individual pacifies the aggression of another by adopting a submissive posture or assuming an inferior social position.
Arthritis Any of a number of diseases of the joints of the skeleton, whose symptoms include inflammation of the joints, swelling, joint pain and stiffness.
Artificial insemination (AI) The injection of sperm into the genital tract of a female animal using a syringe or similar device.
Assay A laboratory procedure that aims to determine the presence of a substance and the amount of that substance in a sample.
Association index A measure of the extent to which two individual animals spend time associating with each other.
Auditory enrichment An enrichment method that employs music or other sound.
AZA Association of Zoos and Aquariums.
Behavioural restriction The phenomenon whereby an animal is motivated to perform a particular behaviour but is unable to do so because it is being restrained or its environment does not provide appropriate stimuli.
BIAZA British and Irish Association of Zoos and Aquariums.
Billboard An American entertainment magazine.
Bond group A social unit made up of several family groups of elephants.
Bull pen An enclosure space specially designed to hold one or more bull elephants. Important for separating adult bulls from a herd to simulate natural social behaviour – adult bulls do not live within mixed-sex social groups – and to confine bulls during musth.
Bullhook *See* ankus.
Calorie (caloric) restriction The restriction of the number of calories consumed by an animal.
Catheter A thin tube which may be inserted into the body for a variety of veterinary purposes, for example to treat diseases or perform surgical procedures.
Cervix The lower, tapered, part of the uterus where it meets the vagina.
Chemical signal A chemical used to pass information from one part of the body to another (e.g. a hormone) or from one individual to another (e.g. a pheromone).
Circadian rhythm Physical and behavioural changes that follow a 24-hour cycle.
CITES Convention on International Trade in Endangered Species of Wild Fauna and Flora 1973. An international law that restricts trade in rare species using a system of import and export permits.
Clade A group of animals that consists of a common ancestor and all of its descendants.
Clan Several bond groups and family groups.
Clinical sign An objective finding or characteristic usually determined by physical examination or investigation (e.g. laboratory test or X-ray) that may be indicative of a particular condition, abnormality or disease.
Concentrate In relation to nutrition, food that has been reduced in volume by the removal of liquid.
Condition score A numerical value derived from a scale representing the physical condition of an animal.
Confounding factor (variable) A variable other than the independent variable of interest that may affect the dependent variable, resulting in erroneous conclusions.
Conspecific An individual of the same species; in the context of this book, another elephant of the same species.
Contrafreeloading Preferring to work for food when it is already freely available at no cost.
Corpus luteum Yellow body. The remnants of the follicle after it ruptures during ovulation. It produces progesterone.
Cortisol A steroid hormone produced by the adrenal glands and involved in the immune response. Monitored as an indicator of stress levels.
Crate training Training an animal to enter a crate to familiarise it with this environment prior to transportation.
Crude protein Protein content estimated by measuring the quantity of nitrogen present.
Cryptic speciation An evolutionary process that produces different species whose individual members look morphologically identical but cannot interbreed.
Dam Mother.
Demography The study of the characteristics of a population, for example its size, growth rate, age structure, distribution, etc.
Digestibility A measure of the proportion of a food that can be digested by an animal.
Digestible energy The gross energy of a food minus the energy lost in the faeces therefrom.
Distress A motivational state in an animal caused by stress that it is unable to deal with.

DNA fingerprinting A molecular technique used to identify genomes by comparison with a known standard or to compare DNA from different sources. May be used to distinguish between subspecies, examine the relatedness of individuals and identify animal parts and products from rare species. It may be used to identifying an individual animal and establish its parentage.

Dominance hierarchy Any of a number of social systems whereby individuals are ranked according to relative status.

Dyad In social network analysis, a pair of individuals; a group of two.

Dystocia An abnormal or difficult labour and birth.

EAZA European Association of Zoos and Aquaria.

EEP Formerly European Endangered species Programme; now European Association of Zoos and Aquaria Ex-situ Programme.

Effective population size A measure of how well a population maintains genetic diversity from one generation to the next. It is usually lower than the actual population size because, for example, the sex ratio may be unequal and some individuals may be infertile.

Elephant sanctuary A place where former zoo, circus or other working elephants are kept after they have become too old or sick to work or when a zoo decides its facilities compromise elephant welfare due to its small enclosure size or the absence of companions.

Elephant transport crate A specially built metal container for transporting a live elephant safely and with high welfare standards. May be connected to an HVAC unit.

Endocrine monitoring The process of measuring and recording the levels of various hormones, or their metabolites, in blood, urine, faeces or saliva, often for the purposes of studying reproductive condition and functioning.

Endocrinology The scientific study of hormones and their functions.

Environmental enrichment A husbandry principle – a device or management practice – that seeks to enhance the quality of animal care in captivity by identifying and providing the environmental stimuli necessary for optimal psychological and physiological well-being.

Epinephrine *See* adrenaline.

Episiotomy A surgical incision of the perineum and posterior vaginal wall to enlarge the opening for the calf to pass through during birth.

Evolutionarily significant unit (ESU) A group of organisms that is distinct for conservation purposes, for example a species or a subspecies. Usually determined by DNA analysis.

Faecal cortisol metabolites (FCMs) Breakdown products of cortisol found in faeces, the measurement of which is used to quantify stress.

Family group Females and their young.

Female philopatry A social system in which females remain in their natal group or in the home ranges where they were born.

Feral Derived from a 'domesticated' population.

Fibreoptoscope A flexible optical device used in endoscopy for examining the inside of the body, for example the oesophagus, stomach, upper respiratory tract and female reproductive tract.

First-degree relatives Two individuals who have 50% of their DNA in common: parent and offspring, full siblings.

Focal sampling In focal sampling one individual is selected and studied for a period of time. For example, all feeding behaviour may be recorded over a 20-minute sampling period.

Foetotomy Dissection of a dead foetus in situ so that it may be more easily removed from the uterus.

Follicle-stimulating hormone (FSH) A hormone that stimulates the maturation of Graafian follicles in the ovary in females and promotes sperm formation in the testes in males.

Follicular phase A part of the oestrous cycle during which a follicle matures in the ovary, ending in the release of an ovum during the process of ovulation.

Force platform An instrument that measures the ground reaction forces generated when a body stands on, or moves across, it. Used for studying gait, movement, balance, etc.

Formol-saline A general fixative used in preparing tissue samples for microscopic examination.

Free contact A method of managing elephants whereby keepers enter enclosures with the animals and engage in physical contact with them.

Gait The pattern of movement of the limbs of an animal during locomotion over a solid substrate, for example during walking or running.

Genetic bottleneck A dramatic reduction in size of a population caused by environmental events or human activity resulting in a loss of genetic diversity.

Genetic drift A loss of genetic variation within a population due to a reduction in the frequency (and sometimes loss) of certain variants as a result of random events.

Gestation The period of time between conception and birth.

Goad *See* ankus.

Gonadotropin Any of a group of hormones that regulate reproduction by stimulating the gonads to carry out their functions.

GPS Global positioning system.

GPS collar A collar worn around an elephant's neck that carries a GPS device so that the animal's position, movement, etc. may be monitored and recorded.

Graviportal Describing an animal whose body is physically adapted to support a great weight while moving slowly over land.

Greeting ceremony A behaviour exhibited by an elephant and directed towards another elephant (or keeper) – usually after a period of separation – consisting of vocalisations, ear-flapping, urination and defaecation.

Gross energy In nutrition studies, the amount of energy in the food as measured by bomb calorimetry, that is, when it undergoes complete combustion in oxygen.

Helping behaviour A behaviour performed for the benefit of another.

Herpesvirus A virus belonging to the family Herpesviridae. Pathogenic in a wide variety of animals.

Heterozygosity The possession by an individual of two different alleles for a particular gene.

Homozygosity The possession by an individual of two identical alleles for a particular gene.

Howdah A carriage that is fixed to the back of an elephant for the purpose of carrying people.

Human chorionic gonadotropin A hormone that promotes the maintenance of the corpus luteum in the ovary at the beginning of pregnancy.

HVAC unit Heating, ventilation and air-conditioning unit.

Hybrid An individual produced by a mating between individuals from two different genera, species or subspecies.

Hybridisation The process of producing hybrids.

Hyperkeratosis An abnormal thickening of the outer layer of the skin caused by excessive production of keratin.

Immunoassay A method of detecting or measuring a specific protein or other molecule by using its ability to bind to an antibody or, sometimes, an antigen.

Inbreeding Breeding between close relatives.

Infrared thermography A technique that converts invisible infrared radiation to visible light. Used to detect heat loss.

Infrasonic communication The passing of information between individuals using low-frequency sound.

Inhibin A hormone that inhibits the production of follicle-stimulating hormone (FSH) by the pituitary gland.

Interdigital glands Glands located between the digits (toes in elephants).

Intramuscular Located in, or administered into, a muscle.

Intravenous Located in, or administered into, a vein.

ISIS International Species Information System. A database of information on animals living in zoos that was the forerunner of ZIMS.

IUCN International Union for the Conservation of Nature and Natural Resources.

Jacobson's organ The vomeronasal organ located in the roof of the mouth of an elephant used to detect chemicals in urine, etc.

Khedda, kheddah A stockade used to capture wild Asian elephants.

Khonkie, koomkie A tame elephant used in the capture of wild elephants.

Life expectancy A statistical measure of the expected time until death from a specified age.

Life table A table of data for a population of animals that is used to calculate mortality rates between age classes and life expectancy.

Locus (pl. loci) A fixed position on a chromosome, for example the location of a gene or genetic marker.

Longevity The time between birth and death.

Luteal phase The early part of the oestrous cycle beginning with the formation of the corpus luteum and ending — if conception occurs — in pregnancy. Circulating progesterone levels are higher during this phase.

Luteinising hormone (LH) A hormone produced by the anterior pituitary gland, an acute rise in which triggers ovulation and the development of the corpus luteum. In elephants there are two LH surges three weeks apart; only the second induces ovulation.

l_x In a life table, the proportion of the individuals in a particular age class that survive to the next age class.

Mahout An individual, usually a man, who cares for, trains and works a captive elephant, especially in an Asian timber camp.

Mating pandemonium A social behaviour sometimes exhibited by a herd of elephants following a mating between a bull and a cow whereby excited individuals gather around the pair, vocalise loudly, flap their ears, urinate and defaecate.

Matrilineal hierarchy A dominance hierarchy within the females in a society.

Mean calving interval Average time between consecutive births to the same dam.

Mesoherbivore A medium-sized herbivore.

Metabolite A starting material in, intermediate in, or end product of, metabolism.

Mitochondrial DNA The DNA that occurs in the mitochondria of eukaryotic cells and is inherited through the female line.

Morphometric Relating to body form (size and shape).

Mother hypothesis As mothers age the cost of reproduction increases. This hypothesis suggests that as she ages a mother's energy would be better spent helping her existing offspring to survive than risking producing more offspring.

Mother–offspring unit Female and her offspring.

Musth A condition, similar to rutting, whereby a bull elephant exhibits increased aggression and unpredictable behaviour associated with an increase in testosterone level.

Naturalistic exhibit A zoo exhibit that mimics natural conditions.

Nulliparous Describing a female that has not given birth.

Obesity A condition whereby excess fat in the body has accumulated to such an extent that it compromises health.

Obstetrics The branch of veterinary (and human) medicine and surgery concerned with pregnancy and birth.

Oestrogens A group of steroid hormones produced by the ovaries which control ovulation by causing a surge in luteinising hormone.

Oestrous cycle The reproductive cycle that occurs in most placental mammals in which there are recurring periods of fertility (oestrus) and infertility (anoestrus) in females.

Oestrus A period of sexual receptivity and fertility in the female.

Oestrus synchrony The synchronisation of the oestrous cycles of females living in close proximity.

Outbreeding The breeding of unrelated animals or distant relatives.

Ovulation The release of an egg from the ovary.

Oxytocin A hormone produced by the posterior pituitary gland which induces uterine contraction during labour and stimulates milk flow.

PCR assay Polymerase chain reaction assay; a technique that amplifies small quantities of DNA and analyses it to determine its structure.

Personality The characteristic of animals that recognises the existence of differences in behaviours, or suites of related behaviours, between individuals of the same species which are consistent over time and driven by differences in how individuals respond to information about their environment.

Phylogeny The evolutionary relationships within and between groups of organisms.

Phylogeography The scientific study of the principles and processes that have led to the geographical distributions of genealogical lineages — especially within and among closely related species — by examining the geographical distribution of DNA sequence variants.

Picket A stake driven into the ground to which an animal is tethered.

Picketing The act of tethering an animal to a picket.

Polymorphic locus A locus that has two or more alleles.

Positive reinforcement The occurrence of a desirable event or stimulus (a reward) presented as a consequence of a behaviour.

Prepubertal Prior to reaching puberty.
Primiparous Giving, or having given, birth for the first time.
Progesterone A hormone produced by the corpus luteum which maintains the uterine endometrium during pregnancy.
Prolactin A hormone produced by the anterior pituitary gland that stimulates lactation. Levels may increase in response to stress.
Prolactinaemia A deficiency or excess of prolactin.
Protected contact A method of husbandry whereby elephants are managed by keepers from behind a fence so that there is no uncontrolled physical contact.
Pubertal Relating to puberty, the physical changes that the body undergoes when an individual reaches sexual maturity and becomes capable of sexual reproduction.
Range state A country (state) where a species naturally occurs in the wild.
Recumbent sleep Sleep performed lying down.
Relatedness Being genetically related; a measure of the closeness of the genetic relationship between individuals.
Relative quantity judgement A dichotomous judgement of unequal quantities ordered in magnitude.
REM sleep Rapid eye movement (D-state) sleep. Associated with dreaming.
Restricted contact Direct contact with an elephant in restraints allowed within a protected contact system.
Retinol Vitamin A.
Retirement The release of an elephant from work (e.g. in forestry or a circus), usually to a sanctuary.
Rewilding The process of recreating lost ecosystems using extant species similar to (or the same as) those that have become (locally) extinct.
Rotational exhibit A zoo exhibit consisting of several discrete compartments that allows animals access to different areas at different times.
Roughage Fibrous material in the diet that is difficult to digest.
Route-tracing Repeatedly walking the same path through an enclosure. In extreme cases this may involve placing each foot in footprints made during an earlier circuit.
RSPCA Royal Society for the Prevention of Cruelty to Animals.
Rumble A deep, resonant vocalisation.
Salivary cortisol The cortisol found in saliva which is a reliable indicator of the circulating level of cortisol in the blood.
Scan sampling A method of studying behaviour by making recordings at set time intervals, for example every 5 minutes.
Secondary positive reinforcement Primary positive reinforcement fulfils a biological need, for example food. Secondary reinforcement is learned and works via association with primary reinforcers. This allows the trainer to continue delivering reinforcement (e.g. food) even if the animal has no immediate biological need for it.
Seismic communication A method of conveying information using mechanical vibrations of the substrate.
Self-recognition The ability to separate the 'self' from others. Often tested by studying the response of an individual to seeing its image in a mirror.
Senescence The deterioration of the body and its functions with age; the process of growing old.
Sexual dimorphism Differences in characteristics between the sexes of a species beyond those conferred by their sex organs (e.g. size, shape, presence and absence of tusks, etc.).
Sire Father.
Social facilitation The phenomenon whereby an individual is more likely to exhibit a behaviour if it is exhibited by other individuals in a group.
Social rank Position in a hierarchy.
Sociogram A diagram that shows and quantifies the relationships between individuals in a group.
Species Groups of interbreeding natural populations that are reproductively isolated from other such groups.
Sperm sex-sorting A sex-selecting technique that involves separating the spermatozoa carrying an X chromosome from those carrying a Y chromosome.
SSC Species Survival Commission of the IUCN.
SSP Species Survival Plan Program. A cooperative breeding programme for a rare species managed by the AZA.
Stereotypic behaviour A repetitive behaviour that has no obvious purpose.

Stress A state of anxiety resulting from specific internal or external stimuli. Associated with changes in hormone levels, especially an increase in cortisol.
Stressor A stimulus capable of causing stress.
Subspecies A subdivision of a species.
TAG Taxon advisory group.
Tail flicking A repeated movement of the tail under the body by a female elephant. It may disperse chemical signals and an increase in its frequency appears to indicate forthcoming ovulation.
Testosterone A steroid hormone produced by the testes which is responsible for the development of the secondary sexual characteristics. In females it is an intermediate in the synthesis of oestrogen.
Tool use The use of an object by an animal as an implement for performing or facilitating a mechanical operation, for example reaching food or scratching an inaccessible part of the body.
Transrectal ultrasound scan A technique used for examining reproductive and other organs in the pelvis using an ultrasound probe placed in the rectum to create images.
Trunk wash A method of obtaining a sample from the trunk by passing liquid into it with a syringe and then collecting it when it drains out.
Tuberculosis (TB) An infection caused by a Mycobacterium contracted by inhaling droplets produced by an infected individual. Causes lesions, especially in the lungs and associated lymph nodes.
Type I error A false positive, for example the apparent detection of a relationship between two variables that does not really exist.
Type II error A false negative, for example the failure to detect a relationship that exists between two variables.
Ultrasonography A diagnostic imaging technique that uses ultrasound to produce images of internal organs.
Urinary cortisol *See* cortisol.
Vaginal vestibulotomy A surgical procedure that involves incising the vertical part of the urinogenital tract to remove a foetus.
Ventral oedema Swelling along the ventral wall of the abdomen.
Visual acuity The clarity of vision.
Well-being The state of being comfortable, healthy and contented.
ZIMS Zoological Management Information System. A sophisticated database of information on animals living in zoos. An important source of data for zoo managers and researchers.

References

Abdul, J.B., Daim, M.S.B., Othman, N.B., Ibrahim, M.A.B., 1996. Elephant Management in Peninsular Malaysia (Unpublished manuscript). Department of Wildlife and National Parks, Kuala Lumpur.

Ahamed, A.M.R., 2015. Activity time budget of the Asian elephant (*Elephas maximus* Linn.) in the wild. Trends in Biosciences 8 (12), ISSN 0974-8, 3024-3028, 2015.

Alcock, J., 1972. The evolution of the use of tools by feeding animals. Evolution 26, 464–473.

Alward, L., 2008. Why circuses are unsuited to elephants. In: Wemmer, C., Christen, C.A. (Eds.), Elephants and Ethics. Towards a Morality of Coexistence. The Johns Hopkins University Press, Baltimore, MD, pp. 205–224.

Amrein, M., Heistermann, M., Weingrill, T., 2014. The effect of fission-fusion housing on hormonal and behavioural indicators of stress in Bornean orangutans (*Pongo pygmaeus*). International Journal of Primatology 35, 509–528.

Andrews, J., Mecklenborg, A., Bercovitch, F.B., 2005. Milk intake and development in a newborn captive African elephant (*Loxodonta africana*). Zoo Biology 24, 275–281.

Anon., 1882. Jumbo, the big African elephant at the Zoological Gardens. The Graphic, 25 February 1882, London, pp. 170–180.

Anon., 1976. Golden Days. Historic Photographs of the London Zoo. Gerald Duckworth & Co. Ltd, London.

Anon., 1992. Asian elephant (*Elephas maximus*) EEP annual report 1991. In: Brouwer, K., Smits, S., de Boer, L.E.M. (Eds.), EEP Yearbook 1991/92 Including the Proceedings of the 9th EEP Conference, Edinburgh, 6–8 July 1992. EAZA/EEP Executive Office, Amsterdam, p. 200.

Anon., 1996. Ein Beitrag zum Workshop 'Elefant und Mensch', Münster, 7–9 September 1996. European Elephant Group, Karl Wenschow, München.

Anon., 1998. International news. Emmen Zoo, Netherlands. International Zoo News 45/1 (282).

Anon., 2000. Asian elephant (*Elephas maximus*) EEP annual report 1998. In: Rietkerk, F., Hiddinga, B., Brouwer, K., Smits, S. (Eds.), EEP Yearbook 1998/99 Including Proceedings of the 16th EAZA Conference, Basel, 7–12 September 1999. EAZA Executive Office, Amsterdam, pp. 234–236.

Anon., 2006a. ISIS Abstracts: *Loxodonta africana*. International Species Information System. <https://app.isis.org/abstracts/abs.asp> (accessed 27.10.06.).

Anon., 2006b. ISIS Abstracts: *Elephas maximus*. International Species Information System. <https://app.isis.org/abstracts/abs.asp> (accessed 27.10.06.).

Anon., 2006c. Maryland Zoo in Baltimore Elephant Facilities Project (Baltimore City). D06E021E. Board of Public Works, Analysis of the FY 2007 Maryland Executive Budget, 2006.

Anon., 2011. 'Cruel' elephant polo match cancelled in Jaipur. The Telegraph. <https://www.telegraph.co.uk/news/worldnews/asia/india/8714257/Cruel-elephant-polo-match-cancelled-in-Jaipur.html> (accessed 12.04.18.).

Anon., 2012a. ASPCA Pays $9.3 Million in Landmark Ringling Bros. and Barnum & Bailey Circus Settlement. Press Release. Feld Entertainment. <https://www.feldentertainment.com/PressRoom/DisplayPressRelease/62237/> (accessed 16.08.19.).

Anon., 2012b. Anne the elephant court case – pre-trial hearing. <http://www.ad-international.org/about_us/go.php?id=2543> (accessed 09.06.18.).

Anon., 2013. Innocent Prisoner: The Plight of Elephants Living in Solitary Confinement in Europe. Born Free Foundation. <https://www.bornfree.org.uk/storage/media/content/files/Publications/INNOCENT-PRISONER_1.pdf> (accessed 29.07.19.).

Anon., 2015. Statement on Swaziland. <https://www.elephanttrust.org/index.php/articles/item/statement-on-swaziland> (accessed 05.06.19.).

Anon., 2016a. Girl, 7, dies after being hit by rock thrown by elephant in Morocco zoo. The Guardian, 28 July 2016. <https://www.theguardian.com/world/2016/jul/28/girl-dies-rabat-zoo-elephant-morocco> (accessed 25.04.16.).

Anon., 2016b. Jumbo the Elephant. The Life and Legacy of History's Most Famous Circus Animal. Charles River Editors.

Anon., 2016c. 5 Year Report. British and Irish Association of Zoos and Aquariums. Elephant Welfare Group, BIAZA, Regent's Park, London.

Anon., 2016d. Thomas Cook have stopped promoting elephant rides and shows. <https://www.worldanimalprotection.org/news/thomas-cook-have-stopped-promoting-elephant-rides-and-shows> (accessed 09.05.19.).

Anon., 2018a. <http://www.liverpoolmuseums.org.uk/wml/collections/blitz/casualties/item-627534.aspx> (accessed 12.03.19.).

Anon., 2018b. Elephants loose on Spanish motorway after circus lorry overturns. Daily Telegraph. <https://www.telegraph.co.uk/news/2018/04/02/elephants-loose-spanish-motorway-circus-lorry-overturns/> (accessed 21.04.19.).

Anon., 2018c. Working Elephants Now Have Oasis for Rehab and Retirement. Press release from *Four Paws*. <http://www.four-paws.us/news-2/press-releases/working-elephants-now-have-oasis-for-rehab-and-retirement-/> (accessed 16.06.18.).

Anon., 2018d. Our elephants are on the move. <https://twycrosszoo.org/animals/asian-elephants/> (accessed 13.07.18.).

Anon., 2018e. <https://www.african-parks.org/largest-elephant-translocation-history-concludes-malawi> (accessed 20.01.18.).

Anon., 2019a. The Elephant Ethogram. ElephantVoices. <https://elephantvoices.org/studies-a-projects/the-elephant-ethogram.html> (accessed 26.03.19.).

Anon., 2019b. Sheldrick Wildlife Trust. <https://www.sheldrickwildlifetrust.org/about/mission-history> (accessed 03.05.19.).

Anon., 2019c. Our Mission. <https://www.newquayzoo.org.uk/support-us/our-mission> (accessed 26.02.19.).

Anon., 2019d. <https://www.elephantvoices.org/studies-a-projects/santuario-de-elefantes-brasil.html> (accessed 19.06.19.).

Anon., 2019e. <https://elephantaidinternational.org/projects/elephant-refuge-north-america/> (accessed 19.06.19.).

Anon., 2019f. <https://www.elephanthaven.com/en/home> (accessed 19.06.19.).

Appayya, M.K., 1995. Elephants in Karnataka (India): a status report. In: Daniel, J.C., Datye, H. (Eds.), A Week with Elephants. Proceedings of the International Seminar on the Conservation of Asian Elephants. Bombay Natural History Society, Oxford University Press, Oxford, pp. 88−93.

Archie, E.A., Morrison, T.A., Foley, C.A.H., Moss, C.J., Alberts, S.C., 2006. Dominance rank relationships among wild female African elephants, *Loxodonta africana*. Animal Behaviour 71, 117−127.

AsERSM, 2017. Asian Elephants Range States Meeting. Final Report, 18−20 April 2017. Jakarta, Indonesia. <https://elephantconservation.org/iefImages/2018/03/AsERSM-2017_Final-Report.pdf> (accessed 10.07.19.).

AZA, 2001. AZA Standards for Elephant Management and Care. Association of Zoos and Aquariums, Silver Spring, MD.

AZA, 2007a. Top Ten AZA Elephant Success Stories. <http://www.aza.org/Newrsroom/PR_Top10ElephStories/> (accessed 23.04.07.).

AZA, 2007b. Zoo Elephants Thriving. <http://www.aza.org/ElephantConservation/PR_ThrivingZoos/> (accessed 23.04.07.).

AZA, 2009. <www.aza.org/behavior-advisory-group/> (accessed 15.06.09.).

AZA, 2012. AZA Standards for Elephant Management and Care. Associations of Zoos and Aquariums, Silver Spring, MD.

AZA, 2013. AZA Elephant Profile Form. Association of Zoos and Aquariums, Silver Spring, MD. <http://www.elephanttag.org/Professional/AZA_Standardized_Elephant_Profile_2013.pdf> (accessed 04.07.18.).

AZA, 2015. AZA's statement on Pittsburgh Zoo & PPG Aquarium's decision to forfeit AZA accreditation Monday, August 17, 2015. <https://www.aza.org/aza-news-releases/posts/azas-statement-on-pittsburgh-zoo--ppg-aquariums-decision-to-forfeit-aza-accreditation> (accessed 27.06.19.).

AZA, 2019a. Standards for Elephant Management and Care, The Accreditation Standards and Related Policies, 2019 ed. Association of Zoos and Aquariums, Silver Spring, MD, pp. 38−66.

AZA, 2019b. Elephant FAQ. <https://www.aza.org/elephant-faq> (accessed 26.12.17.).

AZA, 2019c. Elephant Conservation. <https://www.aza.org/elephant-conservation/> (accessed 30.07.19.).

AZA, 2020. Elephant Conservation. <https://www.aza.org/elephant-conservation/> (accessed 1.06.20.).

Baker, C.M.A., Manwell, C., 1983. Man and elephant the 'dare theory' of domestication and the origin of breeds. Zeitschrift für Tierzüchtung und Züchtungsbiologie 100, 55–75. <https://doi.org/10.1111/j.1439-0388.1983.tb00712.x>.

Baldrey, F.S.H., 1930. Tuberculosis in an elephant. The Journal of the Royal Army Veterinary Corps 1, 252.

Balke, J.M.E., Barker, I.K., Hackenberger, M.K., McManamon, R., Boever, W.J., 1988. Reproductive anatomy of three nulliparous female Asian elephants: the development of artificial breeding techniques. Zoo Biology 7, 99–113.

Ballou, J.D., Foose, T.J., 1996. Demographic and genetic management of captive populations. In: Kleiman, D.G., Allen, M.E., Thompson, K.V., Lumpkin, S. (Eds.), Wild Mammals in Captivity. Principles and Techniques. University of Chicago Press, Chicago, IL, pp. 263–283.

Balmford, A., 2000. Separating fact from artifact in analyses of zoo visitor preferences. Conservation Biology 14, 1193–1195.

Balmford, A., Leader-Williams, N., Mace, G.M., Manica, A., Walter, O., West, C., et al., 2007. Message received? Quantifying the impact of informal conservation education on adults visiting UK zoos. In: Zimmerman, A., Dickie, L., West, C. (Eds.), Zoos in the 21st Century. Catalysts for Conservation. Cambridge University Press, Cambridge, pp. 120–136.

Bandara, R., Tisdell, C., 2004. The net benefit of saving the Asian elephant: a policy and contingent valuation study. Ecological Economics 48, 93–107.

Barandongo, Z.R., Mfune, J.K.E., Turner, W.C., 2018. Bathing behaviors of African herbivores and the potential risk of inhalation anthrax. Journal of Wildlife Diseases 54, 34–44. <https://doi.org/10.7589/2017-04-069>.

Baratay, E., Hardouin-Fugier, E., 2002. Zoo: A History of Zoological Gardens in the West. Reaktion Books Ltd, London.

Barnaby, D., 1988. The Elephant Who Walked to Manchester. Bassett Publications, North Hill, Plymouth.

Barnes, R.F.W., 1995. Elephants. In: Macdonald, D. (Ed.), The Encyclopedia of Mammals. Andromeda Oxford Ltd, Oxford, pp. 452–461.

Barongi, R., Fisken, F.A., Parker, M., Gusset, M. (Eds.), 2015. Committing to Conservation: the World Zoo and Aquarium Conservation Strategy. WAZA Executive Office, Gland.

Barua, P., 1995. Managing a problem population of elephants. In: Daniel, J.C., Datye, H. (Eds.), A Week with Elephants. Proceedings of the International Seminar on the Conservation of Asian Elephant. Bombay Natural History Society, Oxford University Press, Oxford, pp. 150–161.

Barua, M., Tamuly, J., Ahmed, R.A., 2010. Mutiny or clear sailing? Examining the role of the Asian elephant as a flagship species. Human Dimensions of Wildlife 15, 145–160. <https://doi.org/10.1080/10871200903536176>.

Baskaran, N., Das, S., Sukumar, R., 2009. Population, Reproduction and Management of Captive Asian Elephants (*Elephas maximus*) in Jaldapara Wildlife Sanctuary, West Bengal, India. Indian Forester, pp. 1545–1555.

Bates, L.A., Byrne, R.W., 2007. Creative or created: using anecdotes to investigate animal cognition. Methods, Special Issue 'Neurocognitive Mechanisms Creativity: A Toolkit' 42, 12–21.

Bax, N.P., Sheldrick, D.L.W., 1963. Some preliminary observations on the food of elephants in the Tsavo Royal National Park (East) of Kenya. East African Wildlife Journal 1, 40–53.

BBC, 2012. Twycross Zoo arrests: elephant abuse police hold staff. <https://www.bbc.co.uk/news/uk-england-leicestershire-20080453> 25.10.2012 (accessed 28.06.19.).

BBC, 2019a. Singapore seizes elephant ivory and pangolin scales in record $48m haul. <https://www.bbc.co.uk/news/world-asia-49079720> (accessed 29.07.19.).

BBC, 2019b. Denmark buys last circus elephants so they can retire. <https://www.bbc.co.uk/newsround/49607433> (accessed 10.09.19.).

Beach, F.A., 1944. Relative effects of androgen upon the mating behavior of male rats subjected to forebrain injury or castration. The Journal of Experimental Zoology 97, 249–295.

Beach, F.A., 1947. A review of physiological and psychological studies of sexual behavior in mammals. Physiological Reviews 27, 240–307.

Beach, F.A., 1967. Cerebral and hormonal control of reflex mechanisms involved in copulatory behaviour. Physiological Reviews 47, 289–316.

Bechert, U.S., Brown, J.L., Dierenfeld, E.S., Ling, P.D., Molter, C.M., Schulte, B.A., 2019. Zoo elephant research: contributions to conservation of captive and free-ranging species. International Zoo Yearbook 53, 1–27.

Beck, B.B., 1980. Animal Tool Behavior: The Use and Manufacture of Tools by Animals. Garland STPM Press, New York.

Benedict, F.G., 1936. The Physiology of the Elephant. Publication No. 474, Carnegie Institution, Washington, DC.

Benirschke, K., Roocroft, A., 1992. Elephant inflicted injuries. Symposium on Erkrankungen der Zoo-und Wildtiere 34, 239–247.

Bercovitch, F.B., Andrews, J., 2010. Developmental milestones among African elephant calves on their first day of life. Zoo Biology 29, 120–126.

Berger, J., Cunningham, C., 2006. Behavioural ecology in managed reserves: gender-based asymmetries in interspecific dominance in African elephants and rhinos. Animal Conservation 1, 33–38.

Bertone, J.J., 2006. Excessive drowsiness secondary to recumbent sleep deprivation in two horses. Veterinary Clinics of North America: Equine Practice 22, 157–162.

Birkett, A., Stevens-Wood, B., 2005. Effect of low rainfall and browsing by large herbivores on an enclosed savannah habitat in Kenya. African Journal of Ecology 43, 123–130.

Bischof, L.L., Duffield, D.A., 1994. Relatedness estimation of captive Asian elephants (*Elephas maximus*) by DNA fingerprinting. Zoo Biology 13, 77–82.

Blake, S., Hedges, S., 2004. Sinking the flagship: the case of forest elephants in Asia and Africa. Biological Conservation 18, 1191–1202.

Blanc, J., 2008. *Loxodonta africana*. The IUCN Red List of Threatened Species 2008: e.T12392A3339343. <https://doi.org/10.2305/IUCN.UK.2008.RLTS.T12392A3339343.en> (accessed 17.04.18.).

Blaszkiewitz, B., 1999. Elephant breeding in Berlin. International Zoo News 46, 470–474.

Blumenbach, J.F., 1797. Handbuch der Naturgeschichte, fifth ed. J. C. Dieterich, Göttingen.

Bond, I., 1994. The importance of sport-hunted African elephants to CAMPFIRE in Zimbabwe. Traffic Bulletin 14, 117–119.

Boorer, M., 1972. Some aspects of stereotyped patterns of movement exhibited by zoo animals. International Zoo Yearbook 12, 164–166.

Bopayya, A.B., 1928. Tuberculosis in an elephant. Indian Forester 54, 500–502.

Borries, C., 1997. Infanticide in seasonally breeding multimale groups of Himalayan langurs (*Presbytis entellus*) in Ramnagar (South Nepal). Behavioral Ecology and Sociobiology 41, 139–150.

Bouley, D.M., Alarcón, C.N., Hildebrandt, T., O'Connell-Rodwell, C.E., 2007. The distribution, density and three-dimensional histomorphology of Pacinian corpuscles in the foot of the Asian elephant (*Elephas maximus*) and their potential role in seismic communication. Journal of Anatomy 211, 428–435. <https://doi.org/10.1111/j.1469-7580.2007.00792.x>.

Bowman, D., 2012. Conservation: bring elephants to Australia? Nature 482, 30. <https://doi.org/10.1038/482030a>.

Bradshaw, G.A., Schore, A.N., Brown, J.L., Poole, J.H., Moss, C.J., 2005. Elephant breakdown. Nature 433, 807. <https://doi.org/10.1038/433807a>.

Brambell, R., 1965. Report of the Technical Committee to Enquire into the Welfare of Animals Kept Under Intensive Livestock Husbandry Systems. Her Majesty's Stationary Office, London.

Brannian, J.D., Griffin, F., Terranova, P.F., 1989. Urinary androstenedione and luteinizing hormone concentrations during musth in a mature African elephant. Zoo Biology 8, 165–170.

Bremner-Harrison, S., Prodohl, P.A., Elwood, R.W., 2004. Behavioural trait assessment as a release criterion: boldness predicts early death in a reintroduction programme of captive-bred swift fox (*Vulpes velox*). Animal Conservation 7, 313–320.

Bridges, T.C., 1937. Wardens of the Wild. George G. Harrap & Co. Ltd, London.

Broad, G., 1996. Visitor profile and evaluation of informal education at Jersey Zoo. Dodo 32, 166–192.

Brockett, R.C., Stoinski, T.S., Black, J., Markowitz, T., Maple, T.L., 1999. Nocturnal behavior in a group of unchained female African elephants. Zoo Biology 18, 101–109.

Broom, D., 1986. Indicators of poor welfare. The British Veterinary Journal 142, 524–526.

Broom, D.M., 1991. Animal welfare: concepts and measurement. Journal of Animal Science 69, 4167–4175.

Broom, D.M., 2013. Can Animals be Moral?, M. Rowlands. Oxford University Press, New York, Book review. Applied Animal Behaviour Science 147, 243–244.

Brown, R.E., 1986. Social and hormonal factors influencing infanticide and its suppression in adult male Long-Evans hooded rats (*Rattus norvegicus*). Journal of Comparative Psychology 100, 155–161.

Brown, J.L., 2000. Reproductive endocrine monitoring in elephants: an essential tool for assisting captive breeding. Zoo Biology 19, 347−367.

Brown, J.L., Lehnhardt, J., 1995. Serum and urinary hormones during pregnancy and the peri- and postpartum period in an Asian elephant (*Elephas maximus*). Zoo Biology 14, 555−564.

Brown, J.L., Hildebrandt, T.B., Theison, W., Neiffer, D.L., 1999. Endocrine and ultrasound evaluation of a noncycling African elephant: identification of an ovarian follicular cyst. Zoo Biology 18, 223−232.

Brown, J.L., Olson, D., Keele, M., Freeman, E.W., 2004a. Survey of the reproductive cyclicity status of Asian and African elephants in North America. Zoo Biology 23, 309−321.

Brown, J.L., Göritz, F., Pratt-Hawkes, N., Hermes, R., Galloway, M., Graham, L.H., et al., 2004b. Successful artificial insemination of an Asian elephant at the National Zoological Park. Zoo Biology 23, 45−63.

Brown, J.L., Paris, S., Prado-Oviedo, N.A., Meehan, C.L., Hogan, J.N., Morfeld, K.A., et al., 2016. Reproductive health assessment of female elephants in North American zoos and association of husbandry practices with reproductive dysfunction in African elephants (*Loxodonta africana*). PLoS ONE 11 (7), e0145673. <https://doi.org/10.1371/journal.pone.0145673>.

Burks, K.D., Mellen, J.D., Miller, G.W., Lehnhardt, J., Weiss, A., Figueredo, A.J., et al., 2004. Comparison of two introduction methods for African elephants (*Loxodonta africana*). Zoo Biology 23, 109−126.

Byrne, R.W., Bates, L.A., 2011. Elephant cognition: what we know about what elephants know. In: Moss, C.J., Croze, H., Lee, P.C. (Eds.), The Amboseli Elephants. A Long-Term Perspective on a Long-Lived Animal. The University of Chicago Press, Chicago and London, pp. 174−182.

Cameron, E.Z., Ryan, S.J., 2016. Welfare at multiple scales: importance of zoo elephant population welfare in a world of declining wild populations. PLoS ONE 11 (7), e0158701. <https://doi.org/10.1371/journal.pone.0158701>.

Campos-Arceiz, A., Lin, T.Z., Htun, W., Takatsuki, S., Leimgruber, P., 2008. Working with mahouts to explore the diet of work elephants in Myanmar (Burma). Ecological Research 23, 1057−1064. <https://doi.org/10.1007/s11284-008-0466-4>.

Cappellini, E., Gentry, A., Palkopoulou, E., Ishida, Y., Cram, D., Roos, A., et al., 2013. Resolution of the type material of the Asian elephant, *Elephas maximus* Linnaeus, 1758 (Proboscidea, Elephantidae). Zoological Journal of the Linnaean Society, 1−10. <https://doi.org/10.1111/zoj.12084>.

Carlstead, K., 1996. Effects of captivity on the behaviour of wild mammals. In: Kleiman, D.G., Allen, M.E., Thompson, K.V., Lumpkin, S. (Eds.), Wild Mammals in Captivity. Principles and Techniques. University of Chicago Press, Chicago, IL, pp. 317−333.

Carlstead, K., Mench, J.A., Meehan, C., Brown, J.L., 2013. An epidemiological approach to welfare research in zoos: the elephant welfare project. Journal of Applied Animal Welfare Science 16 (4), 319−337. <https://doi.org/10.1080/10888705.2013.827915>.

Carlstead, K., Paris, S., Brown, J.L., 2018. Good keeper-elephant relationships in North American Zoos are mutually beneficial to welfare. Applied Animal Behaviour Science 211, 103−111. <https://doi.org/10.1016/j.applanim.2018.11.003>.

Carr, N., 2016. Ideal animals and animal traits for zoos: general public perspectives. Tourism Management 57, 37−44.

Carr, W.J., Loeb, L.S., Wylie, N.R., 1965. Responses to feminine odors in normal and castrated male rats. Journal of Comparative and Physiological Psychology 62, 336−338.

Carrington, R., 1958. Elephants. A Short Account of Their Natural History, Evolution and Influence on Mankind. Chatto and Windus, London.

Carwardine, M., 1995. The Guiness Book of Animal Records. Guiness Publishing Ltd, Enfield, Middlesex.

Casares, M., Silván, G., Carbonell, M.-D., Gerique, C., Martínez-Fernández, L., Cáceres, S., et al., 2016. Circadian rhythm of salivary cortisol secretion in female zoo-kept African elephants (*Loxodonta africana*). Zoo Biology 35, 65−69.

Chadwick, D.H., 1992. The Fate of the Elephant. Penguin Book Ltd, London.

Chadwick, C.L., Williams, E., Asher, L., Yon, L., 2017. Incorporating stakeholder perspectives into the assessment and provision of captive elephant welfare. Animal Welfare 26, 461−472. <https://doi.org/10.7120/09627286.26.4.461>.

Chambers, P., 2007. Jumbo. This Being the True Story of the Greatest Elephant in the World. André Deutsch, London.

Chandon, P., Morwitz, V.G., Reinartz, W.J., 2005. Do intentions really predict behavior? Self-generated validity effects in survey research. Journal of Marketing 69, 1–14.
Chandrapuria, V.P., Shrivastava, A.B., Agrawal, S., Agarwal, S., 2014. Vaginal vestibulotomy in an Asian elephant. Gajah 40, 39–41.
Chandrasekharan, K., Radhakrishnan, K., Cheeran, J.V., Nair, K.N.M., Prabhakaran, T., 1995. Review of the incidence, etiology and control of common diseases of Asian elephants with special reference to Kerala. In: Daniel, J.C., Datye, H. (Eds.), A Week with Elephants. Proceedings of the International Seminar on the Conservation of Asian Elephant. Bombay Natural History Society, Oxford University Press, Oxford, pp. 439–449.
Charles, M.B., 2007. The rise of the Sassanian Elephant Corps: elephants and the later Roman Empire. Iranica Antigua 42, 301–346.
Chartier, L., Zimmerman, A., Ladle, R.J., 2011. Habitat loss and human–elephant conflict in Assam, India: does a critical threshold exist? Oryx 45, 528–533.
Chase, M.J., Schlossberg, S., Griffin, C.R., Bouché, P.J.C., Djene, S.W., Elkan, P.W., et al., 2016. Continent-wide survey reveals massive decline in African savannah elephants. PeerJ 4, e2354. <https://doi.org/10.7717/peerj.2354>.
Chatkupt, T.T., Sollod, A.E., Sarobol, S., 1999. Elephants in Thailand: determinants of health and welfare in working populations. Journal of Applied Animal Welfare Science 2, 187–203.
Cheeran, J.V., Chandrasekharan, K., Radhakrishnan, K., 1995. Principles and practice of fixing dose of drugs for elephants. In: Daniel, J.C., Datye, H. (Eds.), A Week with Elephants. Proceedings of the International Seminar on the Conservation of Asian Elephant. Bombay Natural History Society, Oxford University Press, Oxford, pp. 430–438.
Chelliah, K., Sukumar, R., 2013. The role of tusks, musth and body size in male–male competition among Asian elephants, *Elephas maximus*. Animal Behaviour 86, 1207–1214.
Chevalier-Skolnikoff, S., Liska, J., 1993. Tool use by wild and captive elephants. Animal Behaviour 46, 209–219.
Choudhury, A., Lahiri Choudhury, D.K., Desai, A., Duckworth, J.W., Easa, P.S., Johnsingh, A.J.T., et al., 2008. *Elephas maximus*. The IUCN Red List of Threatened Species 2008: e.T7140A12828813. <https://doi.org/10.2305/IUCN.UK.2008.RLTS.T7140A12828813.en> (accessed 17.04.18.).
Chusyd, D.E., Brown, J.L., Hambly, C., Johnson, M.S., Morfeld, K., Patki, A., et al., 2018. Adiposity and reproductive cycling status in zoo African elephants. Obesity 26, 103–110.
CITES, 2017a. International trade in live elephants. <https://cites.org/eng/news/statement/international_trade_in_live_elephants> (accessed 22.02.19.).
CITES, 2017b. Conf. 11.20 (Rev. CoP17). Definition of the term 'appropriate and acceptable destinations'. <https://www.cites.org/sites/default/files/document/E-Res-11-20-R17.pdf> (accessed 22.02.19.).
CITES, 2019a. <https://www.cites.org/eng/CITES_conference_responds_to_extinction_crisis_by_strengthening_international_trade_regime_for_wildlife_28082019> (accessed 25.09.19.).
CITES, 2019b. <https://cites.org/sites/default/files/document/E-Res-11-20-R18.pdf> (accessed 12.12.19.).
Clauss, M., Loehlein, W., Kienzle, E., Wiesner, H., 2003. Studies on feed digestibilities in captive Asian elephants (*Elephas maximus*). Journal of Animal Physiology and Animal Nutrition 87, 160–173.
Clauss, M., Streich, W.J., Schwarm, A., Ortmann, S., Hummel, J., 2007. The relationship of food intake and ingesta passage predicts feeding ecology in two different megaherbivore groups. Oikos 116, 209–216.
Clemins, P.J., Johnson, M.T., Leong, K.M., Savage, A., 2005. Automatic classification and speaker identification of African elephant (*Loxodonta africana*) vocalizations. Journal of the Acoustical Society of America 117, 956. <https://doi.org/10.1121/1.1847850>.
Clubb, R., Mason, G., 2002. A Review of the Welfare of Zoo Elephants in Europe. A Report Commissioned by the RSPCA. Animal Behaviour Research Group, Department of Zoology, University of Oxford, Oxford.
Clubb, R., Rowcliffe, M., Lee, P., Mar, K.U., Moss, C., Mason, G.J., 2008. Compromised survivorship in zoo elephants. Science 322, 1649.
Coe, M., 1972. Defaecation by African elephants (*Loxodonta africana africana* (Blumenbach)). African Journal of Ecology 10, 165–174.
Cohn, J.P., 2006. Do elephants belong in zoos? BioScience 56, 714–717. <https://doi.org/10.1641/0006-3568(2006)56[714:DEBIZ]2.0.CO;2>.
Coleing, A., 2009. The application of social network theory to animal behaviour. Bioscience Horizons: International Journal of Student Research 2, 32–43. <https://doi.org/10.1093/biohorizons/hzp008>.

Congostrina, A.L., 2019. Barcelona's push for an "animalist" zoo signals end of a 127-year era. El País. <https://elpais.com/elpais/2019/05/07/inenglish/1557243876_053971.html> (accessed 26.06.19.).

Conway, W., 1998. Zoo reserves; a proposal. AZA Annual Conference Proceedings. American Zoo and Aquarium Association, Silver Spring, MD, pp. 54–58.

Cottrell, S., Jensen, J.L., Peck, S.L., 2014. Resuscitation and resurrection: the ethics of cloning cheetahs, mammoths, and Neanderthals. Life Sciences, Society and Policy 10, 3. <https://doi.org/10.1186/2195-7819-10-3>.

Crawley, J.A.H., Lahdenpera, M., Seltmann, M.W., Htut, W., Aung, H.H., Nyein, K., et al., 2019. Investigating changes within the handling system of the largest semi-captive population of Asian elephants. PLoS ONE 14 (1), e0209701. <https://doi.org/10.1371/journal.pone.029701>.

Cresswell, R., 1883. Aristotle's History of Animals in Ten Books (Translated by Richard Cresswell M.A). George Bell & Sons, York Street, Covent Garden, London.

Croke, V., 1997. The Modern Ark. The Story of Zoos: Past, Present and Future. Scribner, New York.

Csuti, B., Sargent, E.L., Bechert, U.S., 2001. The Elephant's Foot: Prevention and Care of Foot Conditions in Captive Asian and African Elephants. Iowa State University Press, Ames, IA.

Dale, R.H.I., 2010. Birth statistics for African (*Loxodonta africana*) and Asian (*Elephas maximus*) elephants in human care: history and implications for elephant welfare. Zoo Biology 29, 87–103.

Daniel, J.C., Datye, H. (Eds.), 1995. A Week with Elephants. Proceedings of the International Seminar on the Conservation of Asian Elephants. Bombay Natural History Society, Oxford University Press, Oxford.

Darwin Initiative, 2019. Projects. <https://www.darwininitiative.org.uk/project/> (accessed 29.07.19.).

Das, A., Smith, M.L., Saini, M., Katole, S., Kullu, S.S., Gupte, B.K., et al., 2015. Effect of concentrates restriction on feed consumption, diet digestibility, and nitrogen utilization in captive Asian elephants (*Elephas maximus*). Zoo Biology 34, 60–70.

Datye, H.S., Bhagwat, A.M., 1995. Man-elephant conflict: a case study of human deaths caused by elephants in parts of central India. In: Daniel, J.C., Datye, H. (Eds.), A Week with Elephants. Proceedings of the International Seminar of Asian Elephants. Bombay Natural History Society. Oxford University Press, Oxford, pp. 340–349.

Davies, J., Foxall, G.R., Pallister, J., 2002. Beyond the intention-behaviour mythology: an integrated model of recycling. Marketing Theory 2, 29. <https://doi.org/10.1177/1470593102002001645>.

Davies, T.E., Wilson, S., Hazarika, N., Chakrabarty, J., Das, D., Hodgson, D.J., et al., 2011. Effectiveness of intervention methods against crop-raiding elephants. Conservation Letters 4, 346–354.

De Beer, Y., van Aarde, R.J., 2008. Do landscape heterogeneity and water distribution explain aspects of elephant home range in southern Africa's arid savannas? Journal of Arid Environments 72, 2017–2025.

De Oliveira, C.A., West, G.D., Houck, R., Leblanc, M., 2004. Control of musth in an Asian elephant (*Elephas maximus*) using leuprolide acetate. Journal of Zoo and Wildlife Medicine 35, 70–76.

De Silva, S., Wittemyer, G., 2012. A comparison of social organization in Asian elephants and African savannah elephants. International Journal of Primatology 33, 1125–1141. <https://doi.org/10.1007/s10764-011-9564-1>.

De Villiers, P.A., Kok, O.B., 1997. Home range, association and related aspects of elephants in the eastern Transvaal Lowveld. African Journal of Ecology 35, 224–236.

De Waal, F., 1996. Good Natured. The Origins of Right and Wrong in Humans and Other Animals. Harvard University Press, Cambridge, MA.

Defra, 2017. Secretary of State's Standards of Modern Zoo Practice. Appendix 8 – Specialist Exhibits, Elephants. Department for Environment, Food and Rural Affairs, London. <https://www.gov.uk/government/publications/secretary-of-state-s-standards-of-modern-zoo-practice#history> (accessed 24.10.18.).

DeGrazia, D., 2002. Animal Rights. A Very Short Introduction. Oxford University Press, Oxford.

Deleu, R., Veenhuizen, R., Nelissen, M., 2003. Evaluation of the mixed-species exhibit of African elephants and Hamadryas baboons in Safari Beekse Bergen, the Netherlands. Primate Reports 65, 5–19.

Delort, R., 1992. The Life and Lore of the Elephant. Thames and Hudson, London.

Dembiec, D.P., Snider, R.J., Zanella, A.J., 2004. The effects of transport stress on tiger physiology and behavior. Zoo Biology 23, 335–346.

Denis, A., 1963. On Safari: The Story of My Life. William Collins, Sons & Co. Ltd, Glasgow.

Derocher, A.E., Wiig, Ø., 1999. Infanticide and cannibalism of juvenile polar bear (*Ursus maritimus*) in Svalbard. Arctic 52, 307–310.

Dey, S.C., 1991. Depredation by wildlife in fringe areas of North Bengal Forests with special reference to elephant damage. Indian Forester 117, 901–908.

Donlan, J., Greene, H.W., Berger, J., Bock, C.E., Bock, J.H., Burney, D.A., et al., 2005. Re-wilding North America. Nature 436, 913–914.

Donlan, C.J., Berger, J., Bock, C.E., Bock, J.H., Burney, D.A., Estes, J.A., et al., 2006. Pleistocene rewilding: an optimistic agenda for twenty-first century conservation. American Naturalist 168, 660–681. <https://doi.org/10.1086/508027>.

Dorning, J., Harris, S., Pickett, H., 2016. The Welfare of Wild Animals in Travelling Circuses. Report for the Welsh Government, <https://gov.wales/welfare-wild-animals-travelling-circuses>.

Dorresteyn, T., Terkel, A., 2000a. Captive elephant breeding: what should we do? In: Rietkerk, F., Hiddinga, B., Brouwer, K., Smits, S. (Eds.), EEP Yearbook 1998/99 Including Proceedings of the 16th EAZA Conference, Basel, 7–12 September 1999, EAZA Executive Office, Amsterdam, pp. 482–483.

Dorresteyn, T., Terkel, A., 2000b. Forward planning and EEP management for elephants in EAZA institutions. In: Rietkerk, F., Hiddinga, B., Brouwer, K., Smits, S. (Eds.), EEP Yearbook 1998/99 Including Proceedings of the 16th EAZA Conference, Basel, 7–12 September 1999, EAZA Executive Office, Amsterdam, pp. 480–481.

Dougall, H.W., Sheldrick, D.L.W., 1964. The chemical composition of a day's diet of an elephant. East African Wildlife Journal 2, 51–59.

Doughty, L.S., Slater, K., Zitzer, H., Avent, T., Thompson, S., 2014. The impact of male contraception on dominance hierarchy and herd association patterns of African elephants (*Loxodonta africana*) in a fenced game reserve. Global Ecology and Conservation 2, 88–96.

Douglas-Hamilton, 1972. On the Ecology and Behaviour of the African Elephant (Unpublished D. Phil. Thesis). Oxford University.

Douglas-Hamilton, I., Douglas-Hamilton, O., 1975. Among the Elephants. Collins, Glasgow.

Dow, T.L., Holásková, I., Brown, J.L., 2011. Results of the third reproductive assessment survey of North American Asian (*Elephas maximus*) and African (*Loxodonta africana*) female elephants. Zoo Biology 30, 699–711.

Doyle, C., 2014. Captive elephants. In: Gruen, L. (Ed.), The Ethics of Captivity. Oxford University Press, Oxford, pp. 38–56.

Doyle, C., Roy, S., 2006. Comments of In Defense of Animals on USDA Docket No. APHIS-2006-0044 "Captive Elephant Welfare." 11 December 2006. <http://citeseerx.ist.psu.edu/viewdoc/download?doi=10.1.1.485.7314&rep=rep1&type=pdf> (accessed 26.03.19.).

Doyle, C., Hancocks, D., Lindsay, K., Poole, J., Roberts, A.M., (n.d.). FAQs: proposed import of wild elephants from Swaziland by U.S. Zoos. <https://conservationaction.co.za/resources/reports/faqs-proposed-import-of-wild-elephants-from-swaziland-by-u-s-zoos/> (accessed 16.07.18.).

Dublin, H.T., 1983. Cooperation and reproductive competition among female African elephants. In: Wasser, S.K. (Ed.), Social Behavior of Female Vertebrates. Academic Press, New York, pp. 291–313.

Duffy, R., Moore, L., 2010. Neoliberalising nature? Elephant-Back tourism in Thailand and Botswana. Antipode 42, 742–766. <https://doi.org/10.1111/j.1467-8330.2010.00771.x>.

Duffy, R., Moore, L., 2011. Global regulations and local practices: the politics and governance of animal welfare in elephant tourism. Journal of Sustainable Tourism 19, 589–604.

Dugatkin, L.A., 1997. Cooperation Among Animals: An Evolutionary Perspective. Oxford University Press, Oxford.

Edwards, J., 2001. The irony of Hannibal's elephants. Latomus 60, Fasc. 4, 900–905.

Eisenberg, J.M., McKay, G.M., Jainudeen, M.R., 1971. Reproductive behavior of the Asiatic elephant (*Elephas maximus maximus* L.). Behaviour 38, 193–225.

Eisentraut, M., Böhme, W., 1989. Gibt es zwei Elefantenarten in Afrika? Zeitschrift des Kölner Zoo 32, 61–68.

ElephantVoices, 2019. <https://elephantvoices.org/elephants-in-captivity-7/in-zoos.html> (accessed 05.06.19.).

Ellicott, C., 2016. Saving Anne the Elephant. John Blake Publishing Ltd, London.

Eltringham, S.K., 1982. Elephants. Blandford Press, Poole.

Eltringham, S.K., 1984. Elephants in zoos. Biologist 31, 108–111.

Elzanowski, A., Sergiel, A., 2006. Stereotypic behavior of a female Asiatic elephant (*Elephas maximus*) in a zoo. Journal of Applied Animal Welfare Science 9, 223–232.

Endres, J., Haufellner, A., Haufellner, B., Schilfarth, J., Schilfarth, M., Schweiger, G., 2004. Documentation 2002: Elephants in European Zoos and Safari Parks. European Elephant Group. Schüling Verlag, Münster.

English, M., Kaplan, G., Rogers, L.J., 2014. Is painting by elephants in zoos as enriching as we are led to believe? PeerJ 2, e471. <https://doi.org/10.7717/peerj.471>.

EPA, 2018. Standard for Zoo Containment Facilities. Environmental Protection Authority (EPA), New Zealand.
Epplett, C., 2013. Roman beast hunts. In: Christen, P., Kyle, D.G. (Eds.), A Companion to Sport and Spectacle in Greek and Roman Antiquity. Wiley-Blackwell, Chichester, West Sussex, pp. 505−519.
Epstein, R., Lanca, R.P., Skinner, B.F., 1981. 'Self-awareness' in the pigeon. Science 212, 695−696.
Esson, M., Moss, A., Pitchford, L., 2014. The 'Thinking big' elephant project. Journal of the International Association of Zoo Educators 50, 14−16.
Estep, D.Q., Dewsbury, D.A., 1996. Mammalian reproductive behavior. In: Kleiman, D.G., Allen, M.E., Thompson, K.V., Lumpkin, S. (Eds.), Wild Mammals in Captivity. Principles and Techniques. University of Chicago Press, Chicago, IL, pp. 379−389.
Estes, R.D., 1991. The Behaviour Guide to African Mammals, Including Hoofed Mammals, Carnivores, Primates. University of California Press, Berkely and Los Angeles, California.
Evans, G.H., 1910. Elephants and Their Diseases. A Treatise on Elephants. Government Press, Rangoon.
Evans, A., 2017. Incredible moment a pair of frantic adult elephants work together to save a calf from drowning after it slips into pool at zoo enclosure. MailOnline. <https://www.dailymail.co.uk/news/article-4618624/Adult-elephants-save-calf-drowning-Korean-zoo.html> (accessed 11.11.18.).
Evans, K., Moore, R.J., Harris, S., 2013. The release of a captive-raised female African elephant (*Loxodonta africana*) in the Okavango Delta, Botswana. Animals 3, 370−385. <https://doi.org/10.3390/ani3020370>.
Ewer, R.F., 1968. Ethology of Mammals. Paul Elek (Scientific Books) Ltd, London.
Fagen, A., Acharya, N., Kaufman, G.E., 2014. Positive reinforcement training for a trunk wash in Nepal's working elephants: demonstrating alternatives to traditional elephant training techniques. Journal of Applied Animal Welfare Science 17, 83−97.
Faith, J.T., Rowan, J., Du, A., Koch, P.L., 2018. Plio-Pleistocene decline of African megaherbivores: no evidence for ancient hominin impacts. Science 362, 938−941.
Fanson, K.V., Lynch, M., Vogelnest, L., Miller, G., Keeley, T., 2013. Response to long-distance relocation in Asian elephants (*Elephas maximus*): monitoring adrenocortical activity via serum, urine, and feces. European Journal of Wildlife Research 59, 655−664. <https://doi.org/10.1007/s10344-013-0718-7>.
FAO, 1998. Secondary Guidelines for Development of National Farm Animal Genetic Resources Management Plans—Management of Small Populations at Risk. Publications Division, Food and Agriculture Organization of the United Nations, Rome <http://www.fao.org/3/a-w9361e.pdf> (accessed 24.07.19.).
FAO, 1999. Forest Harvesting Case-Study 5. Elephants in Logging Operations in Sri Lanka. Publications Division, Food and Agriculture Organization of the United Nations, Rome <http://www.fao.org/3/v9570E/v9570e00.htm#TopOfPage> (accessed 08.03.19.).
Faust, L.J., Thompson, S.D., Earnhardt, J.M., 2006. Is reversing the decline of Asian elephants in North American zoos possible? An individual-based modeling approach. Zoo Biology 25, 201−218.
Fayrer-Hosken, R.A., Brooks, P., Bertschinger, H.J., Kirkpatrick, J.F., Turner, J.W., Liu, I.K.M., 1997. Management of African elephant populations by immunocontraception. Wildlife Society Bulletin 25, 18−21.
Fellers, J.H., Fellers, G.M., 1976. Tool use in a social insect and its implications for competitive interactions. Science 192, 70−72. <https://doi.org/10.1126/science.192.4234.70>.
Fernando, P., Lande, R., 2000. Molecular genetic and behavioural analysis of social organization in the Asian elephant (*Elephas maximus*). Behavioral Ecology and Sociobiology 48, 84−91.
Fernando, P., Vidya, T.N.C., Payne, J., Stuewe, M., Davison, G., Alfred, R.J., et al., 2003. DNA analysis indicates that Asian elephants are native to Borneo and are therefore a high priority for conservation. PLoS Biology 1 (1), e6. <https://doi.org/10.1371/journal.pbio.0000006>.
Fernando, P., Wikramanayake, E.D., Janaka, H.K., Jayasinghe, L.K.A., Gunawardena, M., Kotagama, S.W., et al., 2008. Ranging behavior of the Asian elephant in Sri Lanka. Mammalian Biology 73, 2−13.
Fernando, P., Janaka, H.K., Ekanayaka, S.K.K., Nishantha, H.G., Pastorini, J., 2009. A simple method for assessing elephant body condition. Gajah 31, 29−31.
Ferrier, A.J., 1947. The Care and Management of Elephants in Burma. Steel Brothers & Co. Ltd, London.
Fickel, J., Lieckfeldt, D., Ratanakorn, P., Pitra, C., 2007. Distribution of haplotypes and microsatellite alleles among Asian elephants (*Elephas maximus*) in Thailand. European Journal of Wildlife Research 53, 298−303.
Fifield, K., 2014. Melbourne Zoo Trail of the Elephants. ZooLex Gallery <https://www.zoolex.org/gallery/show/1119/> (accessed 25.04.19.).
Finn, J.K., Tregenza, T., Norman, M.D., 2009. Defensive tool use in a coconut-carrying octopus. Current Biology 19, R1069−R1070.

Flanagan, J., 2019a. Baby elephants torn from mothers and shipped 7,000 miles to China. The Times, 9 February 2019, p43.
Flanagan, J., 2019b. Baby elephants held in pens for a year after sale to Asia. The Times <https://www.thetimes.co.uk/edition/world/baby-elephants-held-in-pens-for-a-year-after-sale-to-asia-k20b3l77t>. 23.09.2019 (accessed 25.09.19.).
Fleischer, R.C., Perry, E.A., Muralidharan, K., Stebvens, E.E., Wemmer, C.M., 2001. Phylogeography of the Asian elephant (*Elephas maximus*) based on mitochondrial DNA. Evolution 55, 1882–1892. <https://doi.org/10.1554/0014-3820(2001)055[1882:POTAEE]2.0.CO;2>.
Foerder, P., Galloway, M.B.T., Moore III, D.E., Reiss, D., 2011. Insightful problem solving in an Asian elephant. PLoS ONE 6 (8), e23251. <https://doi.org/10.1371/journal.pone.0023251>.
Foley, C.A.H., 2002. High incidence of elephant twin births in Tarangire National Park, Tanzania. Pachyderm 32, 62–66.
Foley, C., Pettorelli, N., Foley, L., 2008. Severe drought and calf survival in elephants. Biology Letters 4, 541–544.
Fowler, E.F., Mikota, S.K., 2006. Biology, Medicine, and Surgery of Elephants. Blackwell Publishing Ltd, Oxford.
Fraser, D.J., 2008. How well can captive breeding programs conserve biodiversity? A review of salmonids. Evolutionary Applications 1, 535–586. <https://doi.org/10.1111/j.1752-4571.2008.00036.x>.
Freeman, E.W., Weiss, E., Brown, J.L., 2004. Examination of the interrelationships of behavior, dominance status, and ovarian activity in captive Asian and African elephants. Zoo Biology 23, 431–448.
Freeman, E.W., Guagnano, G., Olson, D., Keele, M., Brown, J.L., 2009. Social factors influence ovarian cyclicity in captive African elephants (*Loxodonta africana*). Zoo Biology 28, 1–15.
Freeman, E.W., Schulte, B.A., Brown, J.L., 2010a. Investigating the impact of rank and ovarian activity on the social behavior of captive female African elephants. Zoo Biology 29, 154–167.
Freeman, E.W., Schulte, B.A., Brown, J.L., 2010b. Using behavioural observations and keeper questionnaires to assess social relationships among captive female African elephants. Zoo Biology 29, 140–153.
French, F., Mancini, C., Sharp, H., 2016. Exploring methods for interaction design with animals: a case-study with Valli. ACI'16: Proceedings of the Third International Conference on Animal-Computer Interaction. ACM, article no. 3.
French, F., Mancini, C., Sharp, H., 2018. High tech cognitive and acoustic enrichment for captive elephants. Journal of Neuroscience Methods 300, 173–183.
Friend, T.H., Parker, M.L., 1999. The effect of penning versus picketing on stereotypic behaviour of circus elephants. Applied Animal Behaviour Science 64, 213–225.
Frost, P.G.H., Bond, I., 2008. The CAMPFIRE programme in Zimbabwe: payments for wildlife services. Ecological Economics 65, 776–787.
Gaalema, D.E., Perdue, B.M., Kelling, A.S., 2011. Food preference, keeper ratings, and reinforcer effectiveness in exotic animals: the value of systematic testing. Journal of Applied Animal Welfare Science 14, 33–41.
Gajaseni, J., 1993. Energy value of elephant labour. Indian Forester 119, 804–806.
Gallup Jr., G.G., 1970. Chimpanzees: self-recognition. Science 167, 86–87.
Gallup Jr., G.G., Povinelli, D.J., Suarez, S.D., Lethmate, J.R., Menzel, E.W., 1995. Further reflections on self-recognition in primates. Animal Behaviour 50, 1525–1532.
Gambino, L., 2015. Pittsburgh zoo ordered to stop stressing its elephants through use of dogs. The Guardian <https://www.theguardian.com/world/2015/feb/02/pittsburgh-zoo-stop-stressing-elephants-dogs> (accessed 15.06.18.).
Ganswindt, A., Heistermann, M., Borragan, S., Hodges, J.K., 2002. Assessment of testicular endocrine function in captive African elephants by measurement of urinary and fecal androgens. Zoo Biology 21, 27–36.
Ganswindt, A., Heistermann, M., Hodges, K., 2005. Physical, physiological, and behavioral correlates of musth in captive African elephants (*Loxodonta africana*). Physiological and Biochemical Zoology 78, 505–514. <https://doi.org/10.1086/430237>.
Garaï, M.E., 1992. Special relationships between female Asian elephants *Elephas maximus* in zoological gardens. Ethology 90, 197–205.
Garrod, A.H., 1875. A report on the Indian elephant which died in the society's gardens on July 7th 1875. Proceedings of the Zoological Society of London 1875, 542.
Garton, G.A., 1963. The composition and biosynthesis of milk lipids. Journal of Lipid Research 4, 237–254.
Genin, J.J., Willems, P.A., Cavagna, G.A., Lair, R., Heglund, N.C., 2010. Biomechanics of locomotion in Asian elephants. Journal of Experimental Biology 213, 694–706. <https://doi.org/10.1242/jeb.035436>.

Gilbert, J., 1994. Elephant feeder balls. Shape of Enrichment 3 (4), 3–5.
Gokula, V., Varadharajan, M., 1996. The status of temple elephant management in Tamil Nadu, Southern India. Rajah 15, 37–40.
Goodall, J., 1971. In the Shadow of Man. Collins, London.
Gore, M., Hutchins, M., Ray, J., 2006. A review of injuries caused by elephants in captivity: an examination of predominant factors. International Zoo Yearbook 40, 51–62.
Gorman, M.L., 1986. The secretion of the temporal gland of the African elephant *Loxodonta africana* as an elephant repellent. Journal of Tropical Ecology 2, 187–190.
Graham, S., 2016. Grace Mugabe pays military debt to China with 35 elephant calves. The Times <https://www.thetimes.co.uk/article/zimbabwark-to-settle-mugabe-debt-vww9ctqrb> 26.12.2016 (accessed 16.07.18.).
Grandin, T. (Ed.), 2014. Livestock Handling and Transport, fourth ed. CABI Publishing, Wallingford, Oxon.
Gray, C.W., Nettashinghe, A.P.W., 1970. A preliminary study of immobilization of the Asiatic elephant (*Elephas maximus*) utilizing etorphine (M-99). Zoologica; Scientific Contributions of the New York Zoological Society 55, 51–56.
Greco, B.J., Meehan, C.L., Hogan, J.N., Leighty, K.A., Mellen, J., Mason, G.J., et al., 2016a. The days and nights of zoo elephants: using epidemiology to better understand stereotypic behavior of African elephants (*Loxodonta africana*) and Asian elephants (*Elephas maximus*) in North American zoos. PLoS ONE 11 (7), p.e.0144276.
Greco, B.J., Meehan, C.L., Miller, L.J., Shepherdson, D.J., Morfeld, K.A., Andrews, J., et al., 2016b. Elephant management in North American zoos: environmental enrichment, feeding, exercise, and training. PLoS ONE 11 (7), e0152490.
Greco, B.J., Meehan, C.L., Heinsius, J.L., Mench, J.A., 2017. Why pace? The influence of social, housing, management, life history, and demographic characteristics on locomotor stereotypy in zoo elephants. Applied Animal Behaviour Science 194, 104–111.
Green, C., 1993. Enriching an elephant's environment. Shape of Enrichment 2 (1), 5–6.
Greenwald, R., Lyashchenko, O., Esfandiari, J., Miller, M., Mikota, S., Olsen, J.H., et al., 2009. Highly accurate antibody assays for early and rapid detection of tuberculosis in African and Asian elephants. Clinical and Vaccine Immunology 16, 605–612.
Gregory, W.K., 1937. Biographical memoir of Henry Fairfield Osborn 1857 – 1935. Biographical Memoirs of National Academy of Sciences of the United States of America XIX, 51–119.
Gröning, K., Saller, M., 1999. Elephants – A Cultural and Natural History. Könemann Verlagsgesellschaft mbH, Cologne.
Grubb, P., Groves, C.P., Dudley, J.P., Shoshani, J., 2000. Living African elephants belong to two species: *Loxodonta africana* (Blumenbach, 1797) and *Loxodonta cyclotis* (Matschie, 1900). Elephant 2 (4), 1–4. <https://doi.org/10.22237/elephant/1521732169>.
Gruber, T.M., Friend, T.H., Gardner, J.M., Packard, J.M., Beaver, B., Bushong, D., 2000. Variation in stereotypic behavior related to restraint in circus elephants. Zoo Biology 19, 209–221.
Gruen, L. (Ed.). 2014. The Ethics of Captivity. Oxford University Press, Oxford.
Gunhold, I., Weissenböck, N.M., Schwammer, H.M., 2006. The dominance hierarchy of the herd of African elephants (*Loxodonta africana*) at the Schoenbrunn Zoo. Der Zoologische Garten 76, 169–177.
Gupfinger, R., Kaltenbrunner, M., 2018. Animals make music: a look at non-human musical expression. Multimodal Technologies and Interaction 2, 51. <https://doi.org/10.3390/mti2030051>.
Gupta, B. K. 2008. Barrier Designs for Zoos. Central Zoo Authority, Ministry of Environment & Forests, India.
Gurusamy, V., Tribe, A., Toukhsati, S., Phillips, C.J.C., 2015. Public attitudes in India and Australia toward Elephants in Zoos. Anthrozoös 28, 87–100.
Guy, P.R., 1975. The daily food intake of the African elephant, *Loxodonta africana*, Blumenbach, in Rhodesia. Arnoldia Rhodesia 7, 1–8.
Guy, P.R., 1976. The feeding behaviour of elephant (*Loxodonta africana*) in the Sengwa area, Rhodesia. South African Journal of Wildlife Research 6, 55–63.
Hackenberger, M.K., 1987. Diet Digestibilities and Ingesta Transit Times of Captive Asian and African Elephants (Unpublished MSc thesis). University of Guelph.
Hacker, C.E., Miller, L.J., 2016. Zoo visitor perceptions, attitudes, and conservation intent after viewing African elephants at the San Diego Zoo Safari Park. Zoo Biology 35, 355–361.
Haight, J., 1993. Playing with their food – ideas for elephants. Shape of Enrichment 2 (2), 9.

Hambrecht, S., Reichler, S., 2013. Group dynamics of young Asian elephant bulls (*Elephas maximus* Linnaeus, 1758) in Heidelberg Zoo — Integration of a newcomer in an established herd. Gruppendynamik bei jungen Asiatischen Elefantenbullen (*Elephas maximus* Linnaeus, 1758) im Zoo Heidelberg — integration eines Neulings in eine bestehende Herde. Der Zoologische Garten 82, 267−292.

Hanby, J.P., 1972. The Sociosexual Nature of Mounting and Related Behaviours in a Confined Troop of Japanese Macaques (*Macaca fuscata*) (Unpublished Ph.D. thesis). University of Oregon.

Hancocks, D., 2008. Most zoos do not deserve elephants. In: Wemmer, C., Christen, C.A. (Eds.), Elephants and Ethics. Towards a Morality of Coexistence. The Johns Hopkins University Press, Baltimore, MD, pp. 259−283.

Harlow, H.F., Harlow, M.K., 1965. The affectional systems. In: Schrier, A.M. (Ed.), Behavior of Nonhuman Primates, vol. 2. Academic, Orlando, FL, pp. 287−334.

Harris, S., Iossa, G., Soulsbury, C.D., 2006. A Review of the Welfare of Wild Animals in Circuses. School of Biological Sciences, University of Bristol, United Kingdom.

Harris, M., Sherwin, C., Harris, S., 2008. The Welfare, Housing and Husbandry of Elephants in UK Zoos (Unpublished report). University of Bristol, Defra, United Kingdom.

Hart, L.A., 1994. The Asian elephants-driver partnership: the drivers' perspective. Applied Animal Behaviour Science 40, 297−312.

Hart, B.L., Hart, L.A., 1994. Fly switching by Asian elephants: tool use to control parasites. Animal Behaviour 48, 35−45.

Hart, L., Sundar, 2000. Family traditions for mahouts of Asian elephants. Anthrozoös 13, 34−42.

Hartl, G.B., Kurt, F., Hemmer, W., Nadlinger, K., 1995. Electrophoretic and chromosomal variation in captive Asian elephants (*Elephas maximus*). Zoo Biology 14, 87−95.

Hartmann, E., Bernstein, J., Wilson, C., 1968. Sleep + dreaming in the elephant. Psychophysiology 4, 389.

Hartnett, G., 1995. Enrich one, empower the other. Shape of Enrichment 4 (3), 5−6.

Harvey, R., Alden, C., Wu, Y.-S., 2017. Speculating a fire sale: options for Chinese authorities in implementing a domestic ivory trade ban. Ecological Economics 141, 22−31.

Harvey, N.D., Daly, C., Clark, N., Ransford, E., Wallace, S., Yon, L., 2018. Social interactions in two groups of zoo-housed adult female Asian elephants (*Elephas maximus*) that differ in relatedness. Animals 8, 132.

Hasenjager, M.J., Bergl, R.A., 2015. Environmental conditions associated with repetitive behavior in a group of African elephants. Zoo Biology 34, 201−210.

Haslam, C., 2018. Thomas Cook calls time on whale shows. The Sunday Times. 29 July 2018, p5.

Haspeslagh, M., Stevens, J.M.G., De Groot, E., Dewulf, J., Kalmar, I.D., Moons, C.P.H., 2013. A survey of foot problems, stereotypic behaviour and floor type in Asian elephants (*Elephas maximus*) in European zoos. Animal Welfare 22, 437−443. < https://doi.org/10.7120/09627286.22.4.437 > .

Haufellner, A., Kurt, F., Schilfarth, J., Schweiger, G., 1993. Elefanten in Zoo und Zirkus. Dokumentation Teil 1: Europa. Karl Wenschow, München.

Haywood, M., 2014. Human places for large non-humans: from London's imperial elephant stables to Copenhagen's postmodern glasshouse. In: Pauknerova, K., Stella, M., Gibas, P., et al., Non-Humans in Social Science: Ontologies, Theories and Case Studies. Pavel Mervart, Červený Kostelec, pp. 201−218.

Hedges, S., Tyson, M.J., Sitompul, A.F., Hammatt, H., 2006. Why inter-country loans will not help Sumatra's elephants. Zoo Biology 25, 235−246.

Heistermann, M., Trohorsch, B., Hodges, J.K., 1997. Assessment of ovarian function in the African elephant (*Loxodonta africana*) by measurement of 5α-reduced progesterone metabolites in serum and urine. Zoo Biology 16, 273−284.

Hejna, P., Zátopková, L., Šafr, M., 2011. A fatal elephant attack. Journal of Forensic Sciences 57, <https://doi.org/10.1111/j.1556-4029.2011.01967>.

Henshaw, J., 1972. Notes on conflict between elephants and some bovids and other interspecific contacts in Yankari Game Reserve, North East Nigeria. East African Wildlife Journal 10, 151−153.

Hermes, R., Saragusty, J., Schaftenaar, W., Göritz, F., Schmitt, D.L., Hildebrandt, T.B., 2008. Obstetrics in elephants. Theriogenology 70, 131−144.

Hildebrandt, T.B., Göritz, F., Pratt, N.C., Brown, J.L., Montali, R.J., Schmitt, D.L., et al., 2000a. Ultrasonography of the urogenital tract in elephants (*Loxodonta africana* and *Elephas maximus*): an important tool for assessing female reproductive function. Zoo Biology 19, 321−332.

Hildebrandt, T.B., Hermes, R., Pratt, N.C., Fritsch, G., Blottner, S., Schmitt, D.L., et al., 2000b. Ultrasonography of the urogenital tract in elephants (*Loxodonta africana* and *Elephas maximus):* an important tool for assessing male reproductive function. Zoo Biology 19, 333−345.

Hildebrandt, T.B., Göritz, F., Hermes, R., Reid, C., Dehnhard, M., Brown, J.L., 2006. Aspects of the reproductive biology and breeding management of Asian and African elephants *Elephas maximus* and *Loxodonta africana*. International Zoo Yearbook 40, 20−40.

Hildebrandt, T.B., Hermes, R., Saragusty, J., Potier, R., Schwammer, H.M., Balfanz, F., et al., 2012. Enriching the captive elephant population genetic pool through artificial insemination with frozen-thawed semen collected in the wild. Theriogenology 78, 1398−1404.

Hile, M.E., Hintz, H.F., Hollis, N.E., 1997. Predicting body weight from body measurements in Asian elephants (*Elephas maximus*). Journal of Zoo and Wildlife Medicine 28, 424−427.

Hill, S.P., Broom, D.M., 2009. Measuring zoo animal welfare: theory and practice. Zoo Biology 28, 531−544.

HMSO, 1977a. The Meteorological Office. Monthly Weather Report. July 1976, Vol. 93, No.7. Her Majesty's Stationery Office, London.

HMSO, 1977b. The Meteorological Office. Monthly Weather Report. January 1977, Vol. 94, No.1. Her Majesty's Stationery Office, London.

Ho, I., Lindsay, K., 2017. Challenges to CITES regulation of the international trade in live, wild-caught African elephants. In: Sixty-ninth Meeting of the Standing Committee Geneva (Switzerland), 27 November−1 December 2017. Convention on International Trade in Endangered Species of Wild Fauna and Flora SC69 Inf. 37.

Hodges, J.K., Heistermann, M., Beard, A., van Aarde, R.J., 1997. Concentrations of progesterone and the 5α-reduced progestins, 5α-pregnane-3,20- dione and 3α-hydroxy-5α-pregnan-20-one, in luteal tissue and circulating blood and their relationship to luteal function in the African elephant, *Loxodonta africana*. Biology of Reproduction 56, 640−646. <https://doi.org/10.1095/biolreprod56.3.640>.

Holdgate, M.R., Meehan, C.L., Hogan, J.N., Miller, L.J., Soltis, J., Andrews, J., et al., 2016a. Walking behavior of zoo elephants: associations between GPS-measured daily walking distances and environmental factors, social factors, and welfare indicators. PLoS ONE 11, p.e.0150331.

Holdgate, M.R., Meehan, C.L., Hogan, J.N., Miller, L.J., Rushen, J., de Passillé, A.M., et al., 2016b. Recumbence behavior in zoo elephants: determination of patterns and frequency of recumbent rest and associated environmental and social factors. PLoS ONE 11, e0153301.

Holzer, D., Scott, D., Bixler, R.D., 1998. Socialization influences on adult zoo visitation. Journal of Applied Recreation Research 23, 43−62.

Horback, K.M., Miller, L.J., Andrews, J.R., Kuczaj, S.A., 2014. Diurnal and nocturnal activity budgets of zoo elephants in an outdoor facility. Zoo Biology 33, 403−410.

Howard, J.G., Bush, M., de Voss, V., Wildt, D.E., 1989. Electroejaculation, semen characteristics and serum testosterone concentration of free ranging African elephants (*Loxodonta africana*). Journal of Reproduction and Fertility 72, 187−195.

Hoyte, J., 1960. Trunk Road for Hannibal: With an Elephant Over the Alps. Geoffrey Bles, London.

Hughes, J.D., 2003. Europe as consumer of exotic biodiversity: Greek and Roman times. Landscape Research 28, 21−31.

Hundley, G., 1923. The breeding of elephants in captivity. Journal of the Bombay Natural History Society 28, 537−539.

Hunt, G.R., 1996. Manufacture and use of hook-tools by New Caledonian crows. Nature 379, 249−251.

Hutchins, M., 2006. Variation in nature; its implications for zoo elephant management. Zoo Biology 25, 161−171.

Hutchins, M., 2007. The animal rights − conservation debate: can zoos and aquariums play a role? In: Zimmerman, A., Hatchwell, M., Dickie, L., West, C. (Eds.), Zoos in the 21st Century. Cambridge University Press, Cambridge, pp. 92−109.

Hutchins, M., Conway, W., 1995. Beyond Noah's ark: the evolving role of modern zoological parks and aquariums in field conservation. International Zoo Yearbook 34, 117−130.

Hutchins, M., Keele, M., 2006. Elephant importation from range countries: ethical considerations for accredited zoos. Zoo Biology 25, 219−233.

Hutchins, M., Paul, E., Bowdoin, J., 1996. Contributions of zoo and aquarium research to wildlife conservation and science. In: Burghardt, G.M., Bielitzski, J.T., Boyce, J.R., Schaffer, D. (Eds.), The Well-Being of Animals in Zoo and Aquarium Research. Scientist's Center for Animal Welfare, Greenbelt, MD, pp. 23−29.

Hutchinson, J.R., Schwerda, D., Famini, D.J., Dale, R.H.I., Fischer, M.S., Kram, R., 2006. The locomotor kinematics of Asian and African elephants: changes with speed and size. The Journal of Experimental Zoology 209, 3812−3827.

Hvilsom, C., Frandsen, P., Børsting, C., Carlsen, F., Sallé, B., Simonsen, B.T., et al., 2013. Understanding geographic origins and history of admixture among chimpanzees in European zoos, with implications for future breeding programmes. Heredity (Edinb.) 110, 586–593.

Ibler, B., Pankow, R., 2012. Daten zur Schlafdauer in der Herde Asiatischer Elefanten (*Elephas maximus*) im Zoologischen Garten Berlin. Data on the sleep in the herd of Asian Elephants (*Elephas maximus*) at Berlin Zoological Garden. Der Zoologische Garten 81, 239–245.

IDA, 2019. An expanding list of worldwide circus bans and restrictions <https://www.stopcircussuffering.com/circus-bans/> (accessed 27.06.19.).

IFAW, 2015. Captured Zimbabwe Elephant Calves Flown to China. <https://www.ifaw.org/united-states/news/captured-zimbabwe-elephant-calves-flown-china> (accessed 16.07.18.).

Illera, J.-C., Silván, G., Cáceres, S., Carbonell, M.-D., Gerique, C., Martínez-Fernández, L., et al., 2014. Assessment of ovarian cycles in the African elephant (*Loxodonta africana*) by measurement of salivary progesterone metabolites. Zoo Biology 33, 245–249.

IMLS, 2018. Institute of Museum and Library Services. Awarded Grants. Honolulu Zoo, LG-54-09-0071-09 and LG-25-10-0033-10. <https://www.imls.gov/grants/awarded-grants?search_api_views_fulltext=elephant%20and%20zoo&sort_by=field_program> (accessed 06.03.18.).

Independent, 2014. <http://www.independent.co.uk/voices/campaigns/elephant-campaign/elephant-campaign-check-how-your-donations-to-the-2013-christmas-appeal-made-a-real-difference-9871312.html> (accessed 26.12.17.).

Irie, N., Hasegawa, T., 2009. Elephant psychology: what we know and what we would like to know. Japanese Psychological Research 51, 177–181.

Irie-Sugimoto, N., Kobayashi, T., Sato, T., Hasegawa, T., 2009. Relative quantity judgment by Asian elephants (*Elephas maximus*). Animal Cognition 12, 193–199.

Isaac, S., 2017. Treasures from the collections. An anatomical account of the elephant accidentally burnt in Dublin, on Fryday, June 17 in the year 1681 – Allen Mullen (1682). Bulletin of the Royal College of Surgeons of England 99, 127. <https://publishing.rcseng.ac.uk/doi/pdfplus/10.1308/rcsbull.2017.127>.

IUCN, 2017a. Statement and resolutions on the role of captive facilities in in-situ African elephant conservation. <https://www.iucn.org/ssc-groups/mammals/african-elephant-specialist-group/afesg-statements/role-captive-facilities> (accessed 28.12.17.).

IUCN, 2017b. IUCN/SSC Asian Elephant Specialist Group Draft Position Statement on Captive Asian Elephants. <http://www.asesg.org/PDFfiles/asesg-Draft%20Position%20Statement%20on%20Captive%20Asian%20Elephants.pdf> (accessed 28.12.17.).

Jachmann, H., Berry, P.S.M., Imae, H., 1995. Tusklessness in African elephants: a future trend. African Journal of Ecology 33, 230–235.

Jachowski, D.S., Slotow, R., Millspaugh, J.J., 2012. Physiological stress and refuge behavior by African elephants. PLoS ONE 7 (2), p.e.31818.

Jachowski, D.S., Slotow, R., Millspaugh, J.J., 2013. Delayed physiological acclimatization by African elephants following reintroduction. Animal Conservation 16, 575–583.

Jakob-Hoff, R., Kingan, M., Fenemore, C., Schmid, G., Cockrem, J.F., Crackle, A., et al., 2019. Potential impact of construction noise on selected zoo animals. Animals 9, 504. <https://doi.org/10.3390/ani9080504>.

Jones, T., Cussack, J.J., Pozo, R.A., Smit, J., Mkuburo, L., Baran, P., et al., 2018. Age structure as an indicator of poaching pressure: insights from rapid assessments of elephant populations across space and time. Ecological Indicators 88, 115–125.

Kahl, M.P., Armstrong, B.D., 2000. Visual and tactile displays in African elephants, *Loxodonta africana*: a progress report (1991-1997). Elephant 2 (4), 19–21.

Kanagaraj, R., Araújo, M.B., Barman, R., Davidar, P., De, R., Digal, D.K., et al., 2019. Predicting range shifts of Asian elephants under global change. Diversity and Distributions. <https://doi.org/10.1111/ddi.12898>.

Kane, L., 2010. A case study of African elephants' journey from Swaziland to US zoos in 2003: a question of commerce and tale of brinkmanship. Journal of Animal Law 6, 51–79.

Katz, B., 2019. A German circus uses stunning holograms instead of live animal performers. <https://www.smithsonianmag.com/smart-news/german-circus-uses-stunning-holograms-instead-live-animal-performers-180972376/> (accessed 10.09.19.).

Keele, M., Lewis, K., Dever, K., 2010. Asian Elephant (*Elephas maximus*) North American Regional Studbook, 17 July 2007 – 31 August 2010. Association of Zoos and Aquariums, Portland, OR.

Kekkonen, J., Brommer, J.E., 2015. Reducing the loss of genetic diversity associated with assisted colonization-like introductions of animals. Current Zoology 61, 827–834.

Kellert, S.R., 1989. Perceptions of animals in America. In: Hoage, R.J. (Ed.), Perceptions of Animals in American Culture. Smithsonian Institution Press, Washington DC, pp. 5–24.

Kellert, S.R., Dunlap, J., 1989. Informal Learning at the Zoo: A Study of Attitude and Knowledge Impacts. Zoological Society of Philadelphia, Philadelphia, PA.

Kemf, E., Jackson, P., 1995. Wanted Alive: Asian Elephants in the Wild. World Wide Fund for Nature, Gland.

Kiley-Worthington, M., 1990a. Animals in Circuses and Zoos. Chiron's World. Aardvark Publishing, Buntingford.

Kiley-Worthington, M., 1990b. Are elephants in zoos and circuses distressed? Applied Animal Behaviour Science 26, 299.

Kiley-Worthington, M., Randle, H.D., 2005. Assessing captive animals' welfare and quality of life. International Zoo News 343, 324.

King, R.L., Burwell, S., White, P.D., 1938. Some notes on the anatomy of the elephant's heart. American Heart Journal 16, 734–742.

Kingdon, J., 2004. The Kingdon Pocket Guide to African Mammals. A & C Black Publishers Ltd, London.

Kioko, J., Muruthi, P., Omondi, P., Chiyo, P.I., 2008. The performance of electric fences as elephant barriers in Amboseli, Kenya. South African Journal of Wildlife Research 38, 52–58. <https://doi.org/10.3957/0379-4369-38.1.52>.

Kipling, R., 1902. The Elephant's Child. Just So Stories. Macmillan, London.

Kistler, J.M., 2007. War Elephants. University of Nebraska Press, Lincoln and London.

Knight, M., 1967. How to Keep an Elephant. Wolfe Publishing Ltd, London.

Knowles, T.G., Warriss, P.D., Vogel, K., 2014. Stress physiology of animals during transport. In: Grandin, T. (Ed.), Livestock Handling and Transport, fourth ed. CABI Publishing, Wallingford, Oxon, pp. 399–420.

Kock, M.D., 1996. Zimbabwe: a model for the sustainable use of wildlife and the development of innovative wildlife management practices. In: Taylor, V.J., Dunstone, N. (Eds.), The Exploitation of Mammal Populations. Springer, Dordrecht, pp. 229–249.

Kohler, W., 2018. The Mentality of Apes. Routledge, New York.

Kontogeorgopoulos, N., 2009. The role of tourism in elephant welfare in Northern Thailand. Journal of Tourism 10, 1–19.

Kowalski, N.L., Dale, R.H.I., Mazur, C.L.H., 2010. A survey of the management and development of captive African elephant (*Loxodonta africana*) calves: birth to three months of age. Zoo Biology 29, 104–119.

Koyama, N., Ueno, Y., Eguchi, Y., Uetake, K., Tanaka, T., 2012. Effects of daily management changes on behavioral patterns of a solitary female African elephant (*Loxodonta africana*) in a zoo. Animal Science Journal 83, 562–570. <https://doi.org/10.1111/j.1740-0929.2011.00992.x>.

Krishnamurthy, V., 1995. Reproductive pattern in captive elephants in the Tamil Nadu Forest Department: India. In: Daniel, J.C., Datye, H. (Eds.), A Week with Elephants. Proceedings of the International Seminar on the Conservation of Asian Elephants. Bombay Natural History Society, Oxford University Press, Oxford, pp. 450–455.

Krishnamurthy, V., Wemmer, C., 1995. Veterinary care of Asian timber elephants in India: historical accounts and current observations. Zoo Biology 14, 123–133.

Kühme, W., 1961. Beobachtungen am Afrikanischen Elefanten (*Loxodonta africana* Blumenbach 1797) in Gefangenschaft. Zeitschrift fur Tierpsychologie 18, 285–296. <https://doi.org/10.1111/j.1439-0310.1961.tb00420.x>.

Kühme, W., 1963. Ethology of the African elephant (*Loxodonta africana* Blumenbach 1797) in captivity. International Zoo Yearbook 4, 113–121.

Kurt, F., 1994. Die Erhaltung des Asiatischen Elefanten in Menschenobhut – ein Vergleich zwischen verschiedenen Haltungssystemen in Südasien und Europa. Zeitschrift des Kölner Zoo 37, 91–113.

Kurt, F., 1995. Asian elephants (*Elephas maximus*) in captivity and the role of captive propagation for maintenance of the species. In: Spooner, N.G., Whitear, J.A. (Eds.), Proceedings of the Eighth UK Elephant Workshop, 21st September 1994. North of England Zoological Society, Chester, pp. 69–96.

Kurt, F., Garaï, M.E., 2007. The Asian Elephant in Captivity. A Field Study. Cambridge University Press, India Pvt. Ltd., New Delhi.

Kurt, F., Hartl, G.B., Tiedemann, R., 1995. Tuskless bulls in Asian elephant *Elephas maximus*. History and population genetics of a man-made phenomenon. Acta Theriologica 40, 125–143.

Lahdenperä, M., Mar, K.U., Lummaa, V., 2014. Reproductive cessation and post-reproductive lifespan in Asian elephants and pre-industrial humans. Frontiers in Zoology 11, 54.

Lahdenperä, M., Mar, K.U., Lummaa, V., 2016. Short-term and delayed effects of mother death on calf mortality in Asian elephants. Behavioral Ecology 27, 166–174.

Lalchandani, N., Chopra, D., 2009. Elephants to be banished from all zoos. The Times of India, 12 November 2009. <https://timesofindia.indiatimes.com/home/environment/flora-fauna/Elephants-to-be-banished-from-all-zoos/articleshow/5221159.cms> (accessed 20.01.18.).

Lang, E.M., 1980. Observations on growth and molar change in the African elephant. African Journal of Ecology 18, 217–234.

Langbauer, W.R., 2000. Elephant communication. Zoo Biology 19, 425–445.

Langman, V.A., Roberts, T.J., Black, J., Maloiy, G.M.O., Heglund, N.C., Weber, J.-M., et al., 1995. Moving cheaply: energetics of walking in the African elephant. Journal of Experimental Biology 198, 629–632.

Langman, V.A., Rowe, M.F., Roberts, T.J., Langman, N.V., Taylor, C.R., 2012. Minimum cost of transport in Asian elephants: do we really need a bigger elephant? Journal of Experimental Biology 215, 1509–1514. <https://doi.org/10.1242/jeb.063032>.

Langman, V.A., Langman, S.L., Ellifrit, N., 2015. Seasonal acclimatization determined by non-invasive measurements of coat insulation. Zoo Biology 34, 368–373.

Laws, R.M., 1966. Age criteria for the African elephant (*Loxodonta a. africana*). East African Wildlife Journal 4, 1–37.

Laws, R.M., 1967. Occurrence of placental scars in the uterus of the African elephant, *Loxodonta africana*. Journal of Reproduction and Fertility 14, 445–449.

Laws, R.M., 1969. Aspects of reproduction in the African elephant *Loxodonta africana*. Journal of Reproduction and Fertility (Suppl.) 6, 193–217.

Laws, R.M., 1970a. Biology of African elephants. Science Progress 58, 251–262.

Laws, R.M., 1970b. Elephants as agents of habitat and landscape change in East Africa. Oikos 21, 1–15.

Laws, R.M., Parker, I.S.C., 1968. Recent studies on elephant populations in East Africa. Symposia of the Zoological Society of London 21, 319–359.

Laws, R.M., Parker, I.S.C., Johnstone, R.C.B., 1975. Elephants and Their Habitats. The Ecology of Elephants in North Bunyoro, Uganda. Clarendon Press, Oxford.

Laws, N., Ganswindt, A., Heistermann, M., Harris, M., Harris, S., Sherwin, S., 2007. A case study: fecal corticosteroid and behavior as indicators of welfare during relocation of an Asian elephant. Journal of Applied Animal Welfare Science 10, 349–358.

Leader-Williams, N., 1990. Black rhinos and African elephants: lessons for conservation funding. Oryx 24, 23–29.

Leakey, R.E., Morell, V., 2001. Wildlife Wars: My Battle to Save Kenya's Elephants. Macmillan, London.

Lee, P.C., 1986. Early social development among African elephants. National Geographic Research 2, 388–401.

Lee, P.C., 1987. Allomothering among African elephants. Animal Behaviour 35, 278–291.

Lee, P.C., 1989. Family structure, communal care and female reproductive effort. In: Standen, V., Foley, R.A. (Eds.), Comparative Socioecology. The Behaviour and Ecology of Humans and Other Mammals. Blackwell, Oxford, pp. 323–340.

Lee, P.C., Lindsay, W.K., Moss, C.J., 2011. Ecological patterns of variability in demographic rates. In: Moss, C.J., Croze, H., Lee, P.C. (Eds.), The Amboseli Elephants. A Long-Term Perspective on a Long-Lived Animal. The University of Chicago Press, Chicago and London, pp. 74–88.

Lehnhardt, J., 1991. Elephant handling: a problem of risk management and resource allocation. AAZPA Regional Conference Proceedings. American Association of Zoological Parks and Aquariums, Wheeling, VA, pp. 569–575.

Leighty, K.A., Soltis, J., Wesolek, C.M., Savage, A., Mellen, J., Lehnhardt, J., 2009. GPS determination of walking rates in captive African elephants (*Loxodonta africana*). Zoo Biology 28, 16–28.

Leighty, K.A., Soltis, J., Savage, A., 2010. GPS assessment of the use of exhibit space and resources by African elephants (*Loxodonta africana*). Zoo Biology 29, 210–220.

Leimgruber, P., Snior, B., Uga Aung, M., Songer, M.A., Mueller, T., Wemmer, C., et al., 2008. Modeling population viability of captive elephants in Myanmar (Burma): implications for wild populations. Animal Conservation 11, 198–205.

Leong, K.M., Burks, K., Rizkalla, C.E., Savage, A., 2005. Effects of reproductive and social context on vocal communication in captive female African elephants (*Loxodonta africana*). Zoo Biology 24, 331−347.

Lewis, G., Fish, B., 1978. I Loved Rogues: The Life of an Elephant Tramp. Superior Publishing Company, Seattle, WA.

Lewis, K.D., Shepherdson, D.J., Owens, T.M., Keele, M., 2010. A survey of elephant husbandry and foot health in North American zoos. Zoo Biology 29, 221−236.

Lincoln, G.A., Ratnasooriya, W.D., 1996. Testosterone secretion, musth behaviour and social dominance in captive male Asian elephants living near the equator. Journal of Reproduction and Fertility 108, 107−113.

Lindsay, K., Chase, M., Landen, K., Nowak, K., 2017. The shared nature of Africa's elephants. Biological Conservation 215, 260−267.

Linnaeus, C., 1758. tenth ed. Systema Naturae per Regna Tria Naturae, Secundum Classes, Ordines, Genera, Species, cum Characteribus, Differentiis, Synonymis, Locis, vol. 1. Impensis Direct. Laurentii Salvii, Holmiae.

Lippé, C., Dumont, P., Bernatchez, L., 2006. High genetic diversity and no inbreeding in the endangered copper redhorse, *Moxostoma hubbsi* (Catostomidae, Pisces); the positive sides of a long generation time. Molecular Ecology 7, 1769−1780.

Lobban, R.A., de Liedekerke, V., 2000. Elephants in Ancient Egypt and Nubia. Anthrozoös 13, 232−244. <https://doi.org/10.2752/089279300786999707>.

Lueders, I., Taya, K., Watanabe, G., Tamamato, Y., Yamamato, T., Kaewmanee, S., et al., 2011. Role of the double luteinizing hormone peak, luteinizing follicles, and the secretion of inhibin for dominant follicle selection in Asian elephants (*Elephas maximus*). Biology of Reproduction 85, 714−720.

Lukacs, D.E., Poulin, M., Besenthal, H., Fad, O.C., Miller, S.P., Atkinson, J.L., et al., 2016. Diurnal and nocturnal activity time budgets of Asian elephants (*Elephas maximus*) in a Zoological Park. Animal Behavior and Cognition 3, 63−77.

Luo, S.-J., Johnson, W.E., Martenson, J., Antunes, A., Martelli, P., Uphyrkina, O., et al., 2008. Subspecies genetic assignments of worldwide captive tigers increase conservation value of captive populations. Current Biology 18, 592−596.

Macdonald, D. (Ed.), 1984. The Encyclopedia of Mammals. Andromeda Oxford Ltd, Abingdon.

Mader, G.J., 2006. Triumphal elephants and political circus at Plutarch, Pomp. 14.6. Classical World 99, 397−403. The Johns Hopkins University Press. Retrieved March 5, 2019, from Project MUSE database.

MAF, 2007. MAF Biosecurity New Zealand Standard 154.03.04: containment facilities for zoo animals (29 January 2007).

Magda, S., Spohn, O., Angkawanish, T., Smith, D.A., Pearl, D.L., 2015. Risk factors for saddle-related skin lesions on elephants used in the tourism industry in Thailand. BMC Veterinary Research 11, 117. <https://doi.org/10.1186/s12917-015-0438-1>.

Makecha, R., Fad, O., Kuczaj II, S.A., 2012. The role of touch in the social interactions of Asian elephants (*Elephas maximus*). International Journal of Comparative Psychology 25, 60−82.

Makhabu, S.W., Skarpe, C., Hytteborn, H., 2006. Elephant impact on shoot distribution on trees and on rebrowsing by smaller browsers. Acta Oecologica 30, 136−146.

Manning, A., 1972. An Introduction to Animal Behaviour, second ed. Edward Arnold (Publishers) Limited, London.

Mar, K.U., 2001. The studbook of timber elephants of Myanmar with special reference to survivorship analysis. In: Baker, I., Kashio, M. (Eds.), Giants on Our Hands: Proceedings of the International Workshop on the Domesticated Asian Elephant. 5-10 February 2001, Bangkok, Thailand. Food and Agriculture Organization of the United Nations, Regional Office for Asia and the Pacific, Bangkok, Thailand. <http://www.fao.org/3/ad031e/ad031e0m.htm#bm22> (accessed 24.07.19.).

Mar, K.U., Maung, M., Thein, M., Khaing, A.T., Tun, W., Nyunt, T., 1995. Electroejaculation and semen characteristics in Myanmar timber elephants. In: Daniel, J.C., Datye, H. (Eds.), A Week with Elephants. Proceedings of the International Seminar on the Conservation of Asian Elephant. Bombay Natural History Society, Oxford University Press, Oxford, pp. 473−482.

Marcilla, A.M., Urios, V., Limiñana, R., 2012. Seasonal rhythms of salivary cortisol secretion in captive Asian elephants (*Elephas maximus*). General and Comparative Endocrinology 176, 259−264.

Markowitz, H., Schmidt, M., Nadal, L., Squier, L., 1975. Do elephants ever forget? Journal of Applied Behavior Analysis 8, 333−335. <https://doi.org/10.1901/jaba.1975.8-333>.

Marten, K., Psarakos, S., 1995. Using self-view television to distinguish between self-examination and social behaviour in the bottlenose dolphin (*Tursiops truncatus*). Consciousness and Cognition 4, 205–224.

Mason, G.J., 1991a. Stereotypies: a critical review. Animal Behaviour 41, 1015–1037.

Mason, G.J., 1991b. Stereotypies and suffering. Behavioural Processes 25, 103–115.

Mason, G.J., Burn, C.C., 2011. Behavioural restriction. In: Appleby, M.C., Mench, J.A., Olsson, I.A.S., Hughes, B.O. (Eds.), Animal Welfare, second ed. CAB International, Wallingford, pp. 98–119.

Mason, G.J., Latham, N.R., 2004. Can't stop, won't stop: is stereotypy a reliable animal welfare indicator? Animal Welfare 13 (Suppl.), 557–569.

Mason, G.J., Veasey, J.S., 2010a. How should the psychological well-being of zoo elephants be objectively investigated? Zoo Biology 29, 237–255.

Mason, G.J., Veasey, J.S., 2010b. What do population-level welfare indices suggest about the well-being of zoo elephants? Zoo Biology 29, 256–273.

Massen, J.J.M., Gallup, A.C., 2017. Why contagious yawning does not (yet) equate to empathy. Neuroscience & Biobehavioral Reviews 80, 573–585.

Matschie, P., 1900. Über geographische Albarten des Afrikanischen elephantens. Sitzungsberichte Gesellschaft naturforschunde Freunde Berlin 8, 189–197.

McClellan, A., Barnum, P.T., 2012. Jumbo the elephant, and the Barnum Museum of Natural History at Tufts University. Journal of the History of Collections 24, 45–62. <https://doi.org/10.1093/jhc/fhr001>.

McComb, K., Moss, C., Durant, S.M., Baker, L., Sayialel, S., 2001. Matriarchs as repositories of social knowledge in African elephants. Science 292, 491–494.

McCullagh, K.G., Widdowson, E.M., 1970. The milk of the African elephant. British Journal of Nutrition 24, 109–117.

McCullagh, K.G., Lincoln, H.B., Southgate, D.A.T., 1969. Fatty acid composition of milk fat of the African elephant. Nature 222, 493–494.

McKay, G.M., 1973. Behavior and ecology of the Asian elephant in Southeastern Ceylon. Smithsonian Contributions to Zoology 125, 1–113.

Meehan, C.L., Hogan, J.N., Bonaparte-Saller, M.K., Mench, J.A., 2016a. Housing and social environments of African (*Loxodonta africana*) and Asian (*Elephas maximus*) elephants in North American zoos. PLoS ONE 11 (7), e0146703. <https://doi.org/10.1371/journal.pone.0146703>.

Meehan, C.L., Mench, J.A., Carlstead, K., Hogan, J.N., 2016b. Determining connections between the daily lives of zoo elephants and their welfare: an epidemiological approach. PLoS ONE 11 (7), e0158124. <https://doi.org/10.1371/journal.pone.0158124>.

Meehan, C., Greco, B., Lynn, B., Morfeld, K., Vicino, G., Orban, D., et al., 2019. The Elephant Welfare Initiative: a model for advancing evidence-based zoo animal welfare monitoring, assessment and enhancement. International Zoo Yearbook <https://doi.org/10.1111/izy.12222>.

Meller, C.L., Croney, C.C., Sheperdson, D., 2007. Effects of rubberized flooring on Asian elephant behaviour in captivity. Zoo Biology 26, 51–61.

Menargues, A., Urios, V., Limiñana, R., Mauri, M., 2012. Circadian rhythm of salivary cortisol in Asian elephants (*Elephas maximus*): a factor to consider during welfare assessment. Journal of Applied Animal Welfare Science 15, 383–390.

Mikota, S.K., 2006. Preventative health care and physical examination. In: Fowler, E.F., Mikota, S.K. (Eds.), Biology, Medicine, and Surgery of Elephants. Blackwell Publishing Ltd, Oxford, pp. 67–73.

Mikota, S.K., Larsen, R.S., Montali, R.J., 2000. Tuberculosis in elephants in North America. Zoo Biology 19, 393–403.

Miller, W., Drautz, D.I., Ratan, A., Pusey, B., Qi, J., Lesk, A.M., et al., 2008. Sequencing the nuclear genome of the extinct woolly mammoth. Nature 456, 387–390.

Miller, L.J., Andrews, J., Anderson, M., 2012. Validating methods to determine walking rates of elephants within a zoological institution. Animal Welfare 21, 577–582.

Miller, A.K., Hensman, M.C., Hensman, S., Schultz, K., Reid, P., Shore, M., et al., 2015. African elephants (*Loxodonta africana*) can detect TNT using olfaction: implications for biosensor application. Applied Animal Behaviour Science 171, 177–183.

Miller, L.J., Chase, M.J., Hacker, C.E., 2016. A comparison of walking rates between wild and zoo African elephants. Journal of Applied Animal Welfare Science 19, 271–279.

Miller, L.J., Luebke, J.F., Matiasek, J., 2018. Viewing African and Asian elephants at accredited zoological institutions: conservation intent and perceptions of animal welfare. Zoo Biology 37, 466–477.
Milman, O., 2016. US zoos secretly fly 18 elephants out of Swaziland ahead of court challenge. The Guardian <https://www.theguardian.com/world/2016/mar/09/us-zoos-secretly-fly-elephants-swaziland-dallas-kansas-nebraska> 9.3.2016 (accessed 16.07.18.).
Minion, J., 2001. The purchase and transport of an Asian elephant (*Elephas maximus*). Ratel 28, 38–45.
Moberg, G.P., 2000. Biological response to stress: implications for animal welfare. In: Moberg, G.P., Mench, J.,A. (Eds.), The Biology of Animal Stress. CAB International, Wallingford, pp. 1–21.
Mohapatra, K.K., Patra, A.K., Paramanik, D.S., 2013. Food and feeding behaviour of Asiatic elephant (*Elephas maximus* Linn.) in Kuldiha Wild Life Sanctuary, Odisha, India. Journal of Environmental Biology 34, 87–92.
Mondol, S., Moltke, I., Hart, J., Keigwin, M., Brown, L., Stephens, M., et al., 2015. New evidence for hybrid zones of forest and savanna elephants in Central and West Africa. Molecular Ecology 24, 6134–6147.
Montali, R.J., Richman, L.K., Hildebrandt, T.B., 1998. Highly fatal disease of Asian elephants in North America and Europe is attributed to a newly recognised endotheliotropic herpesvirus. Elephant Journal 1 (3–4), 3.
Morfeld, K.A., Brown, J.L., 2014. Ovarian acyclicity in zoo African elephants (*Loxodonta africana*) is associated with high body condition scores and elevated serum insulin and leptin. Reproduction, Fertility, and Development 28, 640–647.
Morfeld, K.A., Ball, R.L., Brown, J.L., 2014a. Recurrence of hyperprolactinemia and continuation of ovarian acyclicity in captive African elephants (*Loxodonta africana*) treated with cabergoline. Journal of Zoo and Wildlife Medicine 45, 569–576.
Morfeld, K.A., Lehnhardt, J., Alligood, C., Bolling, J., Brown, J.L., 2014b. Development of a body condition scoring index for female African elephants validated by ultrasound measurements of subcutaneous fat. PLoS ONE 9 (4), e93802. <https://doi.org/10.1371/journal.pone.0093802>.
Morfeld, K.A., Meehan, C.L., Hogan, J.N., Brown, J.L., 2016. Assessment of body condition in African (*Loxodonta africana*) and Asian (*Elephas maximus*) Elephants in North America zoos and management practices associated with high body condition scores. PLoS ONE 11 (7), e0155146. <https://doi.org/10.1371/journal.pone.0155146>.
Morris, R.C., 1958. Stride of elephant crossing trench. Journal of the Bombay Natural History Society 52, 206.
Morris, D., 1967. The Naked Ape. Jonathon Cape, London.
Morris, R., Morris, D., 1966. Men and Pandas. McGraw-Hill, New York.
Morrison, M.L., 2007. Health benefits of animal-assisted interventions. Journal of Evidence-Based Integrative Medicine 12, 51–62.
Moss, C.J., 1983. Oestrous behaviour and female choice in the African elephant. Behaviour 86, 167–195.
Moss, C., 1988. Elephant Memories. Thirteen Years in the Life of an Elephant Family. William Collins Sons & Co. Ltd, Glasgow.
Moss, C.J., 2001. The demography of an African elephant (*Loxodonta africana*) population in Amboseli, Kenya. Journal of Zoology 255, 145–156. <https://doi.org/10.1017/S095283690100121259>.
Moss, A., Esson, M., 2010. Visitor interest in zoo animals and the implications for collection planning and zoo education programmes. Zoo Biology 29, 715–731. <https://doi.org/10.1002/zoo.20316>.
Moss, A., Esson, M., 2013. The Educational claims of zoos: where do we go from here? Zoo Biology 32, 13–18.
Moss, A., Francis, D., Esson, M., 2008. The relationship between viewing area size and visitor behavior in an immersive Asian elephant exhibit. Visitor Studies 11, 26–40.
Moss, C.J., Croze, H., Lee, P.C., 2011. The Amboseli Elephants. A Long-Term Perspective on a Long-Lived Mammal. University of Chicago Press, Chicago, IL.
Moss, A., Jensen, E., Gusset, M., 2017. Impact of a global biodiversity education campaign on zoo and aquarium visitors, Frontiers in Ecology and the Environment, 15. pp. 243–247.
Mullen, A., 1682. An Anatomical Account of the Elephant Accidentally Burnt in Dublin, on Fryday, June 17. In the Year 1681: Sent in a letter to Sir Will. Petty, Fellow of the Royal Society. Together with a Relation of New Anatomical Observations in the Eyes of Animals: Communicated in Another Letter to the Honourable R. Boyle, Esq; Fellow of the same Society. Printed for Sam. Smith, bookseller, at the Prince's Arms in St. Paul's Church-Yard, London.
National Research Council, 2008. Recognition and Alleviation of Distress in Laboratory Animals. The National Academies Press, Washington, DC. <https://doi.org/10.17226/11931>.

Neupane, D., Kunwar, S., Bohara, A.K., Risch, T.S., Johnson, R.L., 2017. Willingness to pay for mitigating human-elephant conflict by residents of Nepal. Journal for Nature Conservation 36, 65–76.

NEZS, 1977. Report of the Council and Statement of Accounts 1977. North of England Zoological Society, Chester Zoo, Cheshire.

NEZS, 2019. Our mission. <https://www.chesterzoo.org/global/about-us/mission>. (accessed 26.02.19.).

Nicholls, H., 2008. Darwin 200: let's make a mammoth. Nature 456, 310–314.

Niemuller, C.A., Liptrap, R.M., 1991. Altered androstenedione to testosterone ratios and LH concentrations during musth in the captive male Asian elephant (*Elephas maximus*). Journal of Reproduction and Fertility 91, 139–146.

Niemullar, C.A., Shaw, H.J., Hodges, J.K., 1997. Pregnancy determination in the Asian elephant (*Elephas maximus*): a change in the plasma progesterone to 17α hydroxyprogesterone ratio. Zoo Biology 16, 415–426.

Nijman, V., Shepherd, C.R., 2014. Emergence of Mong La on the Myanmar-China border as a global hub for the international trade in ivory and elephant parts. Biological Conservation 179, 17–22.

Nissani, M., 2008. Elephant cognition: a review of recent experiments. Gajah 28, 44–52.

Nissani, M., Hoefler-Nissani, D., Lay, U.T., Htun, U.W., 2005. Simultaneous visual discrimination in Asian elephants. Journal of the Experimental Analysis of Behavior 83, 15–29.

Nobbe, G., 1992. Going into orbit. Wildlife Conservation 95, 62–64.

Nowak, R.M., 1999. sixth ed. Walker's Mammals of the World, vol. II. The Johns Hopkins University Press, Baltimore and London.

O'Connell-Rodwell, C.E., 2007. Keeping an "ear" to the ground: seismic communication in elephants. Physiology 22, 287–294.

O'Connell-Rodwell, C.E., 2011. Exploring the use of acoustics as a tool in male elephant/human conflict mitigation. Journal of the Acoustical Society of America 130, 2459. <https://doi.org/10.1121/1.3654877>.

O'Connell-Rodwell, C.E., Wood, J.D., Kinzley, C., Rodwell, T.C., Alarcón, C., Wasser, S.K., et al., 2011. Male African elephants (*Loxodonta africana*) queue when the stakes are high. Ethology Ecology & Evolution 23, 388–397.

Odberg, F.O., 1978. Abnormal behaviour: (stereotypies). Proceedings of the First World Congress of Ethology Applied to Zootechnics. Industrias Grafices Espana, Madrid, pp. 475–480.

Oerke, A.-K., Heistermann, M., Hodges, K., 2006. Duration of pregnancy and its relation to sex of calf and age of cow in the European population of Asian and African elephants. In: Proceedings of the International Elephant Conservation and Research Symposium, 2006, pp. 125–131.

Olson, D., Wiese, R.J., 2000. State of the North American African elephant population and projections for the future. Zoo Biology 19, 311–320.

Omondi, P., Wambwa, E., Gakuya, F., Bitok, E., Ndeere, D., Manyibe, T., et al., 2002. Recent translocation of elephant family units from Sweetwaters Rhino Sanctuary to Meru National Park, Kenya. Pacyderm 32, 39–48.

Ortolani, A., Leong, K., Graham, L., Savage, A., 2005. Behavioral indices of estrus in a group of captive African elephants (*Loxodonta africana*). Zoo Biology 24, 311–329.

Osborn, H.F., 1936. Proboscidea, Vol. 1. The American Museum of Natural History. New York.

Osborn, H.F., 1942. Proboscidea, Vol. 2. The American Museum of Natural History, New York. (Published posthumously).

Panagiotopoulou, O., Pataky, T.C., Hill, Z., Hutchinson, J.R., 2012. Statistical parametric mapping of the regional distribution and ontogenetic scaling of foot pressures during walking in Asian elephants (*Elephas maximus*). Journal of Experimental Biology 215, 1584–1593. <https://doi.org/10.1242/jeb.065862>.

Panagiotopoulou, O., Pataky, T.C., Day, M., Hensman, M.C., Hensman, S., Hutchinson, J.R., et al., 2016. Foot pressure distributions during walking in African elephants (*Loxodonta africana*). Royal Society Open Science 3, 160203. <https://doi.org/10.1098/rsos.160203>.

Pavlov, I.P., 1906. The scientific investigation of the psychical faculties or processes in the higher animals. Science 24, 613–619.

PAWS, 2019. PAWS Wildlife Sanctuaries. Performing Animals Welfare Society. Available from: www.pawsweb.org/care_and_management.html (accessed 28.07.19.).

Payne, K., 2003. Sources of social complexity in the three elephant species. In: de Waal, F.B.M., Tyack, P.L. (Eds.), Animal Social Complexity. Intelligence, Culture and Individualized Societies. Harvard University Press, Cambridge, Massachusetts and London, pp. 57–85.

Payne, K.B., Langbauer Jr., W.R., 1992. Elephant communication. In: Shoshani, J. (Ed.), Elephants. Weldon Owen, San Francisco, CA, pp. 116–123.

Pearce, E.A., Smith, C.G., 2000. The Hutchinson World Weather Guide. Helicon Publishing Limited, Oxford.

Pendlebury, C., Odongo, N.E., Renjifo, A., Naelitz, J., Valdes, E.V., McBride, B.W., 2005. Acid-insoluble ash as a measure of dry matter digestibility in captive African elephants (Loxodonta africana). Zoo Biology 24, 261–265.

Perdue, B.M., Talbot, C.F., Stone, A.M., Beran, M.J., 2012. Putting the elephant back in the herd: elephant relative quantity judgments match those of other species. Animal Cognition 15, 955–961.

PETA, 2014. Pittsburgh Zoo Under Fire for Ordering Dogs to Harass Elephants. <https://www.peta.org/media/news-releases/pittsburgh-zoo-fire-ordering-dogs-harass-elephants/> (accessed 27.06.19.).

PETA, 2015. Update: Animal Trainer Who Starved Elephant Denied Another Exhibitor License. <https://www.peta.org/blog/man-starved-elephant-charged-violating-animal-welfare-act/> (accessed 05.05.19.).

PETA, 2019a. Elephant Incidents in North America. People for the Ethical Treatment of Animals (PETA), Norfolk, VA. <https://www.mediapeta.com/peta/pdf/Elephant-Incident-List-US-only.pdf> (accessed 05.09.19.).

PETA, 2019b. Elephant-Free Zoos. People for the Ethical Treatment of Animals (PETA), Norfolk, VA. <https://www.peta.org/issues/animals-in-entertainment/zoos/elephant-free-zoos/> (accessed 03.09.19.).

Petrides, G.A., Swank, W.G., 1966 Estimating the productivity and energy relations of an African elephant population. In: Proceedings of the Ninth International Grassland Congress, San Paulo, Brazil, pp. 831–842.

Phillips, P.K., Heath, J.E., 1992. Heat exchange by the pinna of the African elephant (Loxodonta africana). Comparative Biochemistry and Physiology Part A: Physiology 101, 693–699.

Phillipson, J., 1975. Rainfall, primary production and 'carrying capacity' of Tsavo National Park (East), Kenya. East African Wildlife Journal 13, 171–201.

Pimm, S.L., van Aarde, R.J., 2001. African elephants and contraception. Nature 411, 766.

Pina-Aguilar, R.E., Lopez-Saucedo, J., Sheffield, R., Ruiz-Galaz, L.I., De Barroso-Padilla, J.J., Gutierrez-Gutierrez, A., 2009. Revival of extinct species using nuclear transfer: hope for the mammoth, true for the Pyrenean ibex, but is it time for "Conservation Cloning"? Cloning and Stem Cells 11, 341–346

Pinter-Wollman, N., Isbell, L.A., Hart, L.A., 2009. Assessing translocation outcome: comparing behavioral and physiological aspects of translocated and resident African elephants (Loxodonta africana). Biological Conservation 142, 1116–1124.

Platek, S.M., Mohamed, F.B., Gallup Jr., G.G., 2005. Contagious yawning and the brain. Cognitive Brain Research 23, 448–452.

Plotnik, J.M., de Waal, F.B.M., Reiss, D., 2006. Self-recognition in an Asian elephant. Proceedings of the National Academy of Sciences 103 (45), 17053–17057. <https://doi.org/10.1073/pnas.0608062103>.

Plotnik, J.M., de Waal, F.B.M., Moore, D., Reiss, D., 2010. Self-recognition in the Asian elephant and future directions for cognitive research with elephants in zoological settings. Zoo Biology 29, 179–191.

Plotnik, J.M., Lair, R., Suphachoksahakun, W., de Waal, F.B.M., 2011. Elephants know when they need a helping trunk in a cooperative task. Proceedings of the National Academy of Sciences 108, 5116–5121. <https://doi.org/10.1073/pnas.1101765108>.

Poole, J.H., 1989. Announcing intent: the aggressive state of musth in African elephants. Animal Behaviour 37, 140–152.

Poole, J.H., 1999. Signals and assessment in African elephants: evidence from playback experiments. Animal Behaviour 58, 185–193.

Poole, J.H., 2010. Open letter to the Director of Zimbabwe National Parks and Wildlife Authority (ZNPWA) regarding elephant capture. Elephant Voices <https://elephantvoices.org/multimedia-resources/statements-a-testimonies.html> (accessed 29.07.19.).

Poole, J.H., Moss, C.J., 1981. Musth in the African elephant, Loxodonta africana. Nature 292, 830–831.

Poole, T.B., Taylor, V.J., Fernando, S.B.U., Ratnasooriya, W.D., Ratnayeke, A., Lincoln, G., et al., 1997. Social behaviour and breeding physiology of a group of Asian elephants Elephas maximus at the Pinnawala Elephant Orphanage, Sri Lanka. International Zoo Yearbook 35, 297–331.

Posta, B., 2011. The Effects of Housing and Enrichment on Zoo Elephant Behaviour (Unpublished MSc thesis). Bowling Green State University, Bowling Green, OH.

Posta, B., Huber, R., Moore, D.E., 2013a. The effects of housing on zoo elephant behavior: a quantitative case study of diurnal and seasonal variation. International Journal of Comparative Psychology 26, 37–52.

Posta, B., Leighty, K.A., Alligood, C., Carlstead, K., 2013b. Using science to understand elephant welfare. Journal of Applied Animal Welfare Science 16, 395–396.

Povinelli, D.J., 1989. Failure to find self-recognition in Asian elephants (*Elephas maximus*) in contrast to their use of mirror cues to discover hidden food. Journal of Comparative Psychology 103, 122–131.

Povinelli, D.J., Preuss, T.M., 1995. Theory of mind: evolutionary history of a cognitive specialization. Trends in Neurosciences 18, 418–424.

Powell, D.M., Vitale, C., 2016. Behavioral changes in female Asian elephants when given access to an outdoor yard overnight. Zoo Biology 35, 298–303.

Pradhan, N.M.B., Williams, A.C., Dhakal, M., 2011. Current status of Asian elephants in Nepal. Gajah 35, 87–92.

Prado-Oviedo, N.A., Bonaparte-Saller, M.K., Malloy, E.J., Meehan, C.L., Mench, J.A., Carlstead, K., et al., 2016. Evaluation of demographics and social life events of Asian (*Elephas maximus*) and African elephants (*Loxodonta africana*) in North American zoos. PLoS ONE 11 (7), e0154750. <https://doi.org/10.1371/journal.pone.0154750>.

Proctor, C.M., Brown, J.L., 2015. A preliminary analysis of the influence of handling method on adrenal activity in zoo African and Asian elephants. Journal of Zoo and Aquarium Research 3, 1–5. <https://doi.org/10.19227/jzar.v3i1.100>.

Proctor, C.M., Freeman, E.W., Brown, J.L., 2010a. Results of a second survey to assess the reproductive status of female Asian and African elephants in North America. Zoo Biology 29, 127–139.

Proctor, C.M., Freeman, E.W., Brown, J.L., 2010b. Influence of dominance status on adrenal activity and ovarian cyclicity status in captive African elephants. Zoo Biology 29, 168–178.

Prusty, B.C., Singh, L.A.K., 1995. Male-male aggression in Asian elephant observed in Similipal Tiger Reserve, Orissa. Indian Forester 121, 902–908.

Pushpakumara, P.G.A., Rajapakse, R.C., Perera, B.M.A.O., Brown, J.L., 2016. Reproductive performance of the largest captive Asian elephant (*Elephas maximus*) population in Sri Lanka. Animal Reproduction Science 174, 93–99.

Rachels, J., 1986. The End of Life. Oxford University Press, New York.

Radford, M., 2007. Wild Animals in Travelling Circuses. The Report of the Chairman of the Circus Working Group. Department for Environment, Food and Rural Affairs (Defra), London. <https://webarchive.nationalarchives.gov.uk/20130403081434/http://archive.defra.gov.uk/foodfarm/farmanimal/welfare/documents/circus-report.pdf> (accessed 21.03.19.).

Ramanathan, A., Mallapur, A., 2008. A visual health assessment of captive Asian elephants (*Elephas maximus*) housed in India. Journal of Zoo and Wildlife Medicine 39, 148–154. <https://doi.org/10.1638/2007-0008R1.1>.

Ramesh, T., Sankar, K., Qureshi, Q., Kalle, R., 2011. Assessment of wild Asiatic elephant (*Elephas maximus indicus*) body condition by simple scoring method in a tropical deciduous forest of Western Ghats, Southern India. Wildlife Biology in Practice 7, 47–54.

Rapaport, L., Haight, J., 1987. Some observations regarding allomaternal caretaking among captive Asian elephants (*Elephas maximus*). Journal of Mammalogy 68, 438–442.

Rasmussen, L.E.L., Krishnamurthy, V., 2000. How chemical signals integrate Asian elephant society: the known and the unknown. Zoo Biology 19, 405–423.

Rasmussen, L.E.L., Munger, B., 1987. The sensorimotor specializations of the trunk tip of the Asian elephant, *Elephas maximus*. Anatomical Record 246, 127–134.

Rasmussen, L.E.L., Schulte, B.A., 1998. Chemical signals in the reproduction of Asian (*Elephas maximus*) and African (*Loxodonta africana*) elephants. Animal Reproduction Science 53, 19–34.

Rees, P.A., 1977. Some Aspects of the Feeding Ecology of the African Elephant (*Loxodonta africana africana*, Blumenbach 1797) in Captivity (Unpublished BSc dissertation). University of Liverpool.

Rees, P.A., 1979. Bottom biting. Letters. New Scientist, 29 November 1979, p719.

Rees, P.A., 1982. Gross assimilation efficiency and food passage time in the African elephant. African Journal of Ecology 20, 193–198.

Rees, P.A., 1983. Synchronization of defaecation in the African elephant (*Loxodonta africana*). Journal of Zoology 201, 581–585.

Rees, P.A., 2000. The introduction of a captive herd of Asian elephants (*Elephas maximus*) to a novel area. Ratel 27, 120–126.

Rees, P.A., 2001a. A history of the National Elephant Centre, Chester Zoo. International Zoo News 48, 170–183.

Rees, P.A., 2001b. Captive breeding of Asian elephants (*Elephas maximus*): the importance of producing socially competent animals. In: Hosetti, B.B., Venkateshwarlu, M. (Eds.), Trends in Wildlife Biodiversity, Conservation and Management, vol. 1. Daya Publishing House, Delhi, pp. 76–91.

Rees, P.A., 2002a. Asian elephants (*Elephas maximus*) dust bathe in response to an increase in environmental temperature. Journal of Thermal Biology 27, 353–358.

Rees, P.A., 2002b. RSPCA elephant welfare recommendations would compromise zoo breeding programmes. International Zoo News 50, 86–90.

Rees, P.A., 2003a. Early experience of sexual behaviour in Asian elephants (*Elephas maximus*). International Zoo News 50, 200–206.

Rees, P.A., 2003b. Asian elephants in zoos face global extinction: should zoos accept the inevitable? Oryx 37, 20–22.

Rees, P.A., 2004a. Low environmental temperature causes an increase in stereotypic behaviour in captive Asian elephants (*Elephas maximus*). Journal of Thermal Biology 29, 37–43.

Rees, P.A., 2004b. Some preliminary evidence of the social facilitation of mounting behavior in the Asian elephant (*Elephas maximus*). Journal of Applied Animal Welfare Science 7, 49–58.

Rees, P.A., 2004c. Unreported appeasement behaviours in the Asian elephant *(Elephas maximus)*. Journal of the Bombay Natural History Society 101, 71–78.

Rees, P.A., 2005a. Will the EC Zoos Directive increase the conservation value of zoo research? Oryx 39, 128–131.

Rees, P.A., 2005b. The EC Zoos Directive: a lost opportunity to implement the Convention on Biological Diversity. Journal of International Wildlife Law and Policy 8, 51–62.

Rees, P.A., 2009a. The sizes of elephant groups in zoos: implications for elephant welfare. Journal of Applied Animal Welfare Science 12, 44–60.

Rees, P.A., 2009b. Activity budgets and the relationship between feeding and stereotypic behaviors in Asian elephants (*Elephas maximus*) in a zoo. Zoo Biology 28, 79–97.

Rees, P.A., 2011. Introduction to Zoo Biology and Management. Wiley-Blackwell, Chichester, West Sussex.

Rees, P.A., 2013. Dictionary of Zoo Biology and Animal Management. Wiley-Blackwell, Chichester, West Sussex.

Rees, P.A., 2015. Studying Captive Animals: A Workbook of Methods in Behaviour, Welfare and Ecology. Chichester, West Sussex.

Rees, P.A., 2018a. Examining Ecology. Exercises in Environmental Biology and Conservation. Elsevier, London.

Rees, P.A., 2018b. The Laws Protecting Animals and Ecosystems. Wiley-Blackwell, Chichester, West Sussex.

Ren, L., Hutchinson, J.R., 2008. The three-dimensional locomotor dynamics of African (*Loxodonta africana*) and Asian (*Elephas maximus*) elephants reveal a smooth gait transition at moderate speed. Journal of the Royal Society Interface 5 (19), 195–211. <https://doi.org/10.1098/rsif.2007.1095>.

Ren, L., Miller, C.E., Lair, R., Hutchinson, J.R., 2010. Integration of biomechanical compliance, leverage, and power in elephant limbs. Proceedings of the National Academy of Sciences 107 (15), 7078–7082.

Richman, L.K., Montali, R.J., Garber, R.L., Kennedy, M.A., Lehnhardt, J., Hildebrandt, T.B., et al., 1999. Novel endotheliotropic herpesviruses fatal for Asian and African elephants. Science 283, 1171–1176.

Richman, L.K., Montali, R.J., Hayward, G.S., 2000. Review of a newly recognised disease of elephants caused by endotheliotropic herpesviruses. Zoo Biology 19, 383–392.

Richmond, R.D., 1932. Miscellaneous notes VII – Elephants – age to which they live in captivity. Journal of the Bombay Natural History Society 36, 494–496.

Rietkerk, F.E., Hiddingh, H., van Dijk, S., 1993. Hand-rearing an Asian elephant *(Elephas maximus)* at the Noorder Zoo, Emmen. International Zoo Yearbook 32, 244–252.

Ritchie, J., 2009. Fact or fiction?: Elephants never forget. Scientific American <https://www.scientificamerican.com/article/elephants-never-forget/> (accessed 27.06.18.).

Robinson, M.R., Mar, K.U., Lummaa, V., 2012. Senescence and age-specific trade-offs between reproduction and survival in female Asian elephants. Ecology Letters 15, 260–266. <https://doi.org/10.1111/j.1461-0248.2011.01735.x>.

Roca, A.L., Georgiadis, N., O'Brien, S.J., 2004. Cytonuclear genomic dissociation in African elephant species. Nature Genetics 37, 96–100.

Roca, A.L., Ishida, Y., Brandt, A.L., Benjamin, N.R., Zhao, K., Georgiadis, N.J., 2015. Elephant natural history: a genomic perspective. Annual Review of Animal Biosciences 3, 139–167.

Rodgers, L.J., 1998. Minds of Their Own. Thinking and Awareness in Animals. Allen & Unwin, St. Leonards, NSW.

Roehrs, J.M., Brockway, C.R., Ross, D.V., Reichard, T.A., Ullrey, D.E., 1989. Digestibility of timothy hay by African elephants. Zoo Biology 8, 331–337.

Roocroft, A., 2007. Protected contact training of elephants in Europe. International Zoo News 54, 6–26.

Roosevelt, T., 1910. African Game Trails. John Murray, London.

Rossman, Z.T., Hart, B.L., Greco, B.J., Young, D., Padfield, C., Wiedner, L., et al., 2017. When yawning occurs in elephants. Frontiers in Veterinary Science <https://doi.org/10.3389/fvets.2017.00022>.

Rothwell, J., 2016. Seven-year-old girl dies in Moroccan zoo after elephant throws rock at her head. <https://www.telegraph.co.uk/news/2016/07/28/seven-year-old-girl-dies-in-moroccan-zoo-after-elephant-threw-ro/> (accessed 03.07.18.).

Rothwell, E.S., Bercovitch, F.B., Andrews, J.R.M., Anderson, M.J., 2011. Estimating walking distance of captive African elephants using an accelerometer. Zoo Biology 30, 579–591.

Rowe, M.F., Bakken, G.S., Ratliff, J.J., Langman, V.A., 2013. Heat storage in Asian elephants during submaximal exercise: behavioral regulation of thermoregulatory constraints on activity in endothermic gigantotherms. Journal of Experimental Biology 216, 1774–1785.

Roy, K., 2013. Use of Elephants in Indian Warfare (1000BCE – 1943 CE). In: Coetzee, D., Eysturlid, L.W. (Eds.), Philosophers of War: The Evolution of History's Greatest Military Thinkers, Vol. 1: The Ancient to the Premodern World, 3000 BCE-1815 CE. Praeger, Santa Barbara, CA, pp. 390–394.

RSPCA, 2006. Live Hard, Die Young: How Elephants Suffer in Zoos. Royal Society for the Prevention of Cruelty to Animals, Horsham, West Sussex.

Rübel, A., 2014. Kaeng Krachan Elephant Park at Zoo Zürich. Gajah 40, 44–45.

Rushen, J., 1993. The "coping" hypothesis of stereotypic behaviour. Animal Behaviour 45, 613–615.

Russow, L.M., 1994. Why do species matter? In: Westphal, D., Westphal, F. (Eds.), Planet in Peril. Essays in Environmental Ethics. Holt, Reinhart and Winston, Orlando, FL, pp. 149–170.

Ryan, S.J., Thompson, S.D., 2001. Disease risk and inter-institutional transfer of specimens in cooperative breeding programs: herpes and the elephant species survival plans. Zoo Biology 20, 89–101.

Ryder, O., 1990. Saving species in their habitats: the transfer of genetics research technologies. In: AZA Annual Conference Proceedings 1990. Wheeling, WV, pp. 41–45.

RZSS, 2019. Our Mission. <https://www.rzss.org.uk/about-rzss/> (accessed 26.02.19.).

Sach, F., Fitzpatrick, M., Masters, N., Field, D., 2019. Financial planning required to keep elephants in zoos in the United Kingdom in accordance with the Secretary of State's Standards of Modern Zoo Practice for the next 30 years. International Zoo Yearbook. <https://doi.org/10.1111/izy.12213>.

Samarasinghe, W.M.P., Ahamed, A.M.R., 2016. A preliminary study on activity budgets of Asian elephant (*Elephas maximus* Linn.) at elephant orphanage. Bulletin of Environment, Pharmacology and Life Sciences 5, 47–50.

Sandom, C., Faurby, S., Sandel, B., Svenning, J.-C., 2014. Global late Quaternary megafauna extinctions linked to humans, not climate change. Proceedings of the Royal Society B: Biological Sciences 281, 20133254. <https://doi.org/10.1098/rspb.2013.3254>.

Santiapillai, C., Jackson, P., 1990. The Asian Elephant: An Action Plan for its Conservation. IUCN/SSC Asian Elephant Specialist Group. IUCN, Gland.

Saragusty, J., Hermes, R., Göritz, F., Schmitt, D.L., Hildebrandt, T.B., 2009. Skewed birth sex ratio and premature mortality in elephants. Animal Reproduction Science 115, 247–254.

Satiansukpong, N., Pongsaksri, M., Sung-U, S., Vittayakorn, S., Tipprasert, P., Pedugsorn, M., et al., 2008. Thai elephant-assisted therapy program: the feasibility in assisting an individual with autism. World Federation of Occupational Therapists (WFOT) Bulletin 58, 17–26. <https://doi.org/10.1179/otb.2008.58.1.004>.

Satiansukpong, N., Pongsaksri, M., Sasat, D., 2016. Thai elephant-assisted therapy programme in children with Down syndrome. Occupational Therapy International 23, 121–131.

Saunders, W.L., Young, G.D., 1985. An experimental study of the effect of the presence or absence of living visual aids in high school biology classrooms upon attitudes toward science and biology achievement. Journal of Research in Science Teaching 22, 619–629.

Savage, A., Rice, J.M., Brangan, J.M., Martini, D.P., Pugh, J.A., Miller, C.D., 1994. Performance of African elephants (*Loxodonta africana*) and California sea lions (*Zalophus californianus*) on a two-choice object discrimination task. Zoo Biology 13, 69–75.

Schiffmann, C., Hoby, S., Wenker, C., Hård, T., Scholz, R., Clauss, M., et al., 2018. When elephants fall asleep: a literature review on elephant rest with case studies on elephant falling bouts, and practical solutions for zoo elephants. Zoo Biology 37, 133–145.

Schiffmann, C., Clauss, M., Hoby, S., 2019. Impact of a new exhibit on stereotypic behaviour in an elderly captive African elephant *(Loxodonta africana)*. Journal of Zoo and Aquarium Research 7, 37–43.

Schmid, J., 1995. Keeping circus elephants temporarily in paddocks – the effects on their behaviour. Animal Welfare 4, 87–101.

Schmid, J., 1998. Status and reproductive capacity of the Asian elephant in zoos and circuses in Europe. International Zoo News 45, 341–351.

Schmid, J., Zeeb, K., 1994. The introduction of paddocks in circus elephant husbandry. Deutsche Tierärztliche Wochenschrift 101, 50–52.

Schmid, J., Heistermann, M., Glanslober, U., Hodges, J.K., 2001. Introduction of foreign female Asian elephants *(Elephas maximus)* into an existing group: behavioural reactions and changes in cortisol levels. Animal Welfare 10, 357–372.

Schmidt, M.J., 1982. Studies on Asian elephant reproduction at the Washington Park Zoo. Zoo Biology 1, 141–147. <https://doi.org/10.1002/zoo.1430010208>.

Schmidt, H., Kappelhof, J., 2019. Review of the management of the Asian elephant *Elephas maximus* EEP: current challenges and future solutions. International Zoo Yearbook 53, 1–14.

Schmidt, M.J., Mar, K.U., 1996. Reproductive performance of captive Asian elephants in Myanmar. Gajah 16, 23–42.

Schmitt, D.L., 1998. Report of a successful artificial insemination in an Asian elephant. Third International Elephant Research Symposium. Springfield, MO.

Schmitt, D., 2008. View from the big top. Why elephants belong in North American circuses. In: Wemmer, C., Christen, C.A. (Eds.), Elephants and Ethics. Towards a Morality of Coexistence. The Johns Hopkins University Press, Baltimore, MD, pp. 227–234.

Schou, M.F., Loeschcke, V., Bechsgaard, J., Schlötterer, C., Kristensen, T.N., 2017. Unexpected high genetic diversity in small populations suggests maintenance by associative dominance. Molecular Ecology 26, 6510–6523.

Schulte, B.A., 2000. Social structure and helping behaviour in captive elephants. Zoo Biology 19, 447–459.

Schürer, U., 2017. Über waldelefanten. Der Zoologische Garten 86, 108–166.

Scigliano, E., 2002. Love, War and Circuses. The Age-Old Relationship Between Elephants and Humans. Bloomsbury Publishing Plc, London.

Sclater, P.L., 1908. The domestication of the African elephant. Journal of the Society for the Preservation of the Wild Fauna of the Empire IV, 47–50. <https://archive.org/details/journalofsociety1908soci/page/47>.

Searle, D.A., 2018. War elephants and early tanks: a transepochal comparison of ancient and modern warfare. Militärgeschichtliche Zeitschrift 77, 37–77. <https://doi.org/10.1515/mgzs-2018-0002>.

Seddon, P.J., Price, M.S., Launay, F., Maunder, M., Soorae, P., Molur, S., et al., 2011. Frankenstein ecosystems and 21st century conservation agendas: reply to Oliveira-Santos and Fernandez. Conservation Biology 25, 212–213.

Seidensticker, J., 2008. Foreword. In: Wemmer, C., Christen, C.A. (Eds.), Elephants and Ethics. Towards a Morality of Coexistence. The Johns Hopkins University Press, Baltimore, MD, pp. xi–xiii.

Seltmann, M.W., Helle, S., Adams, M.J., Mar, K.U., Mar, Lahdenperä, M., 2018. Evaluating the personality structure of semi-captive Asian elephants living in their natural habitat. Royal Society Open Science 5, 172026. <https://doi.org/10.1098/rsos.172026>.

Seth-Smith, A.M.D., Parker, I.S.C., 1967. A record of twin foetuses in the African elephant. African Journal of Ecology 5, 167.

Sharma, M.C., 2006. Hastyăyurvĕda – a complete treatise on elephants. Bulletin of the Indian Institute of History of Medicine 36, 145–158.

Sheldrick, D., 1977. The Orphans of Tsavo. Fontana, Glasgow.

Shepherdson, D.J., 1998. Introduction. Tracing the path of environmental enrichment in zoos. In: Shepherdson, D.J., Mellen, J.D., Hutchins, M. (Eds.), Second Nature. Environmental Enrichment for Captive Animals. Smithsonian Institution Press, Washington, DC and London, pp. 1–12.

Shettleworth, S.J., 2012. Do animals have insight, and what is insight anyway? Canadian Journal of Experimental Psychology/Revue canadienne de. psychologie expérimentale 66, 217–226.

Short, R.V., 1966. Oestrous behaviour, ovulation and the formation of the corpus luteum in the African elephant, *Loxodonta africana*. African Journal of Ecology 4, 56–68.

Short, R.V., 1969. Notes on the teeth and ovaries of an African elephant (*Loxodonta africana*) of known age. Journal of Zoology 158, 421–425.

Shoshani, J., Eisenberg, J.F., 1982. *Elephas maximus*. Mammalian Species 182, 1–8.

Shyan-Norwalt, M.R., Petersen, J., King, B.,M., Staggs, T.E., Dale, R.H.I., 2010. Initial findings on visual acuity thresholds in an African elephant (*Loxodonta africana*). Zoo Biology 29, 30–35.

Singer, P., 1975. Animal Liberation. A New Ethics of Our Treatment of Animals. Jonathan Cape Ltd, London.

Singer, P., 1995. Animal Liberation, second ed. Pimlico, London.

Sitompul, A.F., Griffin, C.R., Fuller, T.K., 2013. Diurnal activity and food choice of free-foraging captive elephants at the Seblat Elephant Conservation Center, Sumatra, Indonesia. Gajah 38, 19–24.

Slade-Cain, B.E., Rasmussen, L.E.L., Schulte, B.A., 2008. Estrous state influences on investigative, aggressive, and tail flicking behavior in captive female Asian elephants. Zoo Biology 27, 167–180.

Slotow, R., van Dyk, G., Poole, J., Page, B., Klocke, A., 2000. Older bull elephants control young males. Nature 408, 425–426. <https://doi.org/10.1038/35044191>.

Smet, A.F., Byrne, R.W., 2013. African elephants can use human pointing cues to find hidden food. Current Biology 23, 2033–2037.

Smith, T.M., 1963. The Nature of the Beast. Jarrolds Publishers Ltd, London.

Smith, L., Broad, S., 2008. Do zoo visitors attend to conservation messages? A case study of an elephant exhibit. Tourism Review International 11, 225–235.

Smith, B., Hutchins, M., 2000. The value of captive breeding programmes to field conservation: elephants as an example. Pachyderm 28, 101–109.

Soltis, J., 2010. Vocal communication in African elephants (*Loxodonta africana*). Zoo Biology 29, 192–209.

Soltis, J., Leong, K., Savage, A., 2005a. African elephant vocal communication I: Antiphonal calling behaviour among affiliated females. Animal Behaviour 70, 579–587.

Soltis, J., Leong, K., Savage, A., 2005b. African elephant vocal communication II: Rumble variation reflects the individual identity and emotional state of callers. Animal Behaviour 70, 589–599.

Somvanshi, R., 2006. Veterinary medicine and animal keeping in ancient India. Asian Agri-History 10, 133–146.

Spielman, D., Brook, B., Briscoe, D., Frankham, R., 2004. Does inbreeding and loss of genetic diversity decrease disease resistance? Conservation Genetics 5, 439–448.

Spinage, C., 1994. Elephants. T. & A. D. Poysner Ltd, London.

Spinage, C., 2019. Member Profile, The Niche. British Ecological Society, London, March 2019, p. 52.

Spooner, N.G., Whitear, J.A. (Eds.), 1995. Proceedings of the Eighth UK Elephant Workshop, 21 September 1994. North of England Zoological Society, Chester.

SSSMZP, 2012. Secretary of State's Standards of Modern Zoo Practice. Defra, London.

SSSMZP, 2017. Secretary of State's Standards of Modern Zoo Practice. Defra, London.

Stafford, R., Goodenough, A.E., Slater, K., Carpenter, W., Collins, L., Cruickshank, H., et al., 2011. Inferential and visual analysis of ethogram data using multivariate techniques. Animal Behaviour 83, 563–569.

Stansfield, F.J., Chavatte-Palmer, P., 2015. A novel objective method of estimating the age of mandibles from African elephants (*Loxodonta africana africana*). PLoS ONE 10 (5), e0124980.

Steel, J.H., 1885. A Manual of the Diseases of the Elephant and His Management and Uses. W. H. Moore, Lawrence Asylum Press, Madras.

Stevenson, M.F., Walter, O., 2002. Management Guidelines for the Welfare of Zoo Animals Elephants *Loxodonta africana* and *Elephas maximus*. British and Irish Association of Zoos and Aquariums, Regent's Park, London.

Stevenson, M.F., Walter, O., 2006. Management Guidelines for the Welfare of Zoo Animals. (second ed.). Elephants *Loxodonata africana* and *Elephas maximus*. British and Irish Association of Zoos and Aquariums, Regent's Park, London.

Stoeger, A.S., Mietchen, D., Oh, S., Silva, S., Herbst, C.T., Kwon, S., et al., 2012. An Asian elephant imitates human speech. Current Biology 22, 2144–2148.

Stoinski, T.S., Daniel, E., Maple, T.L., 2000. A preliminary study of the behavioral effects of feeding enrichment on African elephants. Zoo Biology 19, 485–493.

Stokes, H., Perera, V., Jayasena, N., Silva-Fletcher, A., 2017. Nocturnal behaviour of orphaned Asian (*Elephas maximus*) calves in Sri Lanka. Zoo Biology 36, 261–272.

Strehlow, H., 1996. Zoos and aquariums of Berlin. In: Hoage, R.J., Deiss, W.A. (Eds.), New Worlds, New Animals. From Menagerie to Zoological Park in the Nineteenth Century. The Johns Hopkins University Press, Baltimore and London, pp. 63–72.
Sukumar, R., 1989. The Asian Elephant: Ecology and Management. Cambridge University Press, Cambridge.
Sukumar, R., 1994. Elephant Days and Nights. Ten Years with the Indian Elephant. Oxford University Press, Oxford.
Sukumar, R., 2003. The Living Elephants. Evolutionary Ecology, Behaviour, and Conservation. Oxford University Press, Oxford.
Sukumar, R., Krishnamurphy, V., Wemmer, C., Rodden, M., 1997. Demography of captive Asian elephants (*Elephas maximus*) in southern India. Zoo Biology 16, 263–272.
Sullivan, K., Kerr, K., Wanty, R., Amaral, B., Olea-Popelka, F., Valdes, E., 2016. Dietary management, husbandry, and body weights of African elephants (*Loxodonta africana*) during successful pregnancies at Disney's Animal Kingdom. Zoo Biology 35, 574–578.
Suter, I.C., Hockings, M., Baxter, G.S., 2013. Changes in elephant ownership and employment in the Lao PDR: implications for the elephant-based logging and tourism industries. Human Dimensions of Wildlife 18, 279–291.
Suzuki, M., Hirako, K., Saito, S., Suzuki, C., Kashiwabara, T., Koie, H., 2008. Usage of high-performance mattresses for transport of Indo-Pacific bottlenose dolphin. Zoo Biology 27, 331–340.
Swanagan, J.S., 2000. Factors influencing zoo visitors' conservation attitudes and behavior. The Journal of Environmental Education 31, 26–31.
Sykes, S.K., 1971. The Natural History of the African Elephant. Weidenfeld and Nicolson, London.
Szdzuy, K., Dehnhard, M., Strauss, G., Eulenburger, K., Hofer, H., 2006. Behavioural and endocrinological parameters of female African and Asian elephants *Loxodonta africana* and *Elephas maximus* in the peripartal period. International Zoo Yearbook 40, 41–50.
Taylor, V.J., Poole, T.B., 1998. Captive breeding and infant mortality in Asian elephants: a comparison between twenty western zoos and three eastern elephant centers. Zoo Biology 17, 311–322.
Terkel, A., 2004. Taking stock of management and welfare of elephants in EAZA. EAZA News 47, 14–16.
Tetley, C.L., O'Hara, S.J., 2012. Ratings of animal personality as a tool for improving the breeding, management and welfare of zoo mammals. Animal Welfare 21, 463–476.
Thitaram, C., 2012. Breeding management of captive Asian elephant *(Elephas maximus)* in range countries and zoos. Japanese Journal of Zoo and Wildlife Medicine 17, 91–96.
Thitaram, C., Pongsopawijit, P., Thongtip, N., Angkavanich, T., Chansittivej, S., Wongkalasin, W., et al., 2006. Dystocia following prolonged retention of a dead fetus in an Asian elephant (*Elephas maximus*). Theriogenology 66, 1284–1291.
Thitaram, C., Dejchaisri, S., Somgird, C., Angkawanish, T., Brown, J., Phumphuay, R., et al., 2015. Social group formation and genetic relatedness in reintroduced Asian elephants (*Elephas maximus*) in Thailand. Applied Animal Behaviour Science 172, 52–57.
Thitaram, C., Matchimakul, P., Pongkan, W., Tangphokhanon, W., Maktrirat, R., Khonmee, J., et al., 2018. Histology of 24 organs from Asian elephant calves (*Elephas maximus*). PeerJ 6, e4947. <https://doi.org/10.7717/peerj.4947>.
Thongtip, N., Saikhun, J., Mahasawangkul, S., Kornkaewrat, K., Pongsopavijtr, P., Songsasen, N., et al., 2008. Potential factors affecting semen quality in the Asian elephant (*Elephas maximus*). Reproductive Biology and Endocrinology 6, 9. <https://doi.org/10.1186/1477-7827-6-9>.
Thongtip, N., Mahasawangkul, S., Thitaram, C., Pongsopavijtr, P., Kornkaewrat, K., Pinyopummin, A., et al., 2009. Successful artificial insemination in the Asian elephant (*Elephas maximus*) using chilled and frozen-thawed semen. Reproductive Biology and Endocrinology 7, 75. <https://doi.org/10.1186/1477-7827-7-75>.
Thouless, C.R., 1995. Long distance movements of elephants in northern Kenya. African Journal of Ecology 33, 321–334.
Thouless, C.R., 1996. Home ranges and social organization of female elephants in northern Kenya. African Journal of Ecology 34, 284–297.
Thouless, C.R., Dublin, H.T., Blanc, J.J., Skinner, D.P., Daniel, T.E., Taylor, R.D., et al., 2016. African Elephant Status Report 2016: An Update From the African Elephant Database. Occasional Paper Series of the IUCN Species Survival Commission, No. 60. IUCN/SSC African Elephant Specialist Group. IUCN, Gland, vi + 309 pp.

Tidière, M., Gaillard, J.-M., Berger, V., Müller, D.W.H., Lackey, L.B., Gimenez, O., et al., 2016. Comparative analyses of longevity and senescence reveal variable survival benefits of living in zoos across mammals. Science Reports 6, 36361. <https://doi.org/10.1038/srep36361>.

Tinbergen, N., 1952. 'Derived' activities; their causation, biological significance, origin and emancipation during evolution. The Quarterly Review of Biology 27, 1–32.

Tobler, I., 1992. Behavioral sleep in the Asian elephant in captivity. Sleep 15, 1–12.

Toscano, G., 1997. Dangerous jobs. Compensation and Working Conditions, Summer 1997, 57–60.

Toscano, M.J., Friend, T.H., Nevill, C.H., 2001. Environmental conditions and body temperature of circus elephants transported during relatively high and low temperature conditions. Journal of the Elephant Managers Association 12, 115–149.

Tresz, H., Wright, H., 2006. Let them be elephants! How Phoenix Zoo integrated three 'problem' animals. International Zoo News 53, 154–160.

Tudge, C., 1991. Last Animals at the Zoo. How Mass Extinction Can Be Stopped. Hutchinson Radius, London.

Tun, Y., 1962. Twin elephant calves and intervals between births of successive elephant calves. Journal of the Bombay Natural History Society 59, 643–644.

Turkalo, A.K., Wrege, P.H., Wittemyer, G., 2017. Slow intrinsic growth rate in forest elephants indicates recovery from poaching will require decades. The Journal of Applied Ecology 54, 153–159.

Tuttle, D., Keele, M., Nunley, L., 1997. Minimum Husbandry Guidelines for Mammals: Elephants. American Zoo and Aquarium Association, Silver Spring, MD.

Uher, J., Asendorpf, J.B., 2008. Personality assessment in the Great Apes: comparing ecologically valid behavior measures, behavior ratings, and adjective ratings. Journal of Research in Personality 42, 821–838.

USDA, 1997. Guidelines for the Control of TB in Elephants. United States Department of Agriculture.

USFWS, 2018a. U.S. Fish and Wildlife Service Division of International Conservation Asian Elephant Conservation Program FY 2018 Summary of Projects. <https://www.fws.gov/international/pdf/project-summaries-asian-elephant-2018.pdf> (accessed 18.06.19.).

USFWS, 2018b U.S. Fish and Wildlife Service Division of International Conservation African Elephant Conservation Program FY 2018 Summary of Projects. <https://www.fws.gov/international/pdf/project-summaries-african-elephant-2018.pdf> (accessed 18.06.19.).

Valeix, R., Fritz, H., Sabatier, R., Murindafomo, F., Cummings, D., Duncan, P., 2011. Elephant-induced structural changes in the vegetation and habitat selection by large herbivores in an African savanna. Biological Conservation 144, 902–912.

Vancuylenberg, B.W.B., 1977. Feeding behaviour of the Asiatic elephant in south-east Sri Lanka in relation to conservation. Biological Conservation 12, 33–54.

Vanitha, V., Baskaran, N., 2010. Seasonal and roofing material influence on the thermoregulation by captive Asian elephants and its implications for captive elephant welfare. Gajah 33, 35–40.

Vanitha, V., Thiyagesan, K., Baskaran, N., 2008. Food and feeding of captive Asian Elephants (*Elephas maximus*) in the three management facilities at Tamil Nadu, South India. Scientific Transactions in Environment and Technovation 2, 87–97.

Vanitha, V., Thiyagesan, K., Baskaran, N., 2010. Status of mahouts and human-captive elephant conflict in three management systems in Tamil Nadu, India. Indian Forester 136, 767–774.

Vanitha, V., Thiyagesan, K., Baskaran, N., 2011. Social life of captive Asian elephants (*Elephas maximus*) in Southern India, implications for elephant welfare. Journal of Applied Animal Welfare Science 14, 42–58.

Varadharajan, V., Krishnamoorthy, T., Nagarajan, B., 2016. Prevalence of stereotypies and its possible causes among captive Asian elephants (*Elephas maximus*) in Tamil Nadu, India. Applied Animal Behaviour Science 174, 137–146.

Varner, G., 2008. Personhood, memory, and elephant management. In: Wemmer, C., Christen, C.A. (Eds.), Elephants and Ethics. Towards a Morality of Coexistence. The Johns Hopkins University Press, Baltimore, MD, pp. 41–67.

Veltre, T., 1996. Menageries, metaphors, and meanings. In: Hoage, R.J., Deiss, W.A. (Eds.), New Worlds, New Animals. From Menagerie to Zoological Park in the Nineteenth Century. The Johns Hopkins University Press, Baltimore and London, pp. 19–29.

Vicino, G.A., Marcacci, E.S., 2015. Intensity of play behaviour as a potential measure of welfare: a novel method for quantifying the integrated intensity of behavior in African elephants. Zoo Biology 34, 492–496.

Von Drückheim, K.E.M., Hoffman, L.C., Leslie, A., Hensman, M.C., Hensman, S., Schultz, K., et al., 2018. African elephants (*Loxodonta africana*) display remarkable olfactory acuity in human scent matching to sample performance. Applied Animal Behaviour Science 200, 123−129.

Vuletic, J., Byard, R.W., 2013. Death due to crushing by an elephant trunk. Forensic Science, Medicine and Pathology 9, 449−451. <https://doi.org/10.1007/s12024-013-9415-9>.

Walls, S.C., 2018. Coping with constraints: achieving effective conservation with limited resources. Frontiers in Ecology and Evolution 6, 24.

Walsh, B., 2017. Asian elephant (*Elephas maximus*) sleep study − long-term quantitative research at Dublin Zoo. Journal of Zoo and Aquarium Research 5, 82−85. <https://doi.org/10.19227/jzar.v5i2.174>.

Walter, O., 2010. Management Guidelines for the Welfare of Zoo Animals: Elephants, third ed. British and Irish Association of Zoos and Aquariums, Regent's Park, London.

Wanghongsa, S., Boonkird, K., Rabiab, S., Ruksat, S., 2006. On the incident of infanticide in wild elephants. Wildlife Yearbook 7, 111−119.

WAP, 2019. Thai Elephant Venue Reopens Without the Cruelty <https://www.worldanimalprotection.org.uk/changchill> (accessed 28.06.19.).

Ward, P.I., Mosberger, N., Kistler, C., Fischer, O., 1998. The relationship between popularity and body size in zoo animals. Conservation Biology 12, 1408−1411. <https://doi.org/10.1111/j.1523-1739.1998.97402.x>.

Watson, M., 1871. Contribution to the anatomy of the Indian elephant. Part I. The thoracic viscera. Journal of Anatomy and Physiology 6, 82−94.

Wehnelt, S., Wilkinson, R., 2005. Research, conservation and zoos: the EC Zoos Directive − a response to Rees. Oryx 39, 132−133.

Weinstein, T.A., Capitanio, J.P., Gosling, S.D., 2008. Personality in animals. Handbook of Personality Theory Research 3, 328−348.

Weissenböck, N.M., Weiss, C.M., Schwammer, H.M., Kratochvil, H., 2010. Thermal windows on the body surface of African elephants (*Loxodonta africana*) studied by infrared thermography. Journal of Thermal Biology 35, 182−188.

Weisz, I., Wuestenhagen, A., Schwammer, H., 2000. Research on nocturnal behaviour of African elephants at Schönbrunn Zoo. International Zoo News 47, 228−233.

Wells, D.L., Irwin, R.M., 2008. Auditory stimulation as enrichment for zoo-housed Asian elephants (*Elephas maximus*). Animal Welfare 17, 335−340.

Wemmer, C., Christen, C.A. (Eds.), 2008. Elephants and Ethics. Towards a Morality of Coexistence. The Johns Hopkins University Press, Baltimore, MD.

Wemmer, C., Krishnamurthy, V., Shrestha, S., Hayek, L.-A., Thant, M., Nanjappa, K.A., 2006. Assessment of body condition in Asian elephants (*Elephas maximus*). Zoo Biology 25, 187−200.

West, L.J., Pierce, C.M., Thomas, W.D., 1962. Lysergic acid diethylamide: its effect on a male Asiatic elephant. Science 138, 1100−1102.

Whilde, J., Marples, N., 2012. Effects of a birth on the behavior of a family group of Asian elephants (*Elephas maximus*) at Dublin Zoo. Zoo Biology 31, 442−452.

White, C., Wyshak, G., 1964. Inheritance in human dizygotic twinning. The New England Journal of Medicine 271, 1003−1005.

Whitehouse, A.M., 2002. Tusklessness in the elephant population of the Addo Elephant National Park, South Africa. Journal of Zoology 257, 249−254.

Whitehouse, A.M., Schoeman, D.S., 2003. Ranging behaviour of elephants within a small, fenced area in Addo Elephant National Park, South Africa. African Zoology 38, 95−108.

Whyte, I.J., Hall-Martin, A., 2018. Growth characteristics of tusks of elephants in Kruger National Park. Pachyderm 59, 31−40.

Wickler, W., Seibt, U., 1997. Aimed object throwing by a wild elephant in an interspecific encounter. Ethology 103, 365−368.

Wiedenmayer, C., 1998. Food hiding and enrichment in captive Asian elephants. Applied Animal Behaviour Science 56, 77−82.

Wiese, R.J., 2000. Asian elephants are not self-sustaining in North America. Zoo Biology 19, 299−309.

Wiese, R.J., Willis, K., 2004. Calculation of longevity and life expectancy in captive elephants. Zoo Biology 23, 365−373.

Wiese, R.J., Willis, K., 2006. Population management of zoo elephants. International Zoo Yearbook 40, 80–87.

Wijesooriya, P.N., Abeykoon, A.M.H.S., Udawatta, L., Punchihewa, A., Nanayakkara, T., 2012. Gait pattern analysis of an Asian elephant. In: ICIAFS 2012 – Proceedings: 2012 IEEE 6th International Conference on Information and Automation for Sustainability, pp. 221–226. 10.1109/ICIAFS.2012.6419908.

Wijeyamohan, S., Treiber, K., Schmitt, D., Santiapillai, C., 2015. A visual system for scoring body condition of Asian elephants (*Elephas maximus*). Zoo Biology 34, 53–59.

Williams, J.H., 1951. Elephant Bill. Rupert Hart-Davis, London.

Williams, G.C., 1957. Pleiotropy, natural selection, and the evolution of senescence. Evolution 11, 398–411.

Williams, T.M., 1990. Heat transfer in elephants: thermal partitioning based on skin temperature profiles. Journal of Zoology 222, 235–245.

Williams, J.L., Friend, T.H., 2003. Behavior of circus elephants during transport. Journal of the Elephant Managers Association 14, 8–11.

Williams, E., Bremner-Harrison, S., Harvey, N., Evison, E., Yon, L., 2015. An investigation into resting behavior in Asian elephants in UK zoos. Zoo Biology 34, 406–417. <https://doi.org/10.1002/zoo.21235>.

Williams, E., Chadwick, C., Yon, L., Asher, L., 2018. A review of current indicators of welfare in captive elephants (*Loxodonta africana* and *Elephas maximus*). Animal Welfare 27, 235–249.

Wilson, J.C.C., 1922. The breeding of elephants in captivity. Journal of the Bombay Natural History Society 28, 1128–1129.

Wilson, H.S., 1944. The baby elephant program an American dream. Prairie Schooner 18, 43–45.

Wilson, E.O., 1980. Sociobiology, The abridged edition The Belknap Press of Harvard University Press, Cambridge, MA.

Wilson, M.L., Bloomsmith, M.A., Maple, T.L., 2004. Stereotypic swaying and serum cortisol concentrations in three captive African elephants (*Loxodonta africana*). Animal Welfare 13, 39–43.

Wilson, M.L., Bashaw, M.J., Fountain, K., Kieschnick, S., Maple, T.L., 2006. Nocturnal behavior in a group of female African elephants. Zoo Biology 25, 173–186.

Wilson, M.L., Perdue, B.M., Bloomsmith, M.A., Maple, T.L., 2015a. Rates of reinforcement and measures of compliance in free and protected contact elephant management systems. Zoo Biology 34, 431–437.

Wilson, S., Davies, T.E., Hazarika, N., Zimmermann, A., 2015b. Understanding spatial and temporal patterns of human–elephant conflict in Assam, India. Oryx 49, 140–149.

Wilson, J.W., Bergl, R.A., Minter, L.J., Loomis, M.R., Kendall, C.J., 2019. The African elephant *Loxodonta* spp conservation programmes of North Carolina Zoo: two decades of using emerging technologies to advance in situ conservation efforts. International Zoo Yearbook <https://doi.org/10.1111/izy.12216>.

Wing, L.D., Buss, I.O., 1970. Elephants and Forests. Wildlife Monographs 19, 3–92.

Wittemyer, G., Getz, W.M., 2007. Hierarchical dominance structure and social organization in African elephants, *Loxodonta africana*. Animal Behaviour 73, 671–681.

Wittemyer, G., Getz, W.M., Vollrath, F., Douglas-Hamilton, I., 2007. Social dominance, seasonal movements, and spatial segregation in African elephants: a contribution to conservation behavior. Behavioral Ecology and Sociobiology 61, 1919–1931.

Woodford, M.H., Trevor, S., 1970. Fostering a baby elephant. East African Wildlife Journal 8, 204–205.

Worsley, H.K., O'Hara, S.J., 2018. Cross-species referential signalling events in domestic dogs (*Canis familiaris*). Animal Cognition 21, 457–465. <https://doi.org/10.1007/s10071-018-1181-3>.

Wright, D.W.M., 2018. Cloning animals for tourism in the year 2070. Futures 95, 58–75.

Wyatt, J.R., Eltringham, S.K., 1974. The daily activity of the elephant in Rwenzori National Park, Uganda. East African Wildlife Journal 12, 273–289.

Wylie, D., 2008. Elephant. Reaktion Books Ltd, London.

Wynne, C.D.L., 2001. Animal Cognition. The Mental Lives of Animals. Palgrave Macmillan, Basingstoke, Hampshire.

Yamamoto, Y., Yuto, N., Yamamoto, T., Kaewmanee, S., Shiina, O., Mouri, Y., et al., 2012. Secretory pattern of inhibin during estrous cycle and pregnancy in African (*Loxodonta africana*) and Asian (*Elephas maximus*) elephants. Zoo Biology 32, 511–522.

Yasui, S., Konno, A., Tanaka, M., Idani, G., Ludwig, A., Lieckfeldt, D., et al., 2013. Personality assessment and its association with genetic factors in captive Asian and African elephants. Zoo Biology 32, 70–78.

Zimmermann, A., Davies, T.E., Hazarika, N., Wilson, S., Chakrabarty, J., Hazarika, B., et al., 2009. Community-based human-elephant conflict management in Assam. Gajah 30, 34–40.

ZIMS, 2018a. Species holding report: *Elephas maximus*/Asian elephant. Zoological Information Management System. Species 360. <https//www.species360.org> (accessed 26.03.18.).
ZIMS 2018b. Species holding report: *Loxodonta africana*/African elephant. Zoological Information Management System. Species 360. <https//www.species360.org> (accessed 26.03.18.).
Zoo Federation, 2003. In Safe Hands – A Response to the RSPCA Report on the Welfare of Elephants in Captivity. Federation of Zoological Gardens of Great Britain and Ireland, Regent's Park, London.
Zoolex, 2018. Zoolex Zoo Design Organization. <https://www.zoolex.org/> (accessed 07.09.19.).
Zoos Forum, 2010. Elephants in UK Zoos. Zoos Forum Review of Issues in Elephant Husbandry in UK Zoos in the Light of the Report by Harris et al (2008). Wildlife Species Conservation Division, Defra, Bristol.
ZSL, n.d. Zoological Society of London's Annual Report and Accounts 2017-18. Zoological Society of London, Regent's Park, London.
Zuckerman, S., 1932. The Social Life of Monkeys and Apes. Routledge, London.

Legislation, case law and other legal documents cited

International laws
Convention on International Trade in Endangered Species of Wild Fauna and Flora (CITES) 1973
United Nations Convention on Biological Diversity 1992
European Union Law
Council Directive 1999/22/EC of 29 March 1999 on the keeping of wild animals in zoos. (The Zoos Directive)
National laws
Australia
New South Wales
Exhibited Animals Protection Regulation, 2005 (New South Wales)
Policy on the Management of Solitary Elephants in New South Wales (pursuant to clause 8(1) of the Exhibited Animals Protection Regulation, 2005)
Queensland
Biosecurity Act 2014
Canada
British North America Act 1867/ The Constitution Act, 1867
India
Central Zoo Authority Letter No. 7-5/2006-CZA (Vol. II) dated 7th November, 2009. Advisory - Banning Elephants from Zoo collections
Performing Animal (Registration) Rules, 2001
Wildlife (Protection) Act 1972
Rajasthan
DB Civil Writ petition PIL DBCW(PIL) NO. 8987/2006 *Naresh Kadyan v/s Chief Secretary, Govt. of Rajasthan & Others*
Kerala
Kerala Captive Elephants (Management and Maintenance) Rules 2003
Kerala Captive Elephants (Management and Maintenance) Rules 2012
Kerala Forests and Wildlife Department. Circular No.01/2019. Alarming rate of death of Captive Elephants in Kerala. Forest.kerala.gov.in/images/Circulars/circ250119pdf
New Zealand
Exhibited Animals Protection Regulation 2005
MAF Biosecurity New Zealand Standard 154.03.04: Containment facilities for zoo animals (29 January 2007)
Standard for Zoo Containment Facilities 2018, Environmental Protection Authority (EPA), New Zealand
Te Awa Tupua (Whanganui River Claims Settlement) Act 2017
Sri Lanka
Fauna and Flora Protection Ordinance 1937
United Kingdom
Animal Welfare Act 2006
Endangered Species (Import and Export) Act 1976
Ivory Act 2018
Planning (Listed Buildings and Conservation Areas) Act 1990

Wild Animals in Travelling Circuses (Scotland) Act 2018
Zoo Licensing Act 1981
United States
Asian Elephant Conservation Act of 1997
African Elephant Conservation Act of 1989
California
California Senate Bill No. 1062 CHAPTER 234 An act to add Section 2128 to the Fish and Game Code, relating to elephants
New York
The Elephant Protection Act 2017
Rhode Island
Rhode Island General Laws § 4-1-43. Use of bullhooks or similar devices on elephants prohibited
United States Case Law
Born Free USA, et al., Plaintiffs, v. Gale Norton, Secretary, Department of the Interior, et al., Defendants, and the Zoological Society of San Diego, et al., Intervenor-Defendants. 278 F. Supp 2d 5 (D.D.C. 2003)
Nonhuman Rights Project, Inc. v. Stanley. 2015 NY Slip Op 25257 [49 Misc 3d 746]

Suggested Further Reading

It is not possible in a book of this size to discuss all of the research conducted to date on elephants under human care. The following is a list of publications to which the reader is directed for further information.

Abbondanza, F.N., Power, M.L., Dickson, M.A., Brown, J., Oftedal, O.T., 2013. Variation in the composition of milk of Asian elephants (*Elephas maximus*). Zoo Biology 32, 291−298.

Adams, J., Berg, J.K., 1980. Behavior of female African elephants (*Loxodonta africana*) in captivity. Applied Animal Ethology 6, 257−276.

Adams, G.P., Plotka, E.D., Asa, C.S., Ginther, O.J., 1991. Feasibility of characterizing reproductive events in large nondomestic species by transrectal ultrasonic imaging. Zoo Biology 10, 247−259.

Allen, W.R., 2006. Ovulation, pregnancy, placentation and husbandry in the African elephant (*Loxodonta africana*). Philosophical Transactions of the Royal Society B 361, 821−834. <https://doi.org/10.1098/rstb.2006.1831>.

Bansiddhi, P., Brown, J.L., Thitaram, C., Punyapornwithaya, V., Somgird, C., Edwards, K.L., et al., 2018. Changing trends in elephant camp management in northern Thailand and implications for welfare. PeerJ 6, e5996. <https://doi.org/10.7717/peerj.5996>.

Bansiddhi, P., Brown, J.L., Thitaram, C., Punyapornwithaya, V., Nganvongpanit, K., 2019. Elephant tourism in Thailand: a review of animal welfare practices and needs. Journal of Applied Animal Welfare Science <https://doi.org/10.1080/10888705.2019.1569522>.

Bates, L.A., Sayialel, K.N., Njiraini, N.W., Poole, J.H., Moss, C.J., Byrne, R.W., 2008. African elephants have expectations about the location of out-of-sight family members. Biology Letters 4, 34−36. <https://doi.org/10.1098/rsbl.2007.0529>.

Benedict, F.G., 1936. Nutrition of the elephant. Problemi Kibernetiki 5, 7−13.

Benedict, F.G., Lee, R.C., 1936. The heart rate of the elephant. Proceedings of the American Philosophical Society 76, 335−341.

Bennett, C.F., 1957. A brief history of trained African elephants in the Belgian Congo. Journal of Geography 56, 168−172.

Berg, J.K., 1983. Vocalizations and associated behaviors of the African elephant (*Loxodonta africana*) in captivity. Zeitschrift fur Tierpsychologie 63, 63−79.

Berg, J.K., 1987. Developmental behavior of three African elephant (*Loxodonta africana*) calves in captivity. Der Zoologische Garten 57, 191−196.

Bist, S.S., 2010. No room for elephants in zoos? Zoos' Print 25, 7−10.

Bonaparte-Saller, M., Mench, J.A., 2018. Assessing the dyadic social relationships of female African (*Loxodonta africana*) and Asian (*Elephas maximus*) zoo elephants using proximity, tactile contact, and keeper surveys. Applied Animal Behaviour Science 199, 45−51.

Boyle, S.A., Roberts, B., Pope, B.M., Blake, M.R., Leavelle, S.E., Marshall, J.J., et al., 2015. Assessment of flooring renovations on African elephant (*Loxodonta africana*) behavior and glucocorticoid response. PLoS ONE <https://doi.org/10.1371/journal.pone.0141009>.

Brown, C.E., 1934. Captive pigmy elephants in America. Journal of Mammalogy 15, 248–250. <https://doi.org/10.1093/jmammal/15.3.248>.

Brown, J.L., 2014. Comparative reproductive biology of elephants. In: Holt, W., Brown, J., Comizzoli, P. (Eds.), Reproductive Sciences in Animal Conservation. Springer, New York, pp. 135–169.

Brown, J.L., Lehnhardt, J., 1997. Secretory patterns of serum prolactin in Asian (*Elephas maximus*) and African (*Loxodonta africana*) elephants during different reproductive states: comparison with concentrations in a noncycling African elephant. Zoo Biology 16, 149–159.

Brown, J.L., Wemmer, C.M., Lehnhardt, J., 1995. Urinary cortisol analysis for monitoring adrenal activity in elephants. Zoo Biology 14, 533–542.

Brown, J.L., Somerville, M., Riddle, H.S., Keele, M., Duer, C.K., Freeman, E.W., 2007. Comparative endocrinology of testicular, adrenal and thyroid function in captive Asian and African elephant bulls. General and Comparative Endocrinology 151, 153–162.

Brown, J.L., Wielebnowski, N., Cheeran, J.V., 2008. Pain, stress, and suffering in elephants. In: Wemmer, C., Christen, C.A. (Eds.), Elephants and Ethics. Towards a Morality of Coexistence. The Johns Hopkins University Press, Baltimore, MD, pp. 121–145.

Brown, J.L., Kersey, D.C., Freeman, E.W., Wagener, T., 2010. Assessment of diurnal urinary cortisol excretion in Asian and African elephants using different endocrine methods. Zoo Biology 29, 274–283.

Cheeran, J.V., Poole, T.B., 1996. The exploitation of Asian elephants. In: Taylor, V.J., Dunstone, N. (Eds.), The Exploitation of Mammal Populations. Springer, Dordrecht, pp. 302–309.

Clauss, M., Hatt, J.M., 2006. Feeding Asian and African elephants *Elephas maximus* and *Loxodonta africana* in captivity. International Zoo Yearbook 40, 88–95.

Clubb, R., Rowcliffe, M., Lee, P., Mar, K.U., Moss, C., Mason, G.J., 2009. Fecundity and population viability in female zoo elephants: problems and possible solutions. Animal Welfare 18, 237–247.

Cooper, K.A., Harder, J.D., Clawson, D.H., Frederick, D.L., Lodge, G.A., Peachey, H.C., et al., 1990. Serum testosterone and musth in captive male African and Asian elephants. Zoo Biology 9, 297–306.

Czekala, N.M., Roocroft, A., Bates, M., Allen, J., Lasley, B.L., 1992. Estrogen metabolism in the Asian elephant (*Elephas maximus*). Zoo Biology 11, 75–80.

Czekala, N.M., MacDonald, E.A., Steinman, K., Walker, S., Garrigues, N.W., Olson, D., et al., 2003. Estrogen and LH dynamics during the follicular phase of the estrous cycle in the Asian elephant. Zoo Biology 22, 443–454.

Dahl, N.J., Olson, D., Schmitt, D.L., Blasko, D.R., Kristipati, R., Roser, J.F., 2004. Development of an enzyme-linked immunosorbent assay (ELISA) for luteinizing hormone (LH) in the elephant (*Loxodonta africana* and *Elephas maximus*). Zoo Biology 23, 65–78.

Dathe, H.H., Kuckelkorn, B., Minnemann, D., 1992. Salivary cortisol assessment for stress detection in the Asian elephant (*Elephas maximus*): a pilot study. Zoo Biology 11, 285–289.

De Mel, R.K., Weerakoon, D.K., Ratnasooriya, W.D., 2013. A comparison of stereotypic behaviour in Asian elephants at three different institutions in Sri Lanka. Gajah 38, 25–29.

Dickson, P., Adams, W.M., 2009. Science and uncertainty in South Africa's elephant culling debate. Environment and Planning C: Politics and Space 27, 110–123.

Dierenfeld, E.S., Dolensek, E.P., 1988. Circulating levels of vitamin E in captive Asian elephants (*Elephas maximus*). Zoo Biology 7, 165–172.

Dittrich, L., 1966. Breeding Indian elephants *Elephas maximus* at Hanover Zoo. International Zoo Yearbook 6, 193–196.

Dunkin, R.C., Wilson, D., Way, N., Johnson, K., Williams, T.M., 2013. Climate influences thermal balance and water use in African and Asian elephants: physiology can predict drivers of elephant distribution. Journal of Experimental Biology 216, 2939–2952.

Edwards, K.L., Trotter, J., Jones, M., Brown, J.L., Steinmetz, H.W., 2016. Investigating temporary acyclicity in a captive group of Asian elephants (*Elephas maximus*): relationship between management, adrenal activity and social factors. General and Comparative Endocrinology 225, 104–116.

Forthman, D.L., Kane, L.F., Hancocks, D., Waldau, P.F., 2008. An Elephant in the Room: The Science and Well-Being of Elephants in Captivity. Tufts University Cummings School of Veterinary Medicine's Center for Animals and Public Policy, North Grafton, MA.

Fowler, M.E., 1993. Foot care in elephants. In: Fowler, M.E. (Ed.), Zoo and Wild Animal Medicine, third ed. W. B. Sanders Company, Philadelphia, PA, pp. 448–453.

Friend, T.H., 1999. Behavior of picketed circus elephants. Applied Animal Behaviour Science 62, 73–88.

Gadgil, M., Nair, P.V., 1984. Observations on the social behavior of free ranging groups of tame Asiatic elephant (*Elephas maximus* Linn). Proceedings of the Indian Academy of Sciences Animal Sciences 93, 225–233.

Gentry, P.A., Ross, M.L., Yamada, M., 1996. Blood coagulation profile of the Asian elephant (*Loxodonta africana*). Zoo Biology 15, 413–423.

Ghosal, R., Sukumar, R., Seshagiri, P.B., 2010. Prediction of estrus cyclicity in Asian elephants (*Elephas maximus*) through estimation of fecal progesterone metabolite: development of an enzyme-linked immuno-sorbent assay. Theriogenology 73, 1051–1060.

Ghosal, R., Kalaivanan, N., Sukumar, R., Seshagiri, P.B., 2012a. Assessment of estrus cyclicity in the Asian elephant (*Elephas maximus*) by measurement of fecal progesterone metabolite 5α-P-3OH, using a non-invasive assay. General and Comparative Endocrinology 175, 100–108.

Ghosal, R., Seshagiri, P.B., Sukumar, R., 2012b. Dung as a potential medium for inter-sexual chemical signaling in Asian elephants (*Elephas maximus*). Behavioural Processes 91, 15–21.

Glaeser, S.S., Hunt, K.E., Martin, M.S., Finnegan, M., Brown, J.L., 2012. Investigation of individual and group variability in estrous cycle characteristics in female Asian elephants (*Elephas maximus*) at the Oregon Zoo. Theriogenology 78, 285–296.

Godfrey, A., Kongmuang, C., 2009. Distribution, demography and basic husbandry of the Asian elephant in the tourism industry in Northern Thailand. Gajah 30, 13–18.

Graham, L., Schwarzenberger, F., Möstl, E., Galama, W., Savage, A., 2001. A versatile immunoassay for the determination of progestogens in feces and serum. Zoo Biology 20, 227–236.

Graham, L.H., Bolling, J., Miller, G., Pratt-Hawkes, N., Joseph, S., 2002. Enzyme-immunoassay for the measurement of luteinizing hormone in the serum of African elephants (*Loxodonta africana*). Zoo Biology 21, 403–408.

Grand, A.P., Kuhar, C.W., Leighty, K.A., Bettinger, T.L., Laudenslager, M.L., 2012. Using personality ratings and cortisol to characterize individual differences in African Elephants (*Loxodonta africana*). Applied Animal Behaviour Science 142, 69–75.

Gurusamy, V., Tribe, A., Phillips, C.J.C., 2014. Identification of major welfare issues for captive elephant husbandry by stakeholders. Animal Welfare 23, 11–24.

Hacker, C.E., Horback, K.M., Miller, L.J., 2015. GPS technology as a proxy tool for determining relationships in social animals: an example with African elephants. Applied Animal Behaviour Science 163, 175–182.

Hacker, C.E., Miller, L.J., Schulte, B.A., 2018. Examination of enrichment using space and food for African elephants (*Loxodonta africana*) at the San Diego Zoo Safari Park. Animal Welfare 27, 55–65.

Hart, L.A., 1997. Tourists' effects on the drivers of working Asian elephants. Anthrozoös 10, 47–49.

Hartley, M., 2016. Assessing risk factors for reproductive failure and associated welfare impacts in elephants in European zoos. Journal of Zoo and Aquarium Research 4, 127–138. <https://doi.org/10.19227/jzar.v4i3.162>.

Hartley, M., Stanley, C., 2016. Survey of reproduction and calf rearing in Asian and African elephants in Europe. Journal of Zoo and Aquarium Research 4, 139–146. <https://doi.org/10.19227/jzar.v4i3.161>.

Hatt, J.M., Clauss, M., 2006. Feeding Asian and African elephants. International Zoo Yearbook 40, 88–95.

Heffner, R.S., Heffner, H.E., 1980. Hearing in the elephant (*Elephas maximus*): absolute sensitivity, frequency discrimination, and sound localization. Science 208, 518–520.

Herbst, C.T., Švec, J.G., Lohscheller, J., Frey, R., Gumpenberger, M., Stoeger, A.S., et al., 2013. Complex vibratory patterns in an elephant larynx. Journal of Experimental Biology 216, 4054–4064.

Hermes, R., Olson, D., Göritz, F., Brown, J.L., Schmitt, D.L., Hagan, D., et al., 2000. Ultrasonography of the estrous cycle in female African elephants (*Loxodonta africana*). Zoo Biology 19, 369–382.

Hildebrandt, T.B., Lueders, I., Hermes, R., Goeritz, F., Saragusty, J., 2011. Reproductive cycle of the elephant. Animal Reproduction Science 124, 176–183.

Hollister-Smith, J.A., Alberts, S.C., Rasmussen, L.E.L., 2008. Do male African elephants, *Loxodonta africana*, signal musth via urine dribbling? Animal Behaviour 76, 1829–1841.

Horback, K.M., Miller, L.J., Andrews, J., Kuczaj II, S.A., Anderson, M., 2012. The effects of GPS collars on African elephant (*Loxodonta africana*) behavior at the San Diego Zoo Safari Park. Applied Animal Behaviour Science 142, 76–81.

Horback, K.M., Miller, L.J., Kuczaj II, S.A., 2013. Personality assessment in African elephants (*Loxodonta africana*): comparing the temporal stability of ethological coding versus trait rating. Applied Animal Behaviour Science 149, 55–62.

Iossa, G., Soulsbury, C.D., Harris, S., 2008. Are wild animals suited to a travelling circus life? Animal Welfare 18, 129–140.

Jainudeen, M., McKay, G., Eisenberg, J., 2009. Observations on musth in the domesticated Asiatic elephant (*Elephas maximus*). Mammalia 36, 247–261.

Juniper, P., 2000. The management of Asian elephants (*Elephas maximus*) at Port Lympne Wild Animal Park. Ratel 27, 168–175.

Kajaysri, J., Nokkaew, W., 2013. Assessment of pregnancy status of Asian elephants (*Elephas maximus*) by measurement of progestagen and glucocorticoid and their metabolite concentrations in serum and feces, using enzyme immunoassay (EIA). The Journal of Veterinary Medical Science 13–0103.

Kane, L., Forthman, D., Hancocks, D., 2005. Optimal conditions for captive elephants: a report by the coalition for captive elephant well-being 2005. <http://elephantcare.org/wp-content/uploads/2017/02/Optimal-Conditions-for-Captive-Elephants-2005.pdf>.

Katole, S.B., Das, A., Agarwal, N., Prakash, B., Saha, S.K., Saini, M., et al., 2014. Influence of work on nutrient utilisation in semicaptive Asian elephants (*Elephas maximus*). Journal of Applied Animal Research 42, 380–388.

Katole, S., Das, A., Saini, M., Sharma, A., 2015. Comparative evaluation of wheat-roti or rice-lentil mixture as supplements for growing Asiatic elephants (*Elephas maximus*). Journal of Zoo and Aquarium Research 3, 63–69. <https://doi.org/10.19227/jzar.v3i2.30>.

Kenny, D.E., 2001. Long-term administration of α-tocopherol in captive Asian elephants (*Elephas maximus*). Zoo Biology 20, 245–250.

Kerley, G.I.H., Shrader, A.M., 2007. Elephant contraception: silver bullet or a potentially bitter pill? South African Journal of Science 103, 181–182.

Khawnual, P., Clarke, B., 2002. General care and reproductive management of pregnant and infant elephants at the Ayutthaya Elephant Camp. In: Baker, I., Kashio, M., (Eds.). Giants on Our Hands: Proceedings of the International Workshop on the Domesticated Asian Elephant: February 5–10, 2001; Bangkok, Thailand. Bangkok, Thailand: FAO Regional Office for Asia and the Pacific, pp. 249–56.

Kongsawasdi, S., Mahasawangkul, S., Pongsopawijit, P., Boonprasert, K., Chuatrakoon, B., Thonglorm, N., 2017. Biomechanical parameters of Asian Elephant (*Elephas maximus*) walking gait. Kafkas Universitesi Veteriner Fakultesi Dergisi 23, 357–362.

Kontogeorgopoulos, N., 2009. Wildlife tourism in semi-captive settings: a case study of elephant camps in Northern Thailand. Current Issues in Tourism 12, 429–449.

Kühme, V.W., 1963. Ergänzende Beobachtungen an Afrikanischen Elefanten (*Loxodonta africana* Blumenbach 1979) im Freigehege. Zeitschrift fur Tierpsychologie 20, 66–79.

Kumar, V., Reddy, V.P., Kokkiligadda, A., Shivaji, S., Umapathy, G., 2014. Non-invasive assessment of reproductive status and stress in captive Asian elephants in three south Indian zoos. General and Comparative Endocrinology 201, 37–44.

Kurt, F., Garaï, M., 2001. Stereotypies in captive Asian elephants – a symptom of social isolation. In: Schwammer, H.M., Foose, T.J., Fouraker, M., Olson, D. (Eds.), Proceedings of the International Elephant and Rhino Research Symposium. Schuling Verlag, Vienna.

Kurt, F., Kumarasinghe, J.C., 1998. Remarks on body growth and phenotypes in Asian elephants *Elephas maximus*. Acta Theriologica 5 (Suppl.), 135–153.

Lair, R.C., 1997. Gone Astray: The Care and Management of the Asian Elephant in Domesticity. FAO Regional Office for Asia and the Pacific, Bangkok.

Lamps, L.W., Smoller, B.R., Goodwin, T.E., Rasmussen, L.E.L., 2004. Hormone receptor expression in interdigital glands of the Asian elephant (*Elephas maximus*). Zoo Biology 23, 463–469.

Litchfield, P., 2005. Leaders and matriarchs – a new look at elephant social hierarchies. International Zoo News 52, 338–339.

Locke, P., 2011. The ethnography of captive elephant management in Nepal: a synopsis. Gajah 34, 32–40.

Lohanan R., 2002. The elephant situation in Thailand and a plea for co-operation. In: Baker I, Kashio M., (Eds.). Giants on Our Hands: Proceedings of the International Workshop on the Domesticated Asian Elephant: February 5–10, 2001; Bangkok, Thailand. FAO Regional Office for Asia and the Pacific, Bangkok, Thailand, pp. 231–238.

Loizi, H., Goodwin, T.E., Rasmussen, L.E.L., Whitehouse, A.M., Schulte, B.A., 2009. Sexual dimorphism in the performance of chemosensory investigatory behaviours by African elephants (*Loxodonta africana*). Behaviour 146, 373–392.

Lucas, C., Stanyon, B., 2017. Improving the welfare of African elephants *Loxodonta africana* in zoological institutions through enclosure design and husbandry management: an example from Blair Drummond Safari and Adventure Park. International Zoo Yearbook 51, 248–257.

Mainka, S.A., Lothrop, C.D., 1990. Reproductive and hormonal changes during the estrous cycle and pregnancy in Asian elephants (*Elephas maximus*). Zoo Biology 9, 411–419.

Mainka, S.A., Cooper, R.M., Black, S.R., Dierenfeld, E.S., 1994. Asian elephant (*Elephas maximus*) milk composition during the first 280 days of lactation. Zoo Biology 13, 389–393.

Malhi, G., Keshavan, M., 2008. Thinking about thinking: a self-portrait by a Thai elephant. Acta Neuropsychiatrica 20, 269–270. <https://doi.org/10.1111/j.1601-5215.2008.00325.x>.

Mallapur, A., Ramanathan, A., 2009. Differences in husbandry and management systems across ten facilities housing Asian elephants *Elephas maximus* in India. International Zoo Yearbook 43, 189–197. <https://doi.org/10.1111/j.1748-1090.2008.00077.x>.

Mar, K.U., Lahdenpera, M., Lummaa, V., 2012. Causes and correlates of calf mortality in captive Asian elephants (*Elephas maximus*). PLoS ONE 2012 (7), e32335.

Mellen, J.D., Stevens, V.J., Markowitz, H., 1981. Environmental enrichment for servals, Indian elephants and Canadian otters at Washington Park Zoo, Portland. International Zoo Yearbook 21, 196–201.

Menargues, A., Urios, V., Mauri, M., 2008. Welfare assessment of captive Asian elephants (*Elephas maximus*) and Indian rhinoceros (*Rhinoceros unicornis*) using salivary cortisol measurement. Animal Welfare 17, 305–312.

Mikota, S.K., Maslow, J.N., 2011. Tuberculosis at the human-animal interface: an emerging disease of elephants. Tuberculosis 91, 208–211.

Millspaugh, J.J., Burke, T., Van Dyk, G., Slotow, R., Washburn, B.E., Woods, R.J., 2007. Stress response of working African elephants to transportation and safari adventures. Journal of Wildlife Management 71, 1257–1260.

Morris, M.D., 1986. Large scale deceit: deception by captive elephants. In: Mitchell, R.W., Thompson, N.S. (Eds.), Deception: Perspectives on Human and Nonhuman Deceit. State University of New York, Albany, NY, pp. 183–191.

Mouttham, L.L., Buhr, M., Freeman, E.W., Widowski, T.M., Graham, L.H., Brown, J.L., 2011. Interrelationship of serum testosterone, dominance and ovarian cyclicity status in female African elephants. Animal Reproduction Science 126, 115–121.

Mumby, H.S., Courtiol, A., Mar, K.U., Lummaa, V., 2013. Climatic variation and age-specific survival in Asian elephants from Myanmar. Ecology 94, 1131–1141.

Mumby, H.S., Mar, K.U., Hayward, A.D., Htut, W., Htut-Aung, W., Lummaa, V., 2015a. Elephants born in the high stress season have faster reproductive ageing. Science Reports 5, Article number: 13946 (2015).

Mumby, H.S., Mar, K.U., Thitaram, C., Courtiol, A., Towiboon, P., Min-Oo, Z., et al., 2015b. Stress and body condition are associated with climate and demography in Asian elephants. Conservation Physiology 3 (1), cov030. <https://doi.org/10.1093/conphys/cov030>.

Nissani, M., 2006. Do Asian elephants (*Elephas maximus*) apply causal reasoning to tool-use tasks? Journal of Experimental Psychology. Animal Behavior Processes 32, 91–96. <https://doi.org/10.1037/0097-7403.32.1.91>.

Nogge, G., 2012. Elefantenhaltung in Europa und Indien – ein Vergleich (Elephant keeping in Europe and India – a comparison). Der Zoologische Garten 81, 231–238.

Norkaew, T., Brown, J.L., Bansiddhi, P., Somgird, C., Thitaram, C., Punyapornwithaya, V., et al., 2019. Influence of season, tourist activities and camp management on body condition, testicular and adrenal steroids, lipid profiles, and metabolic status in captive Asian elephant bulls in Thailand. PLoS ONE 14 (3), e0210537.

Owen-Smith, N., Kerley, G.I.H., Page, B., Slotow, R., Van Aarde, R.J., 2006. A scientific perspective on the management of elephants in the Kruger National Park and elsewhere: elephant conservation. South African Journal of Science 102, 389–394.

Payne, K.B., Langbauer Jr., W.R., Thomas, E.M., 1986. Infrasonic calls of the Asian elephant (*Elephas maximus*). Behavioral Ecology and Sociobiology 18, 297–301.

Plotnik, J.M., Shaw, R.C., Brubaker, D.L., Tiller, L.N., Clayton, N.S., 2014. Thinking with their trunks: elephants use smell but not sound to locate food and exclude nonrewarding alternatives. Animal Behaviour 88, 91–98.

Poole, J., Tyack, P.L., Stoeger-Howarth, A.S., Watwood, S., 2005. Elephants are capable of vocal learning. Nature 434, 455–456.

Rahman, H., Dutta, P.K., Dewan, J.N., 1988. Foot and mouth disease in elephant (*Elephas maximus*). Journal of Veterinary Medicine, Series B 35, 70–71. <https://doi.org/10.1111/j.1439-0450.1988.tb00468.x>.

Rasmussen, L.E.L., Hall-Martin, A.J., Hess, D.L., 1996. Chemical profiles of male African elephants (*Loxodonta africana*): physiological and ecological implications. Journal of Mammalogy 77, 422–439.

Rees, P.A., 2000. Are elephant enrichment studies missing the point? International Zoo News 47, 369–371.

Rensch, B., 1957. The intelligence of elephants. Scientific American 196, 44–49.

Rensch, B., Altevogt, R., 1954. Zaehmung und Dressurleistungen indischer Arbeitselefanten (Taming and training performances of Indian work elephants). Zeitschrift fur Tierpsychologie 11, 497–510.

Riddle, H.S., Riddle, S.W., Rasmussen, L.E.L., Goodwin, T.E., 2000. First disclosure and preliminary investigation of a liquid released from the ears of African elephants. Zoo Biology 19, 475–480.

Ringis, R., 1996. Elephants of Thailand in Myth, Art and Reality. Oxford University Press, Kuala Lumpur.

Romano, M., Palombo, M.R., 2017. When legend, history and science rhyme: Hannibal's war elephants as an explanation to large vertebrate skeletons found in Italy. Historical Biology 29, 1106–1124.

Rosen, L.E., Hanyire, T.G., Dawson, J., Foggin, C.M., Michel, A.L., Huyvaert, K.P., et al., 2017. Tuberculosis serosurveillance and management practices of captive African elephants (*Loxodonta africana*) in the Kavango-Zambezi Transfrontier Conservation Area. Transboundary and Emerging Diseases 65, e344–e354. <https://doi.org/10.1111/tbed.12764>.

Rossman, Z.T., Padfield, C., Young, D., Hart, L.A., 2017. Elephant-initiated interactions with humans: individual differences and specific preferences in captive African elephants (*Loxodonta africana*). Frontiers in Veterinary Science 4, 1–10.

Rübel, A., Zingg, R., 2015. Kaeng Krachan Elefantenpark für Asiatische Elefanten (*Elephas maximus* Linnaeus, 1758) im Zoo ZürichKaeng Krachan Elephant Park for Asian elephants (*Elephas maximus* Linnaeus, 1758) at Zürich Zoo. Der Zoologische Garten 84, 1–12.

Saragusty, J., Hildebrandt, T.B., Natan, Y., Hermes, R., Yavin, S., Goeritz, F., et al., 2005. Effect of egg-phosphatidylchlorine on the chilling sensitivity and lipid phase transition of Asian elephant (*Elephas maximus*) spermatozoa. Zoo Biology 24, 233–245.

Schiffmann, C., Clauss, M., Hoby, S., Hatt, J., 2017. Visual body condition scoring in zoo animals – composite, algorithm and overview approaches. Journal of Zoo and Aquarium Research 5, 1–10. <https://doi.org/10.19227/jzar.v5i1.252>.

Schiffmann, C., Knibbs, K., Clauss, M., Merrington, J., Beasley, D., 2018. Unexpected resting behaviour in a geriatric zoo elephant. Gajah 48, 30–33.

Schmidt-Burbach, J., RonFot, D., Srisangiam, R., 2015. Asian elephant (*Elephas maximus*), Pig-Tailed Macaque (*Macaca nemistrina*) and Tiger (*Panthera tigris*) populations at tourism venues in Thailand and aspects of their welfare. PLoS ONE <https://doi.org/10.1371/journal.pone.0139092>.

Schulte, B.A., Feldman, E., Lambert, R., Oliver, R., Hess, D.L., 2000. Temporary ovarian inactivity in elephants: relationship to status and time outside. Physiology & Behavior 71, 123–131.

Schwarzenberger, F., Strauss, G., Hoppen, H.-O., Schaftenaar, W., Dieleman, S.J., Zenker, W., et al., 1997. Evaluation of progesterone and 20-oxo-progestagens in the plasma of Asian (*Elephas maximus*) and African (*Loxodonta africana*) elephants. Zoo Biology 16, 403–413.

Sedgewick, C.J., 1998. The challenge of elephant health care. Zoo Biology 17, 153–155.

Sharma, R., Krishnamurthy, K.V., 1984. Behavior of a neonate elephant (*Elephas maximus*). Applied Animal Behaviour Science 13, 157–161.

Shepherd, N., 2002. How ecotourism can go wrong: the cases of SeaCanoe and Siam Safari, Thailand. Current Issues in Tourism 5, 309–318.

Simonet, P., 2000. Self-recognition in Asian elephants: preliminary findings. Elephant 2, 103.

Stagni, E., Normando, S., de Mori, B., 2007. Distances between individuals in an artificial herd of African elephants (*Loxodonta africana africana*) during resource utilisation in a semi-captive environment. Research in Veterinary Science 113, 122–129.

Stüwe, M., Abdul, J.B., Nor, B.M., Wemmer, C.M., 1998. Tracking the movements of translocated elephants in Malaysia using satellite telemetry. Oryx 32, 68–74.

Taya, K., Komura, H., Kondoh, M., Ogawa, Y., Nakada, K., Watanabe, G., et al., 1991. Concentrations of progesterone, testosterone and estradiol-17β in the serum during the estrous cycle of Asian elephants (*Elephas maximus*). Zoo Biology 10, 299–307.

Thitaram, C., Brown, J.L., 2018. Monitoring and controlling ovarian activity in elephants. Theriogenology 109, 42−47.

Thitaram, C., Chansitthiwet, S., Pongsopawijit, P., Brown, J.L., Wongkalasin, W., Daram, P., et al., 2009. Use of genital inspection and female urine tests to detect oestrus in captive Asian elephants. Animal Reproduction Science 115, 267−278.

Tipprasert, P., 2002. Elephants and ecotourism in Thailand. In: Baker, I., Kashio, M., (Eds.), Giants on Our Hands: Proceedings of the International Workshop on the Domesticated Asian Elephant: February 5−10, 2001; Bangkok, Thailand. FAO Regional Office for Asia and the Pacific, Bangkok, Thailand, pp. 157−171.

Vanitha, V., Thiyagesan, K., Baskaran, N., 2009. Socio-economic status of elephant keepers (mahouts) and human-captive elephant conflict: a case study from the three management systems in Tamil Nadu, Southern India. Gajah 30, 8−12.

Varadharajan, V., Krishnamoorthy, T., Nagarajan, B., 2010. Social life of captive Asian elephants (*Elephas maximus*) in Southern India: implications for elephant welfare. Journal of Applied Animal Welfare Science 14, 42−58. <https://doi.org/10.1080/10888705.2011.527603>.

Veasey, J., 2006. Concepts in the care and welfare of captive elephants. International Zoo Yearbook 40, 63−79.

Vinod, T., Cheeran, J., 1997. Activity time budget of Asian elephants (*Elephas maximus*) in Idukki Wildlife Sanctuary, Kerala, South India. Indian Forester 123, 948−951.

Wang, L., Lin, L., He, Q., Zhang, J., Zhang, L., 2007. Analysis of nutrient components of food for Asian Elephants in the wild and in captivity. Frontiers of Biology in China 2, 351−355.

Watson, P.F., D'Souza, F., 1975. Detection of oestrus in the African elephant (*Loxodonta africana*). Theriogenology 4, 203−209.

Weissenböck, N.M., Schwammer, H.M., Ruf, T., 2009. Estrous synchrony in a group of African elephants (*Loxodonta africana*) under human care. Animal Reproduction Science 113, 322−327.

Weissenböck, N.M., Arnold, W., Ruf, T., 2011. Taking the heat: thermoregulation in Asian elephants under different climatic conditions. Journal of Comparative Physiology B 182, 311−319.

Wisniewska, M., Freeman, E.W., Schulte, B.A., 2015. Behavioural patterns among female African savannah elephants: the role of age, lactational status, and sex of the nursing calf. Behaviour 152, 1719−1744.

Wong, E.P., Yon, L., Purcell, R., Walker, S.L., Othman, N., Saaban, S., et al., 2016. Concentrations of faecal glucocorticoid metabolites in Asian elephant's dung are stable for up to 8 h in a tropical environment. Conservation Physiology 4 (1), cow070. <https://doi.org/10.1093/conphys/cow070>.

Wuestenhagen, A., Weisz, I., Schwammer, H., 2000. Sleeping behaviour of six African elephants (*Loxodonta africana*) in the Schonbrunn Zoological Garden, Vienna, Austria. Der Zoologische Garten 70, 253−261.

Yon, L., Kanchanapangka, S., Chaiyabutr, N., Meepan, S., Lasley, B., 2007. A longitudinal study of LH, gonadal and adrenal steroids in four intact Asian bull elephants (*Elephas maximus*) and one castrate African bull (*Loxodonta africana*) during musth and non-musth periods. General and Comparative Endocrinology 151, 241−245.

Yon, L., Faulkner, B., Kanchanapangka, S., Chaiyabutr, N., Meepan, S., Lasley, B., 2010. A safe method for studying hormone metabolism in an Asian elephant (*Elephas maximus*): accelerator mass spectrometry. Zoo Biology 29, 760−766.

Yon, L., Williams, E., Harvey, N.D., Asher, L., 2019. Development of a behavioural welfare assessment tool for routine use with captive elephants. PLoS ONE 14 (2), <https://doi.org/10.1371/journal.pone.0210783>.

Zaw, U.K., 1997. Utilization of elephants in timber harvesting in Myanmar. Gajah 17, 9−22.

Index

For individual zoos and safari parks referred to in the text see under 'Zoos'. For national parks, reserves, wildlife sanctuaries and elephant sanctuaries referred to in the text see under 'National park', 'Reserve', 'Wildlife sanctuary', and 'Elephant sanctuary', respectively. All named circuses are listed under 'Circus'. For elephants referred to by name in the text, figures, and tables see alphabetical list under 'Elephants named in the text'. This list excludes the main herd at Chester Zoo (*Chang*, *Thi*, *Sheba*, *Maya*, *Kumara*, *Jangoli*, *Upali* and *Sithami*) because of the large number of references to these animals.

Page numbers followed by 'f' indicates figures, 't' indicates tables and 'b' indicates boxes.

A

Accelerometer, 145, 210
Accidents involving elephants, 69, 127, 247
Acclimatisation, 130
Accommodation for elephants, lack of, 322, 323f
Acid-insoluble ash (AIA) marker, 137
Activity budgets, 39–50, 40–42f
Acyclicity, 59, 91, 98, 162
   causes of, 106, 111, 211, 322
   dominance and, 68
   frequency of, 106–108, 107t
   welfare indicator, as, 174t
Adoption. *See* Calf, adoption of
Adventures with Elephants, 122, 126
African Elephant Status Report, 8
African Parks, 325
Age determination, 154
Agonistic behaviour, 68–79, 70f, 72–75f, 73t, 77–78f
Alexander the Great, 11f, 12, 15
Alloparenting, 95, 96–97b, 96t, 97f
Ambassador, elephant as, 14–15
Amboseli Elephant Research Project, 2, 36, 67, 92, 114, 156, 260
American Society for the Prevention of Cruelty to Animals (ASPCA), 27, 272
Anal flap, 176
Anatomical studies, 18–23, 19–20f, 23f
Animal Defenders International, 272, 277
Animal Liberation, 259, 265t
Animal Welfare Board of India, 309
Ankus, 173, 248, 261f, 276f
   early use of, 275
   ban on use of, 276f, 277
   restricted use in England, 277
Anticipatory behaviour, 144, 144f, 193, 202, 206
Appeasement behaviour, 69, 75–79, 77–77b, 77–78f

Aristotle, xv, 15, 24, 207
   observations on elephants by, 16–19b
ARK 2000, 274
Army, British, use of elephants by, 12, 13f
Arthritis, 215–216, 277
Artificial insemination (AI), 109–112, 110f
   advantages and disadvantages of, 111t
   anatomical barriers to, 109
   conceptions produced by, 109
   births produced by, 110f
   frozen semen, use of in, 112
ASPCA. *See* American Society for the Prevention of Cruelty to Animals (ASPCA)
Assam Haathi Project, 306
Assimilation efficiency. *See* Gross assimilation efficiency
Associations between elephants, 59–63, 61–62f
Association index, 60b, 61–62f, 316
Association of Zoos and Aquariums (AZA), 54, 57, 68, 107, 163, 170, 176, 194, 196, 199, 201, 208t, 214–215, 217, 220, 226b, 234t, 235–236, 244, 252, 271, 273b, 278, 287–288, 295, 305, 307–308
Auditory enrichment. *See* Enrichment, auditory
Autism, elephants as therapy for, 320–321
AZA. *See* Association of Zoos and Aquariums (AZA)

B

Baboons and elephants, 148–149
Bark
   damage to, 150, 150f
   enrichment and, 198, 198f
   feeding on, 133–134
   substrate, used as, 236
Barriers for enclosures
   standards for, 225–226b
   types of, 222, 223–224f

Basic needs test, 260
Battles involving elephants, 9–12, 11t, 12f
Behaviour sampling
   methods for. *See* Focal sampling; Scan sampling
   problems with, 36–38, 38f
Benedict, F. G., 2, 25, 135–136, 138
   apparatus used by, 26f
BIAZA. *See* British and Irish Association of Zoos and Aquariums (BIAZA)
Big Game Parks, 278
Billboard, 58
Biographical life, 260
Biomechanics, 208–210, 209f, 216, 216f
Birth rates, 155–157
Blumenbach, J. F., 4–5, 5f
Body condition score, 176–178, 177–178t, 177f
Bond group, 52, 53f
Born Free Foundation, 264, 265t, 277, 279b, 304
Bowman, David, 315
Brambell Report. *See Report of the technical committee to enquire into the welfare of animals kept under intensive livestock husbandry systems*
British Alpine Hannibal Expedition, 12
British and Irish Association of Zoos and Aquariums (BIAZA), 39, 57, 68, 170, 172, 206–207, 234, 234t
Browse. *See* Enrichment, browse as
Bull pen, 226
   Chester Zoo's, 148f
   introduction to new, 147
   size of, 234t
Bullhook. *See* Ankus

C

Calf
   adoption of, 99
   allomothering of. *See* Alloparenting
   development of, 93
   hand-rearing of, 93, 98–99, 100f, 158f, 244f
   height of, 93
   rejection of, 93, 95, 98–99, 100f, 158f, 243, 244f
   stillborn, 92–93, 155, 158f, 161, 175, 295, 309, 324
   weight of, 91, 93
Calf mortality, 155–162
   allomothers effect on, 95
   effect of mother's death on, 161
   law referring to, 273b
   orphanage, in, 309
   welfare indicator, as, 174–175, 261f
Calorific restriction, 211
Calving interval, mean. *See* Mean calving interval
Capabilities approach, 262
Captive breeding in range states, 309
Caregivers as sources of information, 38–39

Case law
   Born Free USA, et al., v. Gale Norton et al., 279b, 304
   Naresh Kadyan v/s Chief Secretary, Govt. of Rajasthan & Others, 277
   Nonhuman Rights Project, Inc. v. Stanley, 275b
Central Zoo Authority (CZA), 225b, 270, 308, 315
Cervix, 109, 112
Chaining of elephants, 18, 44f, 82, 127, 191, 193–194, 205, 205f, 226, 228f, 261f, 270, 275, 277, 307f
Chastisement, 71, 71t, 72f, 74–75b, 74f
Chemical signalling. *See* Communication, chemical
China
   as ivory market, 267, 325
   shipment of elephants to, 278–279, 281b
Chipperfield, Jimmy, 162
Circus
   Barnum and Bailey, 20–21, 21–22b, 22f, 250
   Bobby Roberts' Super, 277
   Carson & Barnes, 118
   Circus Roncalli, 272
   Gottani (Circo Gottani), 250
   Ringling Bros. and Barnum & Bailey, 14f, 262, 272
Circus, justification for keeping elephants in, 363f
CITES. *See* Legislation, Convention on International Trade in Endangered Species of Wild Fauna and Flora 1973 (CITES)
Clade, 6
Clan, 53f, 118, 156
Climate change, 321
Cloning, 320
Cognition, 76, 113–117
Commands used to direct elephants, 128t
Communication
   chemical, 83t, 88, 91, 100, 106, 120–123, 121–122f
   seismic, 123
   tactile, 36, 90, 123, 127
   vocal, 16b, 62, 86b, 87f, 88, 119–120, 120f, 174t
Comparable life test, 260
Condition scoring. *See* Body condition score
Conservation role of elephants, 287–312
Conservation status of elephants, 7–8
Conservation triage, 311
Consumptive use, 324–325
Containment standards for elephants. *See* Barriers for enclosures, standards for
Contrafreeloading, 43, 44f
Cooperation in elephants, 95, 302f. *See also* Alloparenting; Defensive circle/ring
   experiments on, 127–128
Coping hypothesis, 185
Corpus luteum, 81, 90
Cortisol, 90, 108, 171, 181, 185, 191, 244, 250, 258
Courtship. *See* Mating

COVID-19, 327
Crate training. *See* Training, crate

D

Darwin Initiative, 286, 306
Defaecation
   ecological significance of, 139
   rate of, 135
   synchronisation of, 139, 139t
Defensive circle/ring. *See* Protective formation
Defra. *See* Department for Environment, Food and Rural Affairs (Defra)
Department for Environment, Food and Rural Affairs (Defra), 70, 171–172, 270, 283b
Digestibility, 135, 137, 210–211
Digging, 83, 151–153f, 182, 199, 201. *See also* Dusting behaviour
   as displacement behaviour, 83
Discrimination
   between objects and quantities, 116
   visual, 123–124
Displacement behaviour, 62, 83
Dissection of elephant, first in British Isles, 18, 19f
Distress, 170, 180–183, 185
Diurnal behaviour patterns, 41–42f
DNA analysis, 4–6, 164–165, 300, 316
Domestication, 310
   incipient, 9
Dominance hierarchy, 63, 66–69, 145
Donlan, C. J., 315
Drug doses, 176, 212
Dusting behaviour, 45–46, 45f
   temperature, effect on, 132, 132f
Dyad, 60b, 61f, 73f
Dystocia, 109

E

Ear-flapping, 64, 68, 86b, 131
EAZA. *See* European Association of Zoos and Aquaria (EAZA)
Education, use of elephants in, 300–306
EEHV. *See* Herpesvirus
EEP. *See* European Endangered species Programme (EEP); European Association of Zoos and Aquaria Ex-situ Programme (EEP)
Effective population size, 111t, 112, 165–166
Egypt, elephants in, 8–9
Electric fence, 45, 139, 150, 222, 224f, 225–226b, 247, 311, 316
Electrocution of elephant, 26–27
Electroejaculation, 108–109
Elephant Advisory Group, 171
Elephant-assisted therapy, 320–321

Elephant Bill, 2, 13, 82
Elephant centres
   Ringling Bros. Center for Elephant Conservation, 91
   Seblat Elephant Conservation Center, 43, 133
   Thai Elephant Conservation Center, 109, 205, 209–210
Elephant exhibits, new, 237–238, 240t
Elephant-Free Zoos, 264, 265t
Elephant houses. *See also* Enclosure
   architectural styles of, 231t
   history of, 223, 223f, 226–229, 227–229f, 230b, 231f, 231t, 232–233f
   size of, 229, 231, 234, 234t, 235f, 236–237
Elephant inspector, 172
Elephant orphanages. *See* Pinnawala Elephant Orphanage; Sheldrick, Dame Daphne
Elephant polo, 13, 127, 272, 274
Elephant popularity, 261f, 290–294, 294f
Elephant populations
   culling of, in wild, 69, 92, 154, 278, 297–298, 304
   distribution of
      in zoos, 220–221, 220t
      in wild, 7–8
   size of
      in zoos, 220–221, 221t
      in wild, 219–220
   sustainability of, in zoos, 161–162, 296–297
Elephant profile, 208
Elephant rides, 15f, 55f, 109, 126, 180f
   end of holidays promoting, 326
Elephant sanctuaries. *See also* Wildlife sanctuaries
   Elephant Haven European Elephant Sanctuary, 317–318
   Elephant Refuge North America, 317
   Elephant's Lake, 318
   Riddle's Elephant and Wildlife Sanctuary, 91
   Santuário de Elefantes Brasil, 317
   The Elephant Sanctuary, in Tennessee, 118, 278
Elephant species, 3–7
Elephant subspecies, 4–7, 164
   hybrids between, 163–164, 165t
Elephant Welfare Initiative (EWI), 172
Elephants named in the text
   Adega, 134
   Anne, 277–278
   Assam, 32f, 158f
   Billy, 265t
   Brahmaputri, 109
   Cefrose, 139t
   Celeste, 55f
   Chishuru, 122f
   Chistine, 55f, 291f
   Chova, 122f
   Dicksie, 154

Elephants named in the text (*Continued*)
  Don Pedro, 20–21
  Emmett, 110f
  Esha, 63, 64f
  Ganesh Vijay, 110f
  Gertie, 271
  Happy, 115, 147
  Jack, 18–19, 19f
  Jap, 25, 26f, 135
  Jean, 141
  Jenny, 118, 265t
  Jubilee, 54, 56f, 291
  Jumbo (Alps), 12
  Jumbo (London Zoo), 2, 21, 21–22b, 22f, 169
    transportation of, 250, 252, 253f
  Karha, 56t, 93, 99, 158t, 243, 244f
  Kate, 63, 64f
  Koshik, 120, 120f
  Lucky, 265t
  Ma Shwe, 94
  Maharajah, 22–23, 23f
  Manniken, 53
  Maxine, 115, 146
  Mimbu, 63, 64f
  Modoc, 3
  Molly, 53
  Motty, 56f, 164
  Mussina, 122f
  Nandipa, 316
  Nandita, 158f
  Ndovu, 134
  Nobby, 164, 223
  Noojahan, 63, 64f
  Panya, 141
  Patti (Patty), 115, 147
  Peter Pan, 154
  Pickaninny, 259
  Rajah, 55f
  Shirley, 118
  Susi, 265t
  Tara, 63, 64f
  Tissa, 139t
  Topsy, 26–27
  Tusko, 27
  Tuy Hoa, 177
  Valli, 206, 271
  Victor, 44f, 153f
  Wendy, 54–55f, 291f
  Zebi, 55f
ElephantVoices, 36, 306, 314
Eltringham, S. Keith, 2, 99, 219, 313–314
  speech to British Association for the Advancement of Science, 313

Enclosure
  indoor-outdoor preferences, 146–147
  size of. *See* Elephant houses
  use of, 142–146, 143t, 144f
Endocrine monitoring, 89–90, 106–109
Endoscope-guided catheter, 112
Energetics, 140–141, 141f
Enrichment
  auditory, 206
  browse as, 198
  calves as, 206–207
  football as, 197
  fruitsicle as, 195, 196f
  music as, 205–206
  painting as, 205
  social contact as, 207
  toys, interactive, 206
  training as, 207–208, 302–303f
  tyre, 195f
  water as, 202, 204f
Episiotomy, 109
ESU. *See* Evolutionarily significant unit (ESU)
Ethics of keeping elephants in captivity, 259–264, 261f, 263f
Ethogram, 34–39
  aggressive behaviours, for, 71
  comparisons of different types of, 35t
  courtship behaviour, for, 84–86, 85f, 86t
European Association of Zoos and Aquaria (EAZA), 170, 220, 304
European Association of Zoos and Aquaria Ex-situ Programme (EEP), 156, 165t, 166, 217, 220, 296, 322, 323f
European Endangered species Programme (EEP), 111, 156, 164, 220, 295–297
Evolutionarily significant unit (ESU), 6, 163
Exhibits. *See* Elephant exhibits, new; Elephant houses; Enclosure
Experimental archaeology, 12
Explosives, detection of, 122
Eyesight. *See* Visual acuity

F

Family group, 51, 53f
  aggression in, 69
  dominance relationships in, 66–67
  effect on calf production of, 59
Fecundity, 43–45
Feeding behaviour, 43–45, 44f, 134–136
Feeding devices
  feeder ball, 201f
  feeder wall, 199f
  suspended feeders, 117f, 198, 200f

Fibreoptoscope, 109
Five freedoms, 169–170
Flagship species, elephant as, 14
Floor
   health, effect on, 209, 214
   lying rest, effect on, 48
   preferences of elephants, 146–147, 236
   types of, 146, 201, 202f, 236, 237f
Focal sampling, 36
Foetotomy, 109
Follicular cyst. *See* Ovarian cyst
Follicular phase, 89, 91
Food consumption, 134–138
Food passage time, 135, 137–138
Food supplementation, 134, 139–140, 140t
Foot, role in communication, 123
Foot disease, 214–216, 214–215f
Foot pressure patterns, 214, 216
Force platform, 130, 210, 216, 216f
Forest elephant
   taxonomy of, 5–6
   zoo population of, 6
Forest department, use of data from, 158
Four Paws, 318
Free contact, 107, 208, 214f, 225b, 248, 250t, 261f, 314
Friendship among elephants, 59, 61f
Fritz, Stephen. *See* Stephen Fritz Enterprises, Inc.
Fruitsicle. *See* Enrichment, fruitsicle as
Fundraising for in situ projects, 307–308

G

Gait
   characteristics of, 209f
   energetics of, 209
   measurement of, 210
   photographic study of, 209, 209f
   welfare and, 173, 183, 210
Ganesh, 10, 11f
Genetic bottleneck, 318
Genetic diversity, 5–6
Genetic drift, 318
Genetics, 163–167
Gestation, 17, 24, 90–91, 93
Global Positioning System (GPS), 210
Gonadotropin-releasing hormone (GnRH), 108
GPS. *See* Global positioning system (GPS)
GPS collar, 145
Great Elephant Census, 8
Greeting ceremony, 63f, 118, 245f
Gross assimilation efficiency, 135–136, 136f

H

Habeas corpus, 274, 275b
Ha-ha, 32b, 45, 222, 224f

Hannibal, 11t, 12, 12f
Harrods, sale of elephants by, 271
Heat storage, 131
Henry III, 9, 10f
Herpesvirus, 217, 299f
History of Animals, 15, 16–18b
Holotype, 4
Hot wire. *See* Electric fence
Housing. *See* Elephant houses; Enclosure; Floor
Hoyte, John, 12
Human chorionic gonadotropin, 108
Human scent, detection by elephant, 122, 122f
Human speech imitation, 120, 120f
Hybridisation, 56f, 150, 163–164, 165t

I

Identification of individuals, 29–30, 30f
In Defense of Animals, 264
Inbreeding, 6, 164–166
Inbreeding depression, 164
Inbreeding rate, 166
Infanticide, 92
Infrared thermography, 130–131
Infrasonic communication, 119
Inhibin, 90
In situ conservation, financial support for, 284, 284–286b, 286
Insightful behaviour, 116, 117f
Insurance population, 112, 261f, 287, 294–298
Intensive protection zone (IPZ), 324–325
Interbirth interval. *See* Mean calving interval
Interdigital glands, 121
International Species Information System (ISIS), 54, 58t, 220–221, 220–220t
International Union for the Conservation of Nature and Natural Resources (IUCN)
   elephant taxa recognised by, 6, 164, 165t
   elephant status reports of, 7–8
   captive breeding and, 268, 269b
   CITES and, 280–281b
   UN Convention on Biodiversity and, 268
   zoos, cooperation with, 287
Introduction of elephants to non-range states, 315
ISIS. *See* International Species Information System (ISIS)
IUCN. *See* International Union for the Conservation of Nature and Natural Resources (IUCN)
Ivory trade, 5f, 8–9, 266–267, 286b, 298, 325

J

Jacobson's organ, 84b, 121f
Joint family, 52
Journal of the Bombay Natural History Society, 24

K

Keeper-elephant bonds, 243–244, 244f, 245f
Keepers
  cruelty caused by, 275, 277–278
  deaths of, 246–247
  questionnaires completed by. *See* Questionnaires, use of in research
  safety of. *See* Protected contact
Kenya Wildlife Service (KWS), 316
Khedda, kheddah, 310
KWS. *See* Kenya Wildlife Service (KWS)

L

Lameness, 125
Landolt-C visual acuity test, 123–124
Landscape, elephant damage to, 150–153, 150–153f
Lawrence, Lorenzo, 22–23, 23f
Lectotype, 4
Legal cases. *See* Case law
Legal personality, 274
Legislation. *See also* Legal cases
  African Elephant Conservation Act of 1989 (USA), 284–285b
  Animal Welfare Act 2006 (UK), 234, 277–278
  Asian Elephant Conservation Act of 1997 (USA), 284–285b, 295
  Biosecurity Act 2014 (Queensland), 315
  British North America Act 1867 (Canada), 274
  California Senate Bill (Chapter 234 No. 1062), 276b
  Central Zoo Authority Letter No. 7-5/2006-CZA (Vol.II)(India), 270
  Convention on International Trade in Endangered Species of Wild Fauna and Flora (CITES) 1973
    appendices, 7–8, 278, 279t
    appropriate and acceptable destinations, 8, 279t, 280b, 282
    CoP17 (CITES), 8, 279t, 280b
    CoP18 (CITES), 8, 280–281b, 281f
  Council Directive 1999/22/EC of 29 March 1999 on the keeping of wild animals in zoos (The Zoos Directive), 282, 297, 300
  Elephant Protection Act 2017 (New York), 273
  Endangered Species (Import and Export) Act 1976 (United Kingdom), 271
  Exhibited Animals Protection Regulation, 2005 (New South Wales), 282
  Fauna and Flora Protection Ordinance 1937 (Sri Lanka), 271
  Ivory Act 2018 (UK), 267
  Kerala Captive Elephants (Management and Maintenance) Rules 2003 (India), 282
  Kerala Captive Elephants (Management and Maintenance) Rules 2012 (India), 202, 236, 252, 282
  Kerala Forests and Wildlife Department. Circular No.01/2019 (India), 283
  MAF Biosecurity New Zealand Standard 154.03.04: Containment facilities for zoo animals, 225b
  Performing Animal (Registration) Rules, 2001 (India), 274
  Planning (Listed Buildings and Conservation Areas) Act 1990, 230b
  Policy on the Management of Solitary Elephants in New South Wales, 282
  Rhode Island General Laws § 4-1-43. Use of bullhooks or similar devices on elephants prohibited, 275
  Standard for Zoo Containment Facilities 2018, Environmental Protection Authority (EPA), New Zealand, 225
  Te Awa Tupua (Whanganui River Claims Settlement) Act 2017 (New Zealand), 275b
  United Nations Convention on Biological Diversity 1992, 264, 268, 268b, 286, 304, 306, 308
  Wild Animals in Travelling Circuses (Scotland) Act 2018 (UK), 272
  Wildlife (Protection) Act 1972 (India), 282, 308
  Zoo Licensing Act 1981 (UK), 270, 282, 283b
Leibniz Institute for Zoo and Wildlife Research, 112, 123
LH. *See* Luteinising hormone (LH)
Life expectancy, 154–155, 269b
Life tables, 155, 160
Lifespan, 154–155, 161, 262
Linnaeus, Carl, 4, 4f
Logging camps, 13, 24, 123, 127, 128t, 133, 158–162, 159–160t, 179, 211, 246, 307f, 309, 318, 320, 327
Longevity. *See* Lifespan
LSD, death of elephant from, 27
Luteal phase, 90–91
Luteinising hormone (LH), 89
  double peak in, 89, 112

M

Madras Forest Department, 158
Mahouts
  apprenticeship of, 245
  deaths among, 247
  transfer of traditional knowledge in, 245
Mammoth, woolly, resurrection of, 320
Management Guidelines for the Welfare of Zoo Animals: Elephants, 170, 172, 206, 234t

Mating
    attemps at in juveniles, 103–104b, 103t, 104–105f, 104–105t
    early studies of, 81–83
    ethogram of, 84–86, 85–86f
    interference with, 87, 87f
Mating pandemonium, 85f, 86b, 86t, 87, 87f, 101b, 207
Matriarch as repository of knowledge, 59, 118
Mean calving interval, 154, 157–159, 158f, 298, 322
Memory, 117–118
Metabolic heat production, 131
Mirror self-recognition, 115, 115f
Mitochondrial DNA, 6, 164
Mother
    mortality of and calf survival, 161
    oldest in captivity, 157
Mother hypothesis, 161
Mother-offspring unit, 53f
Mounting behaviour. See Mating
Mudumalai Sanctuary Elephant Camp, 159
Mullen, Allen, 18, 19f
Multi-species exhibits, 148–149
Musth
    aggression and, 69, 77b, 77f, 92, 247
    characteristics of, 88, 88f
    control of, 89
    dominance and, 67
    first description of, 24
Myanmar Timber Enterprise, 76, 108, 160, 318

N
National Elephant Institute, Thailand, 108, 112
National park
    Addo Elephant, 69, 298, 318
    Amboseli, 2, 36, 67, 92, 114, 156, 260
    Chitwan, 9, 13
    Etosha, 67, 132, 299
    Hlane Royal, 278
    Hwange, 278–279, 281b
    Kruger, 69, 123, 178, 299, 316, 318, 325
    Lahugala Kitulana, 46
    Lake Manyara, 2, 65, 133, 157, 200f, 260
    Liwonde, 325
    Meru, 21f, 316
    Murchison Falls, 51, 154
    Nagarahole, 244
    Nairobi, 99
    Queen Elizabeth, 135
    Tarangire, 52f, 92, 118, 150f, 156, 298
    Tsavo (East), 99, 133, 135–136, 138–139, 258
    Udawalawi, 49

Nocturnal behaviour, 36, 46–50
Nonhuman Rights Project (NhRP), 274, 275b
Nulliparous females
    anatomical examination of, 109
    as allomothers, 96

O
Obesity, 210–211, 261f, 299f
Oestrous cycle
    behavioural indicators of, 90–91
    irregular. See Acyclicity
    monitoring of, 89–90
    social factors, influence on, 91
    transfer between zoos, effect of, 108
Olfactory sense, 122, 122f
Orphaned elephants. See also Pinnawala Elephant Orphanage
    introduction to wild herd, 99
Osborn, Henry Fairfield, 3–4
Outbreeding depression, 164
Ovarian cyst, 107–108, 111
Ovulation, 25, 89, 91, 108
Ovulatory phase, 90
Owen, Richard, 18–19

P
Pacinian corpuscles, 123
Pandemonium. See Mating pandemonium
Parasites, 126, 132, 213
Parenting, 94–99, 94f. See also Alloparenting
People for the Ethical Treatment of Animals (PETA)
    elephant incidents in USA, list of, 69, 247, 308
    views of, 264
Performing Animal Welfare Society (PAWS), 264, 274, 311
Personhood, 274, 274–275b
Personality
    definition of, 76
    genetics and, 76, 79
    legal. See Legal personality
    measurement of, 76
    stress response and, 79
PETA. See People for the Ethical Treatment of Animals (PETA)
Pheromones. See Chemical signalling
Phylogeography, 6
Pinnawala Elephant Orphanage, 3, 43, 44f, 57f, 204f, 213f, 309
Pleistocene rewilding, 315
Pointing, 116–117, 118f
Polo. See Elephant polo

Popularity of elephants. *See* Elephant popularity
Population level welfare indices. *See* Welfare indices
Populations, elephant. *See* Elephant populations
Positive reinforcement, 208, 248, 248f, 250
Problem solving. *See* Insightful behaviour
Profile, AZA elephant. *See* Elephant profile
Progesterone, 89–90, 107
Prolactin, 90, 181
Protected contact, 69, 248–250, 250t, 277, 314
  cyclicity, effect on, 107
  safety and, 247
  training for, 248f
  wall, 248–249f
Protective formation, 64–66, 65f
Puberty, 17
Pygmy elephant, 5

Q

Quantity judgement. *See* Discrimination
Questionnaires, use of in research, 39, 68, 76, 160, 174, 185, 305

R

Radford Report, 277
Ragi balls, 212, 212f
Ranching of elephants, 313–314
Red spot test, 114
Reintroduction to the wild, relatedness and, 316
Referential gesturing. *See* Pointing
Rejection of calf. *See* Calf, rejection of
Reproductive behaviour. *See* Mating
Report of the technical committee to enquire into the welfare of animals kept under intensive livestock husbandry systems, 169
Reproductive status. *See* Acyclicity
Reserve
  Kanha Tiger Reserve, 109
  Majete Wildlife Reserve, 325
  Nilgiri Biosphere Reserve, 159
  Nkhotakota Wildlife Reserve, 325
  North Za Ma Yi Forest Reserve, 318
  Pongola Nature Reserve, 36
  Samburu-Buffalo Springs Game Reserve, 125
  Shimba Hills National Reserve, 258
Resting behaviour, 46–50, 47f
Retirement of working elephants, 277–278, 282–283, 317
Rewilding. *See* Pleistocene rewilding
Rhinoceros
  farming of, 314
  white, attacks on, 69
Rides, elephant. *See* Elephant rides
Roosevelt, President Theodore, 20, 21f
Rotational exhibit, 149

Royal Society for the Prevention of Cruelty to Animals (RSPCA)
  campaign by, 264
  report on elephants in European zoos, 39, 170–171
RSPCA. *See* Royal Society for the Prevention of Cruelty to Animals (RSPCA)
Rumbles, 119

S

Saddles, injuries caused by, 179, 180t
Safari parks. *See* Zoos
Sanctuaries. *See* Elephant sanctuaries; Wildlife sanctuaries
Sampling problems. *See* Behaviour sampling, problems with
Satellite tracking, 261f, 299, 306, 316
Scan sampling, 36–37, 40
Secretary of State's Standards of Modern Zoo Practice (SSSMZP), 70, 172, 225b, 236, 238, 241, 270, 277, 282, 283b
Seismic communication. *See* Communication, seismic
Self-awareness, 114–115, 115f
Semen quality, 108, 111
Sex ratio, 58, 157, 159, 162, 166, 221t, 323
Sexual behaviour. *See* Mating
Sexual dimorphism, 176
Sexual maturity, 16
Sheldrick, Dame Daphne, 66, 99
Singer, Peter, 259–260
Sleep, duration of, 46–50, 47f
Smell. *See* Olfactory sense
Smith, T. Murray, 24, 99, 113, 219, 310
Social facilitation
  of defaecation, 139, 139t
  of mounting behaviour, 103–105b
Social network analysis, 68
Social rank, 61, 67–68, 70, 91, 119
Sociogram, 61f
Solitary elephants, management of, 282
Space utilisation. *See* Enclosure, use of
Species Survival Commission (SSC), 7–8, 268, 269b
Species Survival Plan (SSP), 39, 107, 163, 217, 220, 295–296
Sperm
  quality of, 106, 108–109, 154
  sex-sorting of, 308
Sport hunting, 320, 325
SSC. *See* Species Survival Commission (SSC)
SSP. *See* Species Survival Plan (SSP)
SSSMZP. *See* Secretary of State's Standards of Modern Zoo Practice (SSSMZP)
Standards for Elephant Management and Care, 68, 170, 196, 199, 201, 208t, 226b, 234t, 235, 252

Stephen Fritz Enterprises, Inc., 252, 254–257b
Stereotypic behaviours, 183–194
    aetiology of, 185, 185f, 188b, 191
    choice, effect of, 50, 194
    definition of, 183
    diurnal variation in, 189f, 192f
    feeding, correlation with, 186, 191, 193f
    restraint, effect of, 193–194
    sampling period, effect of, 37, 38f
    temperature, effect on, 191, 193f
    types of, 183, 184f, 188b, 190f
Stress, 180–181
Stride of elephant, 145, 210, 222
Subspecies of elephants. *See* Elephant subspecies
Substratum. *See* Floor
Sustainability of zoo populations. *See* Elephant populations, sustainability of, in zoos
Syntype, 4

T

Tail flicking, 90–91
Taxonomy of elephants, 3–7
TB. *See* Tuberculosis (TB)
Teeth, age determination from, 154
Temperature regulation
Temple elephants, 10, 37, 131, 131f, 134, 207, 213, 246–248
Testosterone, 89
TETP. *See* Thai Elephant-Assisted Therapy Project (TETP)
Thermal windows, 131
Thai Elephant-Assisted Therapy Project (TETP), 320–321
Thai Elephant Conservation Center. *See* Elephant centres
Timber camps. *See* Logging camps; Myanmar Timber Enterprise
Tool use, 124–127, 124f, 126f, 127f
Tourism, 13–14, 15f, 109, 117, 122, 127, 180f, 205, 246, 288, 309, 315, 320
    objection to use of elephants in, 264, 326
    welfare of elephants used in, 179–180, 180t, 211–213
Tower of London
    elephant at, 9, 10f
    elephant house, 229, 235f, 237
Traditional knowledge of elephant keeping, 89, 160, 170, 244–248, 275, 307
Training
    AZA checklist, 208, 208t
    crate, 253f, 255f, 257
    enrichment and. *See* Enrichment, training as
    protected contact. *See* Protected Contact
Translocation of elephants, 69, 76, 130, 238, 258, 297–298, 316, 323, 325, 327

Transportation of elephants, 250–258
    crates used for, 253–255f
    elephants, effects on, 257–258
    elephant, death during, 250
Transrectal ultrasound, 106, 108, 112, 208t
Trees
    elephant damage in wild, 150f
    preventing damage to in zoos, 150, 151f
Trunk wash, 208, 208t, 217
Tuberculosis (TB), 208, 216–217
Tuskless elephants, 318, 320
Twins
    frequency of occurrence, 91–92, 159
    survival of, 156
    testing for, 111

U

Ultrasonography. *See* Transrectal ultrasound
United States Department of Agriculture (USDA), 256b, 270
United States Fish and Wildlife Service (USFWS), 163, 278, 279b, 285b
Urinogenital-vaginal junction, 109
Using Science to Understand Zoo Elephant Welfare project, 172, 288

V

Vaginal vestibulotomy, 109
Video cameras, collecting data using, 47–50, 62, 68, 130, 182, 186, 193, 210
Visual acuity. *See* Landolt-C visual acuity test

W

Walking, 46. *See also* Gait
War elephants. *See* Battles involving elephants; Army, British
Water. *See* Enrichment, water as
Welfare. *See also* Stereotypic behaviours
    behaviour as indicator of, 181–183
    definition, 173
    elephants use in tourism and 179–180, 180t
    environment, inferred from, 179
    measurement of, 173–175. *See also* Welfare indices
Welfare indices. *See also* Body condition score
    population level, 175–176
Wellbeing. *See* Welfare
White rhinoceros. *See* Rhinoceros, white, attacks on
Wildlife Conservation Society, 284, 285–286b
Wildlife sanctuaries. *See also* Elephant sanctuaries
    Doi Phamuang Wildlife Sanctuary, 316
    Dong Yai Wildlife Sanctuary, 92
    Jaldapara Wildlife Sanctuary, 324

Wildlife sanctuaries (*Continued*)
  Kuldiha Wildlife Sanctuary, 133
  Mudumalai Sanctuary, 159
  Mwaluganii Elephant Sanctuary, 258
  Sublanka Wildlife Sanctuary, 316
  Sweetwaters Rhino Sanctuary, 316
World Animal Protection, 326

Y
Yawning, 49–50

Z
Zimbabwe, export of elephants from, 278–279, 281b, 282
ZIMS. *See* Zoological Information Management System (ZIMS)
Zoos *See also* Appendix
  Adelaide Botanic and Zoological Gardens, South Australia, 223f
  African Lion Safari, Ontario, Canada, 89
  Alaska Zoo, Alaska, USA, 266
  Antwerp Zoo, Belgium, 223
  Auckland Zoo, New Zealand, 182, 227
  Audubon Zoo, New Orleans, Louisiana, USA, 131, 141, 240t
  Barcelona Zoo, Spain, 265t, 324
    enclosure at, 325f
  Basel Zoo, Switzerland. *See* Zoos, Zoo Basel, Switzerland
  Belle Vue Zoo, Manchester, England, 22
  Berlin Zoo, Germany, 49, 195f, 296, 301f, 327
    enclosure at, 44f, 151f, 153f, 223, 224f, 231t, 239f
  Blackpool Zoo, England, 63, 116, 238, 240, 242, 300, 301f, 302–303f, 324
    enclosure at, 64f, 117f, 195f, 199–200f, 203f, 243f, 249f
  Bristol Zoo, England, 55, 266, 291f
    elephant house at, 54f
    history of elephants at, 53, 55f
  Bronx Zoo, New York, USA, 50, 115, 146, 169, 194, 223
    enclosure, 227f
  Brookfield Zoo, Chicago, Illinois, USA, 154
  Busch Gardens Tampa Bay, Florida, USA, 43, 46, 70, 123, 186, 194, 244
  Chehaw Wild Animal Park, Georgia, USA, 266
  Chester Zoo, England. *See also* Assam Haathi Project
    activity budgets at, 39–46
    associations between elephants at, 60–62b, 61f, 62f
    elephant enclosure at, 33f
    elephant herd, composition of, 32f, 32t
    history of elephants at, 56f
  Chimelong Wildlife Safari Park, Guangzho, China, 278
  Colombo Zoo, Sri Lanka, 57f
    elephant shelter at, 205f
  Commerford Zoo, Connecticut, USA, 274
  Copenhagen Zoo, Denmark, 231t, 240t
  Cricket St Thomas Zoo, England, 266
  Dallas Zoo, Texas, USA, 240t, 265t, 278, 289
  Detroit Zoo, Michigan, USA, 266
  Dickerson Park Zoo, Springfield, Missouri, USA, 25, 109
  Disney's Animal Kingdom, Florida, USA, 90–91, 119, 137, 145
  Dublin Zoo, Ireland, 48, 99, 146, 162, 202f, 204f, 206, 237f, 242f, 290, 292f
  Dudley Zoo, England, 55f, 227, 231t, 266
  Durrell. *See* Zoos, Jersey Zoo, Channel Isles
  Edinburgh Zoo, Scotland, 266
  Everland Zoo, South Korea, 120
  Frank Buck Zoo, Texas, USA, 266
  Franklin Park Zoo, Boston, Massachusetts, USA, 227, 231–232f
  Franklin Zoo and Wildlife Sanctuary, Tuaku, New Zealand, 247
  Grand Park Zoo, Seoul, South Korea, 95
  Heidelberg Zoo, Germany, 63, 240t, 309
  Henry Doorly Zoo, Nebraska, USA, 278, 289
  Henry Vilas Zoo, Wisconsin, USA, 265t, 266
  Higashiyama Zoo, Japan, 60, 191
  Honolulu Zoo, Hawaii, USA, 9
  Ichihara Elephant Kingdom, Japan, 116
  Jardin des Plantes, France, 21b
  Jersey Zoo, Channel Isles, 293
  Kansas City Zoological Park, Missouri, USA, 89
  Knowsley Safari Park, England, 43, 45, 134–135, 138–139, 139t, 140f
    elephant house at, 224f, 226, 228f
    landscape damage in, 150–152, 151–152f
  Lincoln Park Zoo, Chicago, Illinois, USA, 27
  London Zoo, England. *See* Zoos, ZSL London Zoo, England
  Longleat Safari Park, England, 278
  Los Angeles Zoo, California, USA, 265t, 311
  Louisiana Purchase Gardens and Zoo, Louisiana, USA, 266
  Lowry Park Zoo, Florida, USA, 163, 279b
  Marine World/Africa USA, California, USA, 125
  Maryland Zoo, USA, 290
  Melbourne Zoo, Australia, 149, 206, 240t, 258, 305
  Mesker Park Zoo, Indiana, USA, 266
  Mong La Zoo, Myanmar, 267
  National Zoological Park, Washington, USA. *See* Zoos, Smithsonian National Zoological Park, Washington, USA
  Newquay Zoo, England, 288
  Noah's Ark Zoo, England, 142, 240, 249
  Noorder Zoo, Emmen, The Netherlands, 99
  North Carolina Zoo, USA, 193, 240, 306, 309

Oregon Zoo, Portland, Oregon, USA (also known as Portland Zoo and Metro Washington Park Zoo), 2, 25, 95, 109, 117, 146, 254b, 295, 308
Philadelphia Zoo, Pennsylvania, USA, 265t
Rabat Zoo, Morocco, 69, 127
Reid Park Zoo, Arizona, 265t, 309
Safari Beekse Berge, The Netherlands, 148
St Louis Zoo, Missouri, USA, 239f, 241f, 241t
San Antonio Zoo, Texas, USA, 265t
San Diego Zoo, San Diego, California, USA, 130, 195, 240t, 254b, 279b, 311
San Diego Zoo Safari Park, San Diego, California, USA, 50, 145, 163, 181, 211, 309, 327
Schönbrunn Zoo, Vienna, Austria, 48, 295
Sedgwick County Zoo, Kansas, USA, 278, 289
Seneca Park Zoo, New York, USA, 309
Smithsonian National Zoological Park (also known as the National Zoological Park), Washington, USA, 112, 115–116, 229, 287, 290, 296, 307, 327
  elephant house at, 227, 233f, 240t
Sorocaba Zoo, Brazil, 127
Stuttgart Zoo, Germany, 222
Taipei Zoo, Taiwan, 155
Taronga Zoo, Australia, 240t, 258, 265t, 327
Turin Zoo, Italy, 12
Twycross Zoo, England, 63, 110f, 224f, 242, 253f, 266, 278, 323–324
  elephant house at, 237f
Ueno Zoo, Japan, 116
Valley Zoo, Edmonton, Canada, 265t
Vincennes Zoo, Paris, France, 157
Washington Park Zoo, Portland, Oregon, USA. *See* Zoos, Oregon Zoo

Welsh Mountain Zoo, Wales, 266
West Midlands Safari Park, West Midlands, England, 210, 224f
Whipsnade Zoo, England. *See* Zoos, ZSL Whipsnade Zoo, England
Woburn Safari Park, Bedfordshire, England, 210, 216
Woodland Park, Seattle, USA, 241t, 265t
Wroclaw Zoo, Poland, 132
Zoo Atlanta, Georgia, USA, 48, 116, 140, 186, 241t, 305
Zoo Basel, Switzerland, 154, 194
ZooParc de Beauval, France, 242
ZSL London Zoo, England, 2, 19, 21–22b, 154, 169, 217, 227, 238, 250, 253f, 265t, 292, 311
  elephant house at, 226, 229f, 230b, 231t
ZSL Whipsnade Zoo, England, 110f, 187–188b, 216, 238, 292, 300, 303f, 308, 311, 327
  elephant house at, 227, 230b, 231–232, 240t, 241–243
Zürich Zoo, Switzerland, 197, 292
Zoo Biology, papers on elephants in, 25
Zoo elephant population. *See* Elephant populations
Zoological Information Management System (ZIMS), 54, 57, 58t, 220–221, 220–221t
Zoological Society of London, 21, 230b, 285b, 286. *See also* Zoos, ZSL London Zoo; ZSL Whipsnade Zoo
Zoos, anonymity of in research, 30
Zoos Directive. *See* Legislation, Council Directive 1999/22/EC of 29 March 1999 on the keeping of wild animals in zoos
Zoos Expert Committee, 317
Zoos Forum. *See* Zoos Expert Committee

CPI Antony Rowe
Eastbourne, UK
October 06, 2020